D0713968

Electric Motor Control

9th Edition

Stephen L. Herman

Electric Motor Control

9th Edition

Stephen L. Herman

DELMAR
CENGAGE Learning

Australia • Canada • Mexico • Singapore • Spain • United Kingdom • United States

Electric Motor Controls, Ninth Edition

Stephen L. Herman

Vice President, Career and Professional
    Editorial: Dave Garza

Director of Learning Solutions: Sandy Clark

Managing Editor: Larry Main

Senior Product Manager: John Fisher

Senior Editorial Assistant:
    Dawn Daugherty

Vice President, Career and Professional
    Marketing: Jennifer McAvey

Marketing Director: Deborah S. Yarnell

Marketing Manager: Jimmy Stephens

Associate Marketing Manager: Mark Pierro

Production Director: Wendy Troeger

Production Manager: Mark Bernard

Content Project Manager:
    Christopher Chien

Senior Art Director: David Arsenault

Technology Project Manager:
    Christopher Catalina

Production Technology Analyst:
    Thomas Stover

© 2010, 2007 Delmar, Cengage Learning

ALL RIGHTS RESERVED. No part of this work covered by the copyright herein may be reproduced, transmitted, stored, or used in any form or by any means graphic, electronic, or mechanical, including but not limited to photocopying, recording, scanning, digitizing, taping, Web distribution, information networks, or information storage and retrieval systems, except as permitted under Section 107 or 108 of the 1976 United States Copyright Act, without the prior written permission of the publisher.

For product information and technology assistance, contact us at
**Professional Group Cengage Learning Customer &
Sales Support, 1-800-354-9706**
For permission to use material from this text or product,
submit all requests online at **cengage.com/permissions.**
Further permissions questions can be e-mailed to
permissionrequest@cengage.com.

Library of Congress Control Number: 2009925275
ISBN-13: 978-1-4354-8575-4
ISBN-10: 1-4354-8575-0

Delmar
5 Maxwell Drive
Clifton Park, NY 12065-2919
USA

Cengage Learning is a leading provider of customized learning solutions with office locations around the globe, including Singapore, the United Kingdom, Australia, Mexico, Brazil and Japan. Locate your local office at: **international.cengage.com/region**

Cengage Learning products are represented in Canada by Nelson Education, Ltd.

For your lifelong learning solutions, visit **delmar.cengage.com**

Visit our corporate website at **cengage.com.**

NOTICE TO THE READER
Publisher does not warrant or guarantee any of the products described herein or perform any independent analysis in connection with any of the product information contained herein. Publisher does not assume, and expressly disclaims, any obligation to obtain and include information other than that provided to it by the manufacturer. The reader is expressly warned to consider and adopt all safety precautions that might be indicated by the activities described herein and to avoid all potential hazards. By following the instructions contained herein, the reader willingly assumes all risks in connection with such instructions. The publisher makes no representations or warranties of any kind, including but not limited to, the warranties of fitness for particular purpose or merchantability, nor are any such representations implied with respect to the material set forth herein, and the publisher takes no responsibility with respect to such material. The publisher shall not be liable for any special, consequential, or exemplary damages resulting, in whole or part, from the readers' use of, or reliance upon, this material.

Printed in the United States of America
2 3 4 5 6 XX 14 13 12 11

Contents

v

Section 10 Methods of Deceleration 333

Section 11 Motor Drives 361

Section 12 Troubleshooting 413

Preface

Electric Motor Control provides beginning students with a practical approach to motor control. The textbook discusses electrical and mechanical components and how they are connected to control different types of motors. Many different types of control circuit and illustrations are discussed. The text contains a wealth of practical information that will apply to almost any industrial application.

PREREQUISITES AND USE

The text assumes that students have knowledge of basic electrical theory and common series and parallel circuits. *Electric Motor Control* has been used successfully for both formal classroom training and self study. It is used extensively in preapprentice and indentured apprentice training programs and in organized journeyman electrician classes. The practical approach to motor control makes this a very useful handbook on the job for installing, monitoring, and maintaining control systems.

MAJOR FEATURES

Electric Motor Control provides a very practical approach to a somewhat difficult subject. The text is written in easy-to-understand language. Each unit of instruction covers a short, concise topic. Expected student learning is outlined in the objectives at the beginning of each unit. The appendices and glossary provide further explanation of terms and servicing to troubleshooting, which the student is encouraged to use. The ninth edition contains expanded information on overload relays and numerous updated illustrations. The text employs a second color to highlight important concepts. The ninth edition provides an update to a textbook that has long been regarded as an outstanding book on the subject of motor control theory and practical application.

PRACTICAL APPROACH TO PROBLEM SOLVING

Electric Motor Control illustrates control systems, starting with the simplest of equipment, and builds on it in a step-by-step fashion to more complex circuits. Students learn to draw and interpret motor control schematics and wiring diagrams. The text helps teach students to think about the process involved in drawing and reading control schematics. This approach leads students in a natural progression into the basic concepts needed to install and troubleshoot control systems.

SUPPLEMENTS TO THIS TEXT

An **Instructor Resource CD** is available for this text. It contains tools and instructional resources that enrich your classroom and make your preparation time shorter. The elements of the instructor resource link directly to the text and tie together to provide a unified instructional system. Features contained in the instructor resource include:

- An **_Instructor Manual_** as a PDF file that contains answers to the end of unit questions, a comprehensive test, and answers to the comprehensive test.

- Unit presentations created in PowerPoint(®): These slides provide the basis for a lecture outline that helps you to present concepts and material.

- Test Questions: More than 250 questions of varying levels of difficulty are provided in true/false and multiple choice formats. These question scan be used to assess student comprehension or can be made available to the student for self-evaluation.

ISBN: 1435485742

ACKNOWLEDGMENTS

The author and staff at Delmar Cengage Learning wish to express their appreciation to the instructors who reviewed the previous edition and ninth edition revision plan and made suggestions for improvements.

Michel Benzer
Bluegrass Community and Technical College
Lexington, KY

Mark Bohnet
Northwest Iowa Community College
Sheldon, IA

Kevin Boiter
Piedmont Technical College
Greenwood, SC

John L. Brown
Portland Community College
Portland, OR

Michael Brumbach
York Technical College
Rock Hill, SC

John Everett
East Central Community College
Decatur, MS

Ivan Maas
North Dakota State College of Science
Wahpeton, ND

Marvin Moak
Hinds Community College
Raymond, MS

Introduction

General Principles of Electric Motor Control

Objectives

After studying this unit, the student should be able to:

- State the purpose and general principles of electric motor control.
- State the difference between manual and remote control.
- List the conditions of starting and stopping, speed control, and protection of electric motors.
- Explain the difference between compensating and definite time delay action.

There are certain conditions that must be considered when selecting, designing, installing, or maintaining electric motor control equipment. The general principles are discussed to help understanding and to motivate students by simplifying the subject of electric motor control.

Motor control was a simple problem when motors were used to drive a common line shaft to which several machines were connected. It was simply necessary to start and stop the motor a few times a day. However, with individual drive, the motor is now almost an integral part of the machine and it is necessary to design the motor controller to fit the needs of the machine to which it is connected. Large installations and the problems of starting motors in these situations may be observed in Figure 1–1 and Figure 1–2.

Motor control is a broad term that means anything from a simple toggle switch to a complex system with components such as relays, timers, and switches. The common function of all controls, however, is to control the operation of an electric motor. As a result, when motor control equipment is selected and installed, many factors must be considered to ensure that the control will function properly for the motor and the machine for which it is selected.

MOTOR CONTROL INSTALLATION CONSIDERATIONS

When choosing a specific device for a particular application, it is important to remember that the motor, machine, and motor controller are interrelated and need to be considered as a package. In general, five basic factors influence the selection and installation of a controller.

1. ELECTRICAL SERVICE
   Establish whether the service is direct (DC) or alternating current (AC). If AC, determine the frequency (hertz) and number of phases in addition to the voltage.

Fig. I–I Five 2000 hp, 1800 rpm induction motors driving water pumps for a Texas oil/water operation. Pumps are used to force water into the ground and "float" oil upward. *(Courtesy of Electric Machinery Company, Inc.)*

Fig. I–2 Horizontal 4000 hp synchronous motor driving a large centrifugal air compressor. *(Courtesy of Electric Machinery Company, Inc.)*

2. MOTOR

The motor should be matched to the electrical service and correctly sized for the machine load in horsepower rating (hp). Other considerations include motor speed and torque. To select proper protection for the motor, its full-load current rating (FLC), service factor (SF), time rating (duty), and other pertinent data—as shown on the motor nameplate—must be used.

3. OPERATING CHARACTERISTICS OF CONTROLLER

The fundamental tasks of a motor controller are to start and stop the motor and to protect the motor, machine, product, and operator. The controller may also be called upon to provide supplementary functions such as reversing, jogging or inching, plugging, or operating at several speeds or at reduced levels of current and motor torque (see Glossary). Section 430 of the *National Electrical Code® (NEC®)* provides requirements concerning the installation of motor circuits. This section is employed to determine the proper conductor size, overload size, and short circuit protection rating for motor installations. In some industries electrical engineers are responsible for determining the requirements for installing a motor or motors. In other industries the electrician is expected to perform this task.

4. ENVIRONMENT

Controller enclosures serve to provide safety protection for operating personnel by preventing accidental contact with live parts. In certain applications, the controller itself must be protected from a variety of environmental conditions, which might include

- Water, rain, snow, or sleet
- Dirt or noncombustible dust
- Cutting oils, coolants, or lubricants

Both personnel and property require protection in environments made hazardous by the presence of explosive gases or combustible dusts.

5. ELECTRICAL CODES AND STANDARDS
Motor control equipment is designed to meet the provisions of the *National Electrical Code® (NEC®)*. Also, local code requirements must be considered and met when installing motors and control devices. Presently, code sections applying to motors, motor circuits, and controllers and industrial control devices are found in Article 430 on motors and motor controllers, Article 440 on air-conditioning and refrigeration equipment, and Article 500 on hazardous locations of the *NEC®*.

The 1970 Occupational Safety and Health Act (OSHA), as amended, requires that each employer furnish employment in an environment free from recognized hazards likely to cause serious harm.

Standards established by the National Electrical Manufacturers Association (NEMA) assist users in the proper selection of control equipment. NEMA standards provide practical information concerning the construction, testing, performance, and manufacture of motor control devices such as starters, relays, and contactors.

One of the organizations that actually tests for conformity to national codes and standards is Underwriters' Laboratories (UL). Equipment that is tested and approved by UL is listed in an annual publication, which is kept current by means of bimonthly supplements to reflect the latest additions and deletions. A UL listing does not mean that a product is approved by the *NEC*. It must be acceptable to the local authority having jurisdiction.

PURPOSE OF CONTROLLER

Some of the complicated and precise automatic applications of electrical control are illustrated in Figure 1–3 and Figures 1–7A and B. Factors to be considered when selecting and installing motor control components for use with particular machines or systems are described in the following paragraphs.

Starting

The motor may be started by connecting it directly across the source of voltage. Slow and gradual starting may be required, not only to

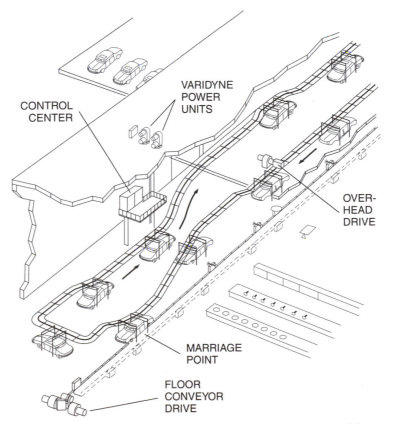

Fig. 1–3 Synchronizing Two Automobile Assembly Systems. *(Courtesy of Emerson Motors)*

Fig. 1–4 Combination fused disconnect switch and motor starter. *(Courtesy of Schneider Electric)*

protect the machine, but also to ensure that the line current inrush on starting is not too great for the power company's system. Some driven machines may be damaged if they are started with a sudden turning effort. The frequency of starting a motor is another factor affecting the controller. A combination fused disconnect switch and motor starter is shown in Figure 1–4.

Stopping

Most controllers allow motors to coast to a standstill. Some impose braking action when the machine must stop quickly. Quick stopping is a vital function of the controller for emergency stops. Controllers assist the stopping action by retarding centrifugal motion of machines and lowering operations of crane hoists.

Reversing

Controllers are required to change the direction of rotation of machines automatically or at the command of an operator at a control station. The reversing action of a controller is a continual process in many industrial applications.

Running

The maintaining of desired operational speeds and characteristics is a prime purpose and function of controllers. They protect motors, operators, machines, and materials while running.

There are many different types of safety circuits and devices to protect people, equipment, and industrial production and processes against possible injury that may occur while the machines are running.

Speed Control

Some controllers can maintain very precise speeds for industrial processes. Other controllers can change the speeds of motors either in steps or gradually through a continuous range of speeds.

Safety of Operator

Many mechanical safeguards have been replaced or aided by electrical means of protection. Electrical control pilot devices in controllers provide a direct means of protecting machine operators from unsafe conditions.

Protection from Damage

Part of the operation of an automatic machine is to protect the machine itself and the manufactured or processed materials it handles. For example, a certain machine control function may be the prevention of conveyor pileups. A machine control can reverse, stop, slow, or do whatever is necessary to protect the machine or processed materials.

Maintenance of Starting Requirements

Once properly installed and adjusted, motor starters will provide reliable operation of starting time, voltages, current, and torques for the benefit of the driven machine and the power system. The *NEC*®, supplemented by local codes, governs the selection of the proper sizes of conductors, starting fuses, circuit breakers, and disconnect switches for specific system requirements.

MANUAL CONTROL

A manual control is one whose operation is accomplished by mechanical means. The effort required to actuate the mechanism is almost always provided by a human operator. The motor may be controlled manually using any one of the following devices.

Toggle Switch

A toggle switch is a manually operated electric switch. Many small motors are started with toggle switches. This means the motor may be started directly without the use of magnetic switches or auxiliary equipment. Motors started with toggle switches are protected by the branch circuit fuse or circuit breaker. These motors generally drive fans, blowers, or other light loads.

Safety Switch

In some cases it is permissible to start a motor directly across the full-line voltage if an externally operated safety switch is used (EXO), Figure 1–5. The motor receives starting and running protection from dual-element, time-delay fuses. The use of a safety switch requires manual operation. A safety switch, therefore, has the same limitations common to most manual starters.

Drum Controller

Drum controllers are rotary, manual switching devices that are often used to reverse motors and to control the speed of AC and DC motors. They are used particularly where frequent start, stop, or reverse operation is required. These controllers may be used without other control components in small motors, generally those with fractional horsepower ratings. Drum controllers are used with magnetic starters in large motors. A drum controller is shown in Figure 1–6.

Fig. 1–6 Drum controller with cover (left) and with cover removed (right). *(Courtesy of Eaton Corporation)*

Faceplate Control

Faceplate controllers have been in use for many years to start DC motors. They have been used for AC induction motor speed control. The faceplate control has multiple switching contacts mounted near a selector arm on the front of an insulated plate. Additional resistors are mounted on the rear to form a complete unit. Technological developments, however, have fewer faceplate controllers being installed.

REMOTE AND AUTOMATIC CONTROL

The motor may be controlled by remote control using push buttons, Figures 1–7A and B. When push-button remote control is used or when automatic devices do not have the electrical capacity to carry the motor starting and running currents, magnetic or electric switches must be included. Magnetic switch control is one whose operation is accomplished by electromagnetic means. The effort required to actuate the electromagnet is supplied by electrical energy rather than by the human operator. If the motor is to be automatically controlled, the following two-wire pilot devices may be used.

Float Switch

The raising or lowering of a float that is mechanically attached to electrical contacts may start motor-driven pumps to empty or fill tanks. Float switches are also used to open or close piping solenoid valves to control fluids,

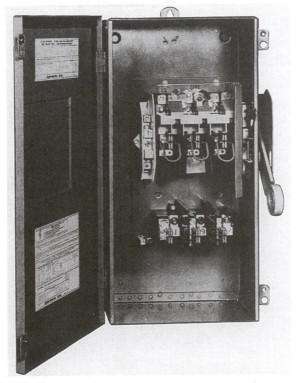

Fig. 1–5 Safety disconnect switch (EXO). *(Courtesy of Eaton Corporation)*

Fig. 1–7(A) Typical cement mill computer console carefully controlling and monitoring motors located remotely. *(Courtesy of Electric Machinery Company, Inc.)*

Fig. 1–7(B) Some typical terminal wiring behind the control panels shown in part A. *(Courtesy of Los Angeles Department of Water and Power)*

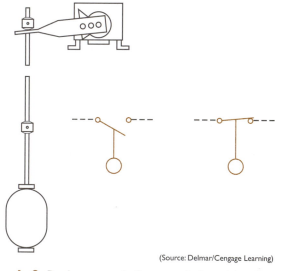

(Source: Delmar/Cengage Learning)

Fig. 1–8 Rod-operated float switch with electrical wiring symbols.

Figure 1–8. There are other similar methods that achieve the same results.

Pressure Switch

Pressure switches are used to control the pressure of liquids and gases (including air) within a desired range, Figure 1–9. Air compressors, for example, are started directly or indirectly on a call for more air by a pressure switch. Electrical wiring symbols are shown as normally closed and normally open in Figure 1–9.

Time Clock

Time clocks can be used when a definite "on and off" period is required and adjustments are

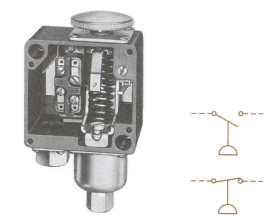

Fig. 1–9 Pressure switches may be necessary to start the motors shown in Figure 1–10. *(Courtesy of Schneider Electric)*

Fig. 1–10 Two 1500 hp vertical induction motors driving pumps. *(Courtesy of Electric Machinery Company, Inc.)*

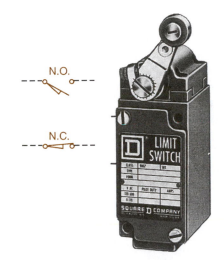

Fig. 1–12 Limit switch shown with electrical wiring symbols. *(Courtesy of Schneider Electric)*

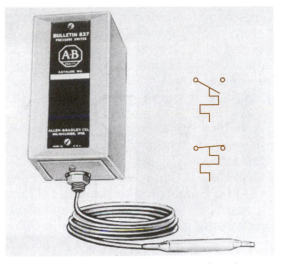

Fig. 1–11 Industrial temperature switch with extension bulb and electrical wiring symbols. *(Courtesy of rockwell Automation)*

not necessary for long periods of time. A typical requirement is a motor that must start every morning at the same time and shut off every night at the same time, or switches the floodlights on and off.

Thermostat

In addition to pilot devices sensitive to liquid levels, gas pressures, and time of day, thermostats sensitive to temperature changes are widely used, Figure 1–11. Thermostats indirectly control large motors in air-conditioning systems and in many industrial applications to maintain the desired temperature range of air, gases, liquids, or solids. There are many types of thermostats and temperature-actuated switches.

Limit Switch

Limit switches, Figure 1–12, are designed to pass an electrical signal only when a predetermined limit is reached. The limit may be a specific position for a machine part or a piece of work, or a certain rotating speed. These devices take the place of a human operator and are often used under conditions where it would be impossible or impractical for the operator to be present or to efficiently direct the machine.

Limit switches are used most frequently as overtravel stops for machines, equipment, and products in process. These devices are used in the control circuits of magnetic starters to govern the starting, stopping, or reversal of electric motors.

Electrical or Mechanical Interlock and Sequence Control

Many of the electrical control devices described in this unit can be connected in an interlocking system so that the final operation of one or more motors depends on the electrical position of each individual control device. For example, a float switch may call for more liquid but will not be satisfied until the prior approval of a pressure switch or time clock is obtained. To design, install, and maintain electrical controls in any electrical or mechanical interlocking system, the electrical technician must understand the total operational system and the function of the individual components. With practice, it is possible to transfer knowledge of circuits and descriptions for an understanding of additional similar controls. It is impossible—in instructional materials—to show all possible combinations of an interlocking control system. However, by understanding the basic functions of control components and their basic circuitry, and by taking the time to trace and draw circuit diagrams, difficult interlocking control systems can become easier to understand.

STARTING AND STOPPING

In starting and stopping a motor and its associated machinery, there are a number of conditions that may affect the motor. A few of them are discussed here.

Frequency of Starting and Stopping

The starting cycle of a controller is an important factor in determining how satisfactorily the controller will perform in a particular application. Magnetic switches, such as motor starters, relays, and contactors, actually beat themselves apart from repeated opening and closing thousands of times. An experienced electrician soon learns to look for this type of component failure when troubleshooting any inoperative control panels. NEMA standards require that the starter size be derated if the frequency of start-stop, jogging, or plugging is more than five times per minute. Therefore, when the frequency of starting the controller is great, the use of heavy-duty controllers and accessories should be considered. For standard-duty controllers, more frequent inspection and maintenance schedules should be followed.

Light- or Heavy-Duty Starting

Some motors may be started with no loads and others must be started with heavy loads. When motors are started, large feeder line disturbances may be created, which can affect the electrical distribution system of the entire industrial plant. The disturbances may even affect the power company's system. As a result, the power companies and electrical inspection agencies place certain limitations on "across-the-line" motor starting.

Fast or Slow Start (Hard or Soft)

To obtain the maximum twisting effort (torque) of the rotor of an AC motor, the best starting condition is to apply full voltage to the motor terminals. The driven machinery, however, may be damaged by the sudden surge of motion.

To prevent this type of damage to machines, equipment, and processed materials, some controllers are designed to start slowly and then increase the motor speed gradually. This type is often recommended by power companies and inspection agencies to avoid electrical line surges.

Smooth Starting

Although reduced electrical and mechanical surges can be obtained with a step-by-step motor starting method, very smooth and gradual starting will require different controlling methods. These are discussed in detail later in the text.

Manual or Automatic Starting and Stopping

Although the manual starting and stopping of machines by an operator is still a common practice, many machines and industrial processes are started and restarted automatically. These automatic devices result in tremendous savings of time and materials. Automatic stopping devices are used in motor control systems for the same reasons. Automatic stopping devices greatly reduce the safety hazards of operating some types of machinery, both for the operator and the materials being processed. An electrically operated mechanical brake is shown in Figure 1–13. Such a brake may be required to stop a machine's motion in a hurry to protect materials being processed or people in the area.

Quick Stop or Slow Stop

Many motors are allowed to coast to a standstill. However, manufacturing requirements and safety considerations often make it necessary to bring machines to as rapid a stop as possible. Automatic controls can retard and brake the speed of a motor and also apply a torque in the

Fig. 1–13 Typical 30-inch brake. (*Courtesy of Eaton Corporation*)

opposite direction of rotation to bring about a rapid stop. This is referred to as "plugging." Plugging can be used only if the driven machine and its load will not be damaged by the reversal of the motor torque. The control of deceleration is one of the important functions of a motor controller.

Accurate Stops

An elevator must stop at precisely the right location so that it is aligned with the floor level. Such accurate stops are possible with the use of the automatic devices interlocked with control systems.

Frequency of Reversals Required

Frequent reversals of the direction of rotation of the motor impose large demands on the controller and the electrical distribution system. Special motors and special starting and running protective devices may be required to meet the conditions of frequent reversals. A heavy-duty drum switch-controller is often used for this purpose.

SPEED CONTROL OF MOTORS

The speed control is concerned not only with starting the motor but also with maintaining or controlling the motor speed while it is running.

There are a number of conditions to be considered for speed control.

Constant Speed

Constant-speed motors are used on water pumps, Figure 1–10 and Figure 1–14. Maintenance of constant speed is essential for motor generator sets under all load conditions. Constant-speed motors with ratings as low as 80 rpm and horsepower ratings up to 5000 hp are used in direct drive units. The simplest method of changing speeds is by gearing. Using gears, almost any "predetermined" speed may be developed by coupling the input gear to the shaft of a squirrel cage induction motor. A speed-reducing gear motor is shown in Figure 1–15.

Varying Speed

A varying speed is usually preferred for cranes and hoists, Figure 1–16. In this type of application, the motor speed slows as the load increases and speeds up as the load decreases.

Adjustable Speed

With adjustable speed controls, an operator can gradually adjust the speed of a motor over a wide range while the motor is running. The speed may be preset, but once it is adjusted it remains essentially constant at any load within the rating of the motor.

Fig. 1–14 Multiple synchronous motors of 3000 hp and 225 rpm driving water pumps. (*Courtesy of Electric Machinery Company, Inc.*)

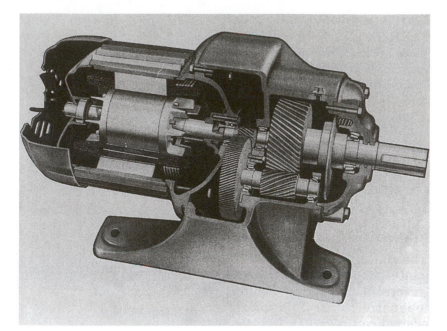

Fig. 1–15 Cutaway view of speed-reducing gear motor (double reduction). *(Courtesy of Emerson Motors)*

Fig. 1–16 Large traveling overhead crane. *(Courtesy of Schneider Electric)*

Multispeed

For multispeed motors, such as the type used on turret lathes in a machine shop, the speed can be set at two or more definite rates. Once the motor is set at a definite speed, the speed will remain practically constant regardless of load changes.

PROTECTIVE FEATURES

The particular application of each motor and control installation must be considered to determine what protective features are required to be installed and maintained.

Overload Protection

Running protection and overload protection refer to the same thing. This protection may be an integral part of the motor or be separate. A controller with electrical overload protection will protect a motor from burning up while allowing the motor to achieve its maximum available power under a range of overload and temperature conditions. An electrical overload on the motor may be caused by mechanical overload on driven machinery, a low line voltage, an open electrical line in a polyphase system resulting in single-phase operation, motor problems such as too badly worn bearings, loose terminal connections, or poor ventilation within the motor.

Open-Field Protection

DC shunt and compound-wound motors can be protected against the loss of field excitation by field loss relays. Other protective arrangements are used with starting equipment for DC and AC synchronous motors. Some sizes of DC motors may race dangerously with the loss of field excitation, whereas other motors may not race due to friction and the fact that they are small.

Open-Phase Protection

Phase failure in a three-phase circuit may be caused by a blown fuse, an open connection, a broken line, or other reasons. If phase failure occurs when the motor is at a standstill during attempts to start, the stator currents will rise to a very high value and will remain there, but the motor will remain stationary (not turn). Because the windings are not properly ventilated while the motor is stationary, the heating produced by the high currents may damage them. Dangerous conditions are also possible while the motor is running. When the motor is running and an open-phase condition occurs, the motor may continue to run. The torque will decrease, possibly to the point of motor "stall"; this condition is called *breakdown torque*.

Reversed-Phase Protection

If two phases of the supply of a three-phase induction motor are interchanged (phase reversal), the motor will reverse its direction of rotation. In elevator operation and industrial applications, this reversal can result in serious damage. Phase failure and phase reversal relays are safety devices used to protect motors, machines, and personnel from the hazards of open-phase or reversed-phase conditions.

Overtravel Protection

Control devices are used in motor starter circuits to govern the starting, stopping, and reversal of electric motors. These devices can be used to control regular machine operation or they can be used as safety emergency switches to prevent the improper functioning of machinery.

Overspeed Protection

Excessive motor speeds can damage a driven machine, materials in the industrial process, or the motor. Overspeed safety protection is provided in control equipment for paper and printing plants, steel mills, processing plants, and the textile industry.

Reversed-Current Protection

Accidental reversal of currents in direct-current controllers can have serious effects. Direct-current controllers used with three-phase alternating-current systems that experience phase failures and phase reversals are also subject to damage. Reversed-current protection is an important provision for battery charging and electroplating equipment.

Fig. 1–17 Spin-on, explosion-proof enclosure for a combination disconnect switch and magnetic motor starter. *(Courtesy of Schneider Electric)*

Mechanical Protection

An enclosure may increase the life span and contribute to the trouble-free operation of a motor and controller. Enclosures with particular ratings such as general purpose, watertight, dustproof, explosion proof, and corrosion resistant are used for specific applications, Figure 1–17. All enclosures must meet the requirements of national and local electrical codes and building codes.

Short-Circuit Protection

For large motors with greater than fractional horsepower ratings, short-circuit and ground fault protection are generally installed in the same enclosure as the motor-disconnecting means. Overcurrent devices (such as fuses and circuit breakers) are used to protect the motor branch circuit conductors, the motor control apparatus, and the motor itself against sustained overcurrent due to short circuits and grounds, and prolonged and excessive starting currents.

CLASSIFICATION OF AUTOMATIC MOTOR STARTING CONTROL SYSTEMS

The numerous types of automatic starting and control systems are grouped into the following classifications: current-limiting acceleration and time-delay acceleration.

Current-Limiting Acceleration

This is also called compensating time. It refers to the amount of current or voltage drop required to open and close magnetic switches when used in a motor accelerating controller. The rise and fall of the current or voltage determines a timing period.

Time-Delay Acceleration

For this classification, *definite time* relays are used to obtain a preset timing period. Once the period is preset, it does not vary regardless of current or voltage changes occurring during motor acceleration. The following timers and timing systems are used for motor acceleration; some are also used in interlocking circuits for automatic control systems.

- Pneumatic timing
- Motor-driven timers
- Capacitor timing
- Electronic timers

Study Questions

1. What is a controller and what is its function? (Use the Glossary and the information from this unit to answer this question.)

2. What is meant by remote control?

3. To what does current limiting, or compensating time, acceleration refer?

4. List at least four devices that are used to control a motor automatically. Briefly describe the purpose of each device.

   Select the *best* answer for each of the following.

5. The general purpose of motor control is
   a. to start the motor
   b. to stop the motor
   c. to reverse the motor
   d. all of the above

6. A motor may be controlled manually by using a
   a. float switch
   b. pressure switch
   c. toggle switch
   d. time clock

7. A motor may be controlled remotely or automatically by using a
   a. drum controller
   b. thermostat
   c. safety switch
   d. faceplate control

8. Conditions that may affect starting and stopping of a motor and the machinery that it drives are
   a. fast or slow starts
   b. starting under light or heavy loads
   c. how often the motor is started and stopped
   d. all of the above

9. Which of the following is *not* considered as motor speed control?
   a. Constant speed
   b. Varying speed
   c. Multispeed
   d. Short-circuit protection

10. Which of the following is not a protective feature of a motor controller?
    a. Overload
    b. Short circuit
    c. Adjustable speed
    d. Mechanical

11. Which function is not a fundamental job of a motor controller?
    a. Start and stop the motor
    b. Protect the motor, machine, and operator
    c. Reverse, inch, jog, speed control
    d. Motor and controller disconnect switch

12. What factors are to be considered when selecting and installing a controller?
    a. Electrical service
    b. Motor
    c. Electrical codes and standards
    d. All of the above

UNIT 2

Fractional and Integral Horsepower Manual Motor Starters

Objectives

After studying this unit, the student should be able to

- Match simple schematic diagrams with the appropriate manual motor starters.

- Connect manual fractional horsepower motor starters for automatic and manual operation.

- Connect integral horsepower manual starters.

- Explain the principles of operation of manual motor starters.

- List common applications of manual starters.

- Read and draw simple schematic diagrams.

- Briefly explain how motors are protected electrically.

FRACTIONAL HORSEPOWER MANUAL MOTOR STARTERS

One of the simplest types of motor starters is an on-off, snap action switch operated by hand on a toggle lever mounted on the front of the starter, Figure 2–1. The motor is connected directly across the matching line voltage when the handle is turned to the START position. This situation usually is not objectionable with motors rated at 1 hp or less. Because a motor may draw up to a 600 percent current surge on starting, larger motors should not be connected directly across the line on startup. Such a connection would result in large line surges that may disrupt power services or cause voltage fluctuations, which impede the normal operation of

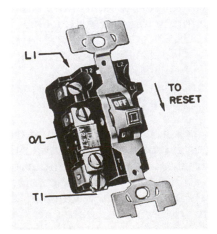

Fig. 2–1 Open-type starter with overload heater. *(Courtesy of Schneider Electric)*

17

other equipment. Proper motor starters for larger motors are discussed later.

Fractional horsepower (FHP) manual motor starters are used whenever it is desired to provide overload protection for a motor as well as "off" and "on" control of small alternating-current single-phase or direct-current motors. Electrical codes require that FHB motors be provided with overload protection whenever they are started automatically or by remote control. Basically, a manual starter is an *on-off* switch with motor overload protection.

Because manual starters are hand-operated mechanical devices (requiring no electrical coil), the contacts remain closed and the lever stays in the ON position in the event of a power failure. As a result, the motor automatically restarts when the power returns. Therefore, low-voltage protection (see Glossary) is not possible with manually operated starters. This automatic restart action is an advantage when the starter is used with motors that run continuously, such as those used on unattended pumps, blowers, fans, and refrigeration processes. This saves the maintenance electrician from running around the plant to restart all these motors after the power returns. On the other hand, the automatic restart feature is a disadvantage on lathes and machines that may be a danger to products, machinery, or people. It is definitely a safety factor to be observed.

The compact construction of the manual starter means that it requires little mounting space and can be installed on the driven machinery and in various other places where the available space is limited. The unenclosed, or open starter, can be mounted in a standard switch or conduit box installed in a wall and can be covered with a standard flush, single-gang switchplate. The ON and OFF positions are clearly marked on the operating lever, which is very similar to a standard lighting toggle switch lever, Figure 2–1.

Application

FHB manual starters have thermal overload protection, Figure 2–1 and Figure 2–2. When an overload occurs, the starter handle automatically moves to the center position to signify that the contacts have opened and the motor is no longer operating. The starter contacts cannot be reclosed until the overload relay is reset manually. The relay is reset by moving the handle to the full OFF position after allowing about 2 minutes for the heater to cool. Should the circuit trip open again, the fault should be located and corrected.

FHB manual starters are provided in several different types of enclosures as well as the open type to be installed in a switch box, flush in the wall, or on the surface. Enclosures are obtained to shield the live starter circuit components from accidental contact, for mounting in machine cavities, to protect the starter from dust and moisture, Figure 2–3, or to prevent the possibility of an explosion when the starter is used in hazardous locations. These different types of enclosures are discussed in more detail in Unit 3.

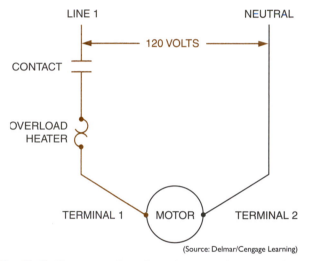

(Source: Delmar/Cengage Learning)

Fig. 2–2 Diagram of single-pole manual starter shown in Figure 2–1.

Fig. 2–3 Watertight and dust-tight manual starter enclosure. *(Courtesy of General Electric Co.)*

AUTOMATIC AND REMOTE OPERATION

Common applications of manual starters provide control of small machine tools, fans, pumps, oil burners, blowers, and unit heaters. Almost any small motor should be controlled with a starter of this type. However, the contact capacity of the starter must be sufficient to make and break the full motor current. Automatic control devices such as pressure switches, float switches, or thermostats rated to carry motor current may be used with FHB manual starters.

The schematic diagram shown in Figure 2–4 illustrates a fractional horsepower an FHB motor controlled automatically by a float switch that is remotely connected in the small motor

Fig. 2–6 FHP manual starter with selector switch mounted in general-purpose enclosure. (*Courtesy of Schneider Electric*)

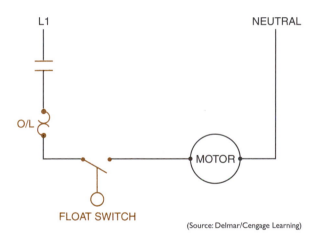

Fig. 2–4 Automatic control with FHP-single-pole, single-phase manual starter for FHP motor without selector switch.

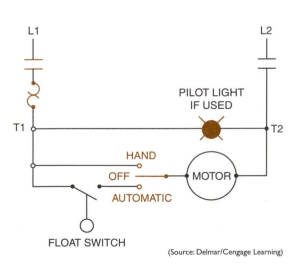

Fig. 2–5 Manual FHP-double-pole, single-phase starter with automatic control for FHP motor using selector switch.

circuit as long as the manual starter contact is closed. When the float is up, the pump motor starts.

In Figure 2–5 and Figure 2–6, the selector switch must be turned to the automatic position if the float switch is to take over an automatic operation, such as sump pumping. A liquid-filled sump raises the float, closes the normally open electrical contact, and starts the motor. When the motor pumps the sump or tank empty, the float lowers and breaks electrical contact with the motor thus stopping the motor. This cycle of events will repeat when the sump fills again, automatically, without a human operator.

Note that a double-pole starter is used in Figure 2–5. This type of starter is required when both lines to the motor must be broken such as for 240 volts, single phase. The double-pole starter is also recommended for heavy-duty applications because of its higher interrupting capacity and longer contact life when using a two-pole motor starter in 120 volts. Normally, the single-pole motor starter is used for 120 volts.

MANUAL PUSH-BUTTON LINE VOLTAGE STARTERS

These are integral horsepower motor starters (not fractional). Generally, manual push-button starters may be used to control single-phase motors rated up to 5 hp, polyphase motors rated up to 10 hp, and DC motors rated up to 2 hp. They are available in two-pole for single-phase and three-pole for polyphase motors. A typical manual three-phase push-button starter and diagram are shown in Figure 2–7 and Figure 2–8.

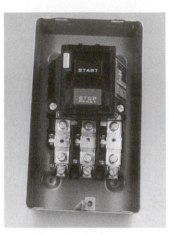

Fig. 2–7 Three-phase line voltage manual starter.

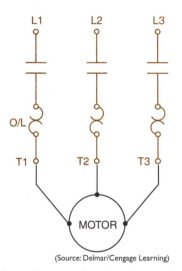

(Source: Delmar/Cengage Learning)

Fig. 2–8 Wiring diagram for the three-pole line voltage push-button manual starter seen in Figure 2–7.

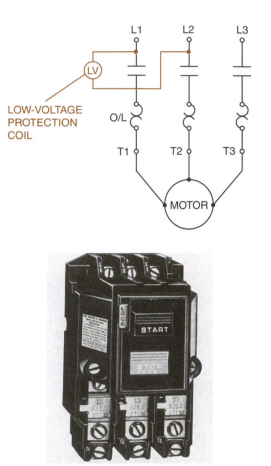

Fig. 2–9 (A) Wiring diagram for integral hp manual starter with low-voltage protection. (B) Manual starter with low-voltage protection. *(Courtesy of Schneider Electric)*

When an overload relay trips, the starter mechanism unlatches, opening the contacts to stop the motor. The contacts cannot be reclosed until the starter mechanism has been reset by pressing the STOP button, after allowing time for the thermal unit to cool.

These starters are designed for infrequent starting of small AC motors. This manual starter provides overload protection also, but it cannot be used where low or undervoltage protection is required or for remote or automatic operation.

Manual Starter with Low-Voltage Protection

Integral horsepower manual starters with low-voltage protection (LVP) prevent automatic startup of motors after a power loss. This is accomplished with a continuous-duty electrical solenoid, which is energized whenever the line-side voltage is present and the start button is pressed, Figure 2–9. If the line voltage is lost or disconnected, the solenoid de-energizes, opening the starter contacts. The contacts will not automatically reclose when the line voltage is restored. To close the contacts to restart the motor again, the device must be manually reset. This manual starter will not function unless the line terminals are energized. This starter should not be confused with magnetic starters described in the next unit. This is *not* a magnetic starter, but a lower cost starter.

Applications

Typical applications include conveyor lines, grinders, metal working machinery, mechanical power presses, mixers, woodworking machinery, and wherever job specifications and standards require low-voltage protection, or wherever machine operator safety could be in jeopardy.

Therefore, this manually operated, push-button starter with low-voltage protection is a method of protecting an operator from injury using automatic restart of a machine on resumption of voltage after a power failure. This is normally accomplished with a magnetic starter with electrical (three-wire) control.

THERMAL OVERLOAD PROTECTION

Thermal overload units are widely used on both the fractional and integral horsepower manual starters for protection of motors from sustained electrical overcurrents that could result from overloading of the driven machine or from excessively low line voltage.

Heater elements that are closely calibrated to the full-load current of the motor are used on the solder-pot and the thermobimetallic types of overload relays. On the solder-pot overload relays, the heating of the element causes alloy elements to melt when there is a motor overload due to excess current in the circuit. When the alloy melts, a spring-loaded ratchet is rotated and trips open a contact, which then opens the supply circuit to the motor; this stops the motor. On the bimetallic overload relays, the heating from the thermal element causes the bimetallic switch to open, opening the supply circuit and thereby stopping the motor.

Normal motor starting currents and momentary overloads will not cause thermal relays to trip because of their inverse-time characteristics (see Glossary). However, continuous overcurrent through the heater unit raises the temperature of the alloy elements. When the melting point is reached, the ratchet is released and the switch mechanism is tripped to open the line or lines to the motor. The switch mechanism is trip-free, which means that it is impossible to hold the contacts closed against an overload.

Only one overload relay is required in either the single-pole or double-pole motor starter, because the starter is intended for use on DC or single-phase AC service. When the line current is excessively high, these relays offer protection against continued operation. Relays with meltable alloy elements are nontemperable and give reliable overload protection. Repeated tripping does not cause deterioration nor does it affect the accuracy of the trip point.

Many types of overload relay heater units are available so that the proper one can be selected on the basis of the actual full-load current rating of the motor. The applicable relay heater units for a particular overload relay are interchangeable and are accessible from the front of the starter. Because the motor current is connected in series with the heater element, the motor will not operate unless the relay unit has the heater installed. Overload units may be changed without disconnecting the wires from the starter or removing the starter from the enclosure. However, the disconnect switch and starter should be turned off first for safety reasons. Additional instruction on motor overload protection is given in Unit 3.

Study Questions

1. If the contacts on a manual starter cannot be closed immediately after a motor over-load has tripped them open, what is the probable reason?

2. If the handle of an installed motor starter is in the center position, what condition does this indicate?

3. How may an automatic operation be achieved when a manual motor starter is used?

4. What does the term *trip-free* mean as it relates to manual starters?

5. If the overload heater element is not installed in a two-pole manual starter, what is the result?

   Select the *best* answer for each of the following.

6. A fractional horsepower manual motor starter
   a. starts and stops motors under 1 hp
   b. has a toggle-switch type of handle and size
   c. cannot offer low-voltage protection
   d. does all of these

7. An automatic operation can be achieved with
   a. an integral horsepower manual starter with LVP coil
   b. a push-button-type manual starter
   c. an FHP manual starter wired in conjunction with a pressure switch
   d. a fractional hp manual starter

8. Low-voltage (or undervoltage) protection (LVP) is available on some
   a. integral hp starters
   b. FHP manual starters
   c. two-pole toggle switches
   d. all of the above

9. LVP is used mainly for
   a. fans
   b. air compressors
   c. pumps
   d. protection of the operator

10. Thermal overload protection
    a. aids motor warm-up
    b. prevents motors from freezing
    c. protects motors and conductors from mechanical damage
    d. provides overcurrent protection for motors

Magnetic Line Voltage Starters

Objectives

After studying this unit, the student should be able to

- Identify common magnetic motor starters and overload relays.
- Describe the construction and operating principles of magnetic switches.
- Describe the operating principle of a solenoid.
- Troubleshoot magnetic switches.
- Select starter protective enclosures for particular applications.

Magnetic control means the use of electro-magnetic energy to close switches. Line voltage (across-the-line) magnetic starters are electro-mechanical devices that provide a safe, convenient, and economical means of full-voltage starting and stopping motors. In addition, these devices can be controlled remotely. They are used when a full-voltage starting torque (see the Glossary and Appendix) may be applied safely to the driven machinery and when the current inrush resulting from across-the-line starting is not objectionable to the power system. Control for these starters is usually provided by pilot devices such as push buttons, float switches, timing relays, and more as discussed in Section 2. Automatic control is obtained from the use of some of these pilot devices.

MAGNETIC STARTERS VERSUS MANUAL

Using manual control, the starter must be mounted so that it is easily within reach of the machine operator. With magnetic control, push-button stations are mounted nearby, but automatic control pilot devices can be mounted almost anywhere on the machine. The push buttons and automatic pilot devices can be connected by control wiring into the coil circuit of a remotely mounted starter, possibly closer to the motor to shorten the power circuit.

Operation

In the construction of a magnetic controller, the armature is mechanically connected to a set of contacts so that when the armature moves to its closed position, the contacts also close. There

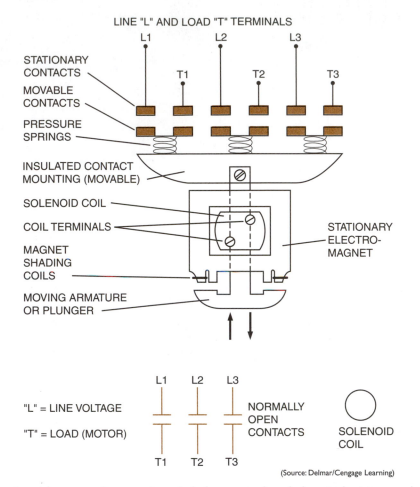

Fig. 3–1 Three-pole, solenoid-operated magnetic switch (contactor) and electrical wiring symbols.

are different variations and positions, but the operating principle is the same.

The simple up-and-down motion of a solenoid-operated, three-pole magnetic switch is shown in Figure 3–1. Not shown are the motor overload relays and the maintaining and auxiliary electrical contacts. Double-break contacts are used on this type of starter to cut the voltage in half on each contact, thus providing high arc rupturing capacity and longer contact life.

STARTER ELECTROMAGNETS

The operating principle that makes a magnetic starter different from a manual starter is the use of an electromagnet. Electrical control equipment makes extensive use of a device called a solenoid. This electromechanical device is used to operate motor starters, contactors, relays, and valves. By placing a coil of many turns of wire around a soft iron core, the magnetic flux set up by the energized coil tends to be concentrated; therefore, the magnetic field effect is strengthened. Because the iron core is the path of least resistance to the magnetic lines of force, magnetic attraction concentrates according to the shape of the magnetic core.

There are several different variations in design of the basic solenoid magnetic core and coil. Figure 3–2 shows a few examples. As shown in the solenoid design of Figure 3–2(C), linkage to the movable contacts assembly is obtained through a hole in the movable plunger. The plunger is shown in the open de-energized position.

The center leg of each of the E-shaped magnet cores in Figure 3–2(B) and (C) is made shorter than the outside legs to prevent the magnetic switch from accidentally staying closed (due to residual magnetism) when power is disconnected.

The major components of a three-phase contactor are shown in Figure 3–3. The armature (moving part of the contactor) contains the movable load contacts. These contacts bridge across the

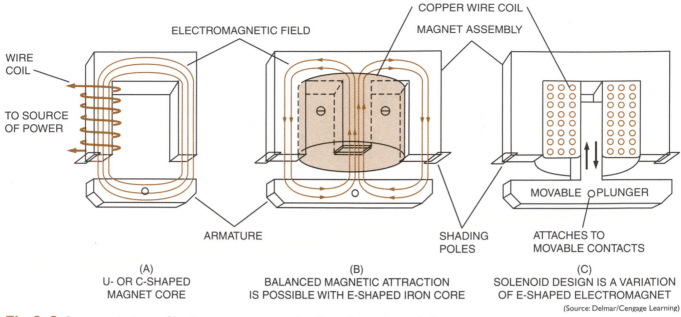

Fig. 3–2 Some variations of basic magnet core and coil configurations of electromagnets.

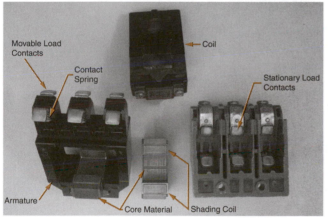

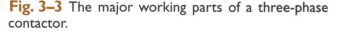

(Source: Delmar/Cengage Learning)

Fig. 3–3 The major working parts of a three-phase contactor.

stationary load contacts when the coil is energized. The core material is placed around the coil to provide a low-reluctance path for magnetic flux, creating a strong magnetic field.

When a magnetic motor starter coil is energized and the armature has sealed in, it is held tightly against the magnetic assembly. A small air gap is always deliberately placed in the center leg, iron circuit. When the coil is de-energized, a small amount of magnetism remains. If it were not for this gap in the iron circuit, the residual magnetism might be enough to hold the movable armature in the sealed-in position. This knowledge can be important to

the electrician when troubleshooting a motor that will not stop.

The OFF or OPEN position is obtained by de-energizing the coil and allowing the force of gravity or spring tension to release the plunger from the magnet body, thereby opening the electrical contacts. The actual contact surfaces of the plunger and core body are machine finished to ensure a high degree of flatness on the contact surfaces so that operation on alternating current is quieter. Improper alignment of the contacting surfaces and foreign matter between the surfaces may cause a noisy hum on alternating-current magnets.

Another source of noise is due to loose laminations. The magnet body and plunger (armature) are made up of thin sheets of iron laminated and riveted together to reduce *eddy currents* and hysteresis, iron losses showing up as heat (see Figure 3–4). Eddy currents are shorted currents induced in the metal by the transformer action of an AC coil. Although these currents are small, they heat up the metal, create an iron loss, and contribute to inefficiency. At one time, laminations in magnets were insulated from each other by a thin, nonmagnetic coating; however, it was found that the normal oxidation of the metallic laminations reduces the effects of eddy currents to a satisfactory degree, thus eliminating the need for a coating.

Eddy currents are induced into solid cores by AC currents. Solid core materials are used with DC circuits.

Laminated cores are used with AC current. Oxide developed between the laminations are resistant to current flow, thus reducing eddy currents.

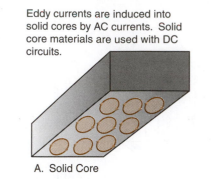

A. Solid Core

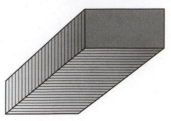

B. Laminated Core

(Source: Delmar/Cengage Learning)

Fig. 3–4 Types of magnet cores.

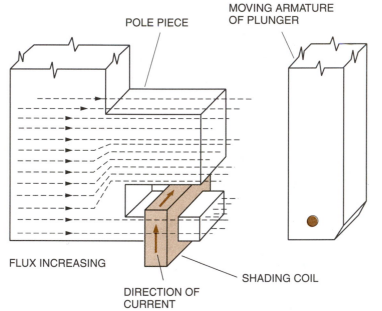

POLE PIECE

MOVING ARMATURE OF PLUNGER

FLUX INCREASING

DIRECTION OF CURRENT

SHADING COIL

(Source: Delmar/Cengage Learning)

Fig. 3–5 Pole face section with shading coil; current is in the clockwise direction for increasing flux (conventional current flow).

SHADED POLE PRINCIPLE

The shaded pole principle is used to provide a time delay in the decay of flux in DC coils and to prevent chatter and wear in the moving parts of AC magnets. Figure 3–5 shows a copper band or short-circuited coil (shading coil) of low resistance connected around a portion of a magnet pole piece. When the flux is increasing in the pole piece from left to right, the induced current in the shading coil is in a clockwise direction.

The magnetic flux produced by the shading coil opposes the direction of the flux of the main field. Therefore, with the shading coil in place, the flux density in the shaded portion of the magnet will be considerably less, and the flux density in the unshaded portion of the magnet will be more than if the shading coil were not in place.

Figure 3–6 shows the magnet pole with the flux direction still from left to right, but now the flux is decreasing in value. The current in the coil is in a counterclockwise direction. As a result, the magnetic flux produced by the coil is in the same direction as the main field flux.

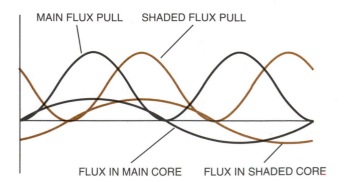

MAIN FLUX PULL   SHADED FLUX PULL

FLUX IN MAIN CORE   FLUX IN SHADED CORE

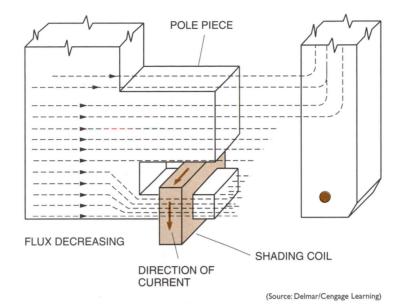

POLE PIECE

FLUX DECREASING

DIRECTION OF
CURRENT

SHADING COIL

(Source: Delmar/Cengage Learning)

Fig. 3–6 Pole face section; current is in the counterclockwise direction for decreasing flux (conventional current flow).

With the shading coil in place, the flux density in the shaded portion of the magnet will be larger and that in the unshaded portion will be less than if the shading coil were not used.

Thus, when the electric circuit of a coil is opened, the current decreases rapidly to zero, but the flux decreases much more slowly because of the action of the shading coil. This produces a more stable magnetic pull on the armature as the AC waveform alternates from maximum to minimum values and helps prevent chatter and AC hum.

Use of the Shading Pole to Prevent Wear and Noise

The attraction of an electromagnet operating on alternating current is pulsating and equals zero twice during each cycle. The pull of the magnet on its armature also drops to zero twice during each cycle. As a result, the sealing surfaces of the magnet tend to separate each time the flux is zero and then contact again as the flux builds up in the opposite direction. This continual making and breaking of contact will result in a noisy starter and wear on the moving parts of the magnet. The noise and wear can be minimized in AC magnets by the use of shaded poles. As shown previously, by shading a pole tip, the flux in the shaded portion lags behind the flux in the unshaded portion. The diagram shows the flux variations with time in both the shaded and unshaded portions of the magnet.

The two flux waves are made as near 90 degrees apart as possible. Pull produced by each flux is also shown. If flux waves are exactly 90 degrees apart, the pulls will be 180 degrees apart and the resultant pull will be constant. However, with fluxes *nearly* 90 degrees apart, the resulting pull varies only a small amount from its average value and never goes through zero. Voltage induced in the shading coil causes flux to exist in the electromagnet, even when the main coil current instantaneously passes through a zero point. As a result, contact between the sealing surfaces of the magnet is not broken and chattering and wear are prevented.

MAGNET COIL

The magnet coil has many turns of insulated copper wire that is tightly wound on a spool. Most coils are protected by a tough epoxy molding that makes them very resistant to mechanical damage, Figure 3–7.

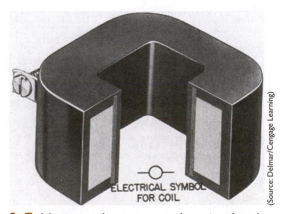

ELECTRICAL SYMBOL
FOR COIL

(Source: Delmar/Cengage Learning)

Fig. 3–7 Magnet coil cutaway to show insulated copper wire wound on a spool and protected by a molding.

Above-Normal Voltage Effects

The manufacturer makes available coils of practically any desired control voltage. Some starters are designed with dual-voltage coils.

NEMA standards require that the magnetic switch operates properly at varying control voltages from a high of 110 percent to a low of 85 percent of the rated coil voltage. This range of required operation is then designed by the manufacturer. It ensures that the coil will withstand elevated temperatures at voltages up to 10 percent over rated voltage and that the armature will pick up and seal in, even though the voltage may drop to 15 percent under the rating. Normally, power company service voltages are very reliable. Plant voltages may vary due to other loaded, operating machines and other reasons affecting the electrical distribution system. If the voltage applied to the coil is too high, the coil will draw too much current. Excessive heat will be produced and may cause the coil insulation to break down and burn out. The magnetic pull will be too high and will cause the armature to slam in with too much force. The magnet pole faces will wear faster, leading to a shortened life for the controller. In addition, reduced contact life may result from excessive contact bounce.

Below-Normal Voltage Effects

Undervoltage produces low coil currents, thereby reducing the magnetic pull. On common starters the magnet may pick up (start to move) but not seal. The armature must seal against the pole faces of the magnet to operate satisfactorily. Without this condition, the coil current will not fall to the sealed value because the magnetic circuit is open, decreasing impedance (AC resistance). Because the coil is not designed to continuously carry a current greater than its sealed current, it will quickly get very hot and burn out. The armature will also chatter. In addition to the noise, there is excessive wear on the magnet pole faces. If the armature does not seal, the contacts may touch but not close with enough pressure, creating another problem. Excessive heat, with arcing and possible welding of the contacts, will occur as the controller attempts to carry a motor-starting current with insufficient contact pressure.

POWER (OR MOTOR) CIRCUIT OF THE MAGNETIC STARTER

The number of poles refers to the number of power contacts determined by the electrical supply service. For example, in a three-phase,

Fig. 3–8 Motor starter with overload relay.

(Source: Delmar/Cengage Learning)

three-wire system, a three-pole starter is required. The power circuit of a starter includes the main stationary and movable contacts and the thermal unit or heater unit of the overload relay assembly. This can be seen in Figure 3–8 (and in Figure 3–1, less the thermal overload relay assembly).

CONTACTS

Contact surfaces are generally a silver alloy for better electrical conductivity. Surfaces also have a mechanical "wiping" action from the point of contact to the sealing position when the starter is activated. This helps to keep the contacts clean. A good operating starter has no contact bounce and has superior resistance to welding and arc erosion. Figure 3–9 shows the 45-degree wedge action design of the contact.

MOTOR OVERHEAT

An electric motor does not know enough to quit when the load gets too much for it. It keeps going until it burns out. If a motor is subjected over a period of time to internal or external heat levels that are high enough to destroy the

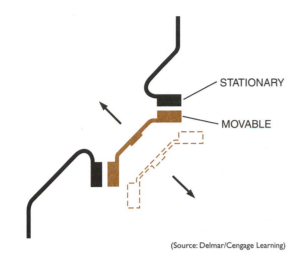

(Source: Delmar/Cengage Learning)

Fig. 3–9 45° wedge action design contacts.

(EFFECTS OF HIGH AMBIENT TEMPERATURE)

insulation on the motor windings, it will fail—burn out.

A solution to this problem *might* be to install a larger motor whose capacity is in excess of the normal horsepower required. This is not very practical because there are other reasons for a motor to overheat besides excess loads. A motor will run cooler in snowy winter weather than in hot tropical summer weather. A high, surrounding air temperature (*ambient temperature*) has the same effect as higher-than-normal current flow through a motor—it tends to deteriorate the insulation on the motor windings.

High ambient temperature is also created by *poor ventilation* of the motor. Motors must get rid of their heat so any obstructions to this must

be avoided. High inrush currents of *excessive starting* create heat within the motor. The same is true with *starting heavy loads*. There are several other related causes that generate heat within a motor such as *voltage unbalance, low voltage,* single phasing. In addition, when the rotating member of the motor will not turn (a condition called *locked rotor*), heat is generated. It must be impossible to design a motor that will adjust itself for all the various changes in total heat that can occur. Some device is needed to protect a motor against expected overheating.

Motor Overload Protection

The ideal overload protection for a motor is an element with current sensing properties very similar to the heating curve of the motor. This would act to open the motor circuit when full load is exceeded. The operation of the protective device is ideal if the motor is allowed to carry small, short, and harmless overloads but is quickly disconnected from the line when an overload has persisted too long. Dual-element, or time-delay, fuses may provide motor overload protection, but they have the disadvantage of being nonrenewable and must be replaced.

An overload relay is added to the magnetic switch that is shown in Figure 3–1. Now it is called a motor starter. The overload relay as sembly is the heart of motor protection. A typical solid-state overload relay is shown in Figure 3–10. The motor can do no more work than the overload relay permits. Like the

Fig. 3–10 Three-phase motor starter with solid-state overload relay. (*Courtesy of Schneider Electric*)

dual-element fuse, the overload relay has characteristics permitting it to hold in during the motor accelerating period when the inrush current is drawn. Nevertheless, it still provides protection on small overloads above full-load current when the motor is running. Unlike the fuse, the overload relay can be reset. It can withstand repeated trip and reset cycles without need of replacement. It is emphasized that the overload relay does *not* provide short-circuit protection. This is the function of overcurrent protective equipment like fuses and circuit breakers, generally located in the disconnecting switch enclosure.

Current draw by a motor is a convenient and accurate measure of the motor load. Motor heating is a combination of the amount of load and the ambient temperature. The overload relay employs some method of sensing motor current. The relay's ability to sense ambient temperature depends on the type of overload relay. Overload relays contain two separate sections; a current sensing section that is connected in series with the motor, and a contact section that contains a set of normally closed contacts connected in series with the coil of the motor starter, Figure 3–11. The contacts of the overload relay are small and cannot carry the load current of a motor. For this reason the contacts are not connected in the motor circuit, but in the control circuit. In the case of a magnetic motor starter, if the motor should become overloaded, the normally closed contacts will open and de-energize the motor starter. The load contacts of the motor starter are utilized to disconnect the motor from the power line. In a manual starter, the overload relay trips a mechanical latch causing the load contacts to open and disconnect the motor from the power line.

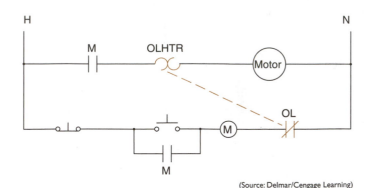

(Source: Delmar/Cengage Learning)

Fig. 3–11 An overload relay has a current sensing section that is connected in series with the motor and a contact section that is connected in series with the motor starter coil.

To provide *overload* or *running protection* to keep a motor from overheating, overload relays are used on starters to limit the amount of current drawn to a predetermined value. The *NEC®* and local electrical codes determine the size of protective overload relays and heating elements, which are properly sized to the motor.

The controller is normally installed in the same room or area as the motor. This makes it subject to the same ambient temperature as the motor. The tripping characteristic of the proper thermal overload relay will then be affected by room temperature exactly as the motor is affected. This is done by selecting a thermal relay element (from a chart provided by the manufacturer) that trips at the danger temperature for the motor windings. When excessive current is drawn, the relay de-energizes the starter and stops the motor.

Overload relays can be classified as being either thermal, electronic, or magnetic. *Magnetic overload relays* react only to current excesses and are not affected by temperature. As the name implies, *thermal overload relays* depend on the rising ambient temperature and temperatures caused by the overload current to trip the overload mechanism.

Thermal overload relays can be further subdivided into two types—melting alloy and bimetallic.

Electronic or *solid-state overload relays*, Figure 3–10, provide the combination of high-speed trip, adjustability, and ease of installation. They are ideal in many precise applications.

Melting Alloy Thermal Units

The melting alloy assembly consisting of a heater element and solder pot is shown in Figure 3–12. The solder pot holds the ratchet wheel in one position. Excessive motor current

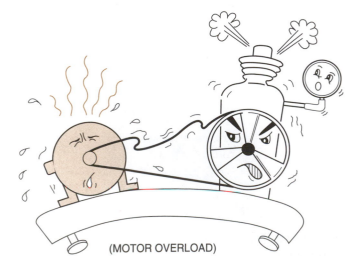

(MOTOR OVERLOAD)

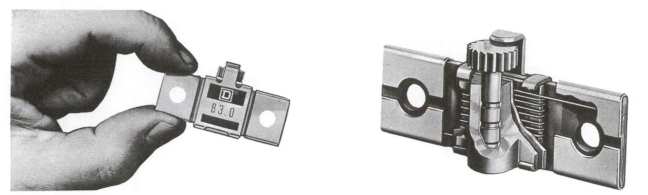

Fig. 3–12 Melting alloy type overload heater. Cutaway view (right) shows construction of heater. *(Courtesy of Schneider Electric)*

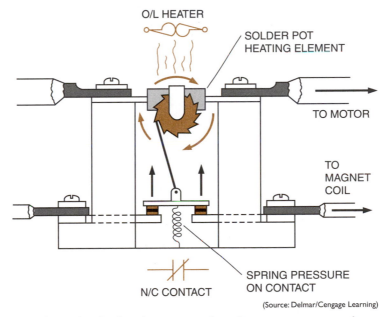

O/L HEATER

SOLDER POT
HEATING ELEMENT

TO MOTOR

TO
MAGNET
COIL

SPRING PRESSURE
ON CONTACT

N/C CONTACT

(Source: Delmar/Cengage Learning)

Fig. 3–13 Melting alloy thermal overload relay. A spring pushes the contact open as heat melts, allowing the ratchet wheel to turn freely. Note the electrical symbols for the overload heater and normally closed control contact.

passes through the heater element and melts an alloy solder pot. Because the ratchet wheel is then free to turn in the molten pool, it trips a set of normally closed contacts that is in the starter control circuit; this stops the motor, Figure 3–13. A cooling-off period is required to allow the solder pot to become solid again before the overload relay can be reset and motor service is restored.

Melting alloy thermal units are interchangeable. They have a one-piece construction that ensures a constant relationship between the heater element and the solder pot. As a result, this unit can be factory calibrated to make it virtually tamperproof in the field. These important features are not possible with any other type of overload relay construction. To obtain appropriate tripping current for motors of different

sizes, a wide selection of interchangeable thermal units (heaters) is available. They give exact overload protection to motors of different full-load current ratings. Thermal units are rated in amperes and are selected on the basis of motor full-load current. For most accurate overload heater selection, the manufacturer publishes a number of rating tables keyed to the controller in which the overload relay is used. The units are easily mounted into the overload relay assembly and held in place with two screws. Being in series with the motor circuit, the motor will not operate without these heating elements installed in the starter. A typical overload heater chart is shown in Figure 3–14. A single-phase solder melting type overload relay is shown in Figure 3–15.

Overload heater selection for NEMA starter sizes 00 - 1. Heaters are calibrated for 115% of motor full load current. For heaters that correspond to 125% of motor full load current use the next size larger heater.

Heater Code	Motor Full Load Current	Heater Code	Motor Full Load Current	Heater Code	Motor Full Load Current
XX01	.25 - .27	XX18	1.35 - 1.47	XX35	6.5 - 7.1
XX02	.28 - .31	XX19	1.48 - 1.62	XX36	7.2 - 7.8
XX03	.32 - .34	XX20	1.63 - 1.78	XX37	7.9 - 8.5
XX04	.35 - .38	XX21	1.79 - 1.95	XX38	8.6 - 9.4
XX05	.39 - .42	XX22	1.96 - 2.15	XX39	9.5 - 10.3
XX06	.43 - .46	XX23	2.16 - 2.35	XX40	10.4 - 11.3
XX07	.47 - .50	XX24	2.36 - 2.58	XX41	11.4 - 12.4
XX08	.51 - .55	XX25	2.59 - 2.83	XX42	12.5 - 13.5
XX09	.56 - .62	XX26	2.84 - 3.11	XX43	13.6 - 14.9
XX10	.63 - .68	XX27	3.12 - 3.42	XX44	15.0 - 16.3
XX11	.69 - .75	XX28	3.43 - 3.73	XX45	16,4 - 18.0
XX12	.76 - .83	XX29	3.74 - 4.07	XX46	18.1 - 19.8
XX13	.84 - .91	XX30	4.08 - 4.39	XX47	19.9 - 21.7
XX14	.92 - 1.00	XX31	4.40 - 4.87	XX48	21,8 - 23.9
XX15	1.01 - 1.11	XX32	4.88 - 5.3	XX49	24.0 - 26.2
XX16	.1.12 - 1.22	XX33	5.4 - 5.9		
XX17	1.23 - 1.34	XX34	6.0 - 6.4		

Fig. 3–14 Typical overload heater chart.

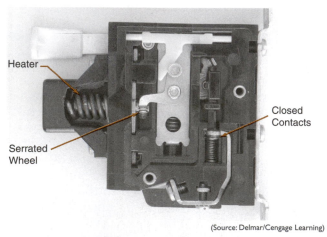

Heater

Serrated
Wheel

Closed
Contacts

(Source: Delmar/Cengage Learning)

Fig. 3–15 Solder melting type overload relay.

Bimetallic Overload Relays

Bimetal strip type overload relays operate on the principle that metals expand and contract with a change of temperature. A bimetal strip is made by bonding two dissimilar metals together, Figure 3–16. Since the metals are dissimilar, they will expand and contract at different rates. When the strip is exposed to a different temperature, it will warp or bend. Bimetal strip type overload relays have a couple of features that are different from the solder pot type. One difference is that the trip current can be adjusted by means of an adjustment knob, Figure 3–17. A second difference is that most bimetal strip type overload relays can be set for automatic or manual reset. In the example shown in Figure 3–18, a spring is moved between manual or automatic position.

In the automatic reset position, the relay contacts, after tripping, will automatically reclose after the relay has cooled down. This is an advantage when the reset button is hard to reach. Automatic reset overload relays are not normally recommended when used with automatic (two-wire) pilot control devices. With

A bimetal strip is made by bonding two dissimilar metals together

A bimetal strip will warp or bend due to a change in temperature.

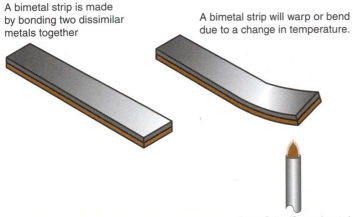

(Source: Delmar/Cengage Learning)

Fig. 3–16 A bimetal strip will bend or warp with a change of temperature.

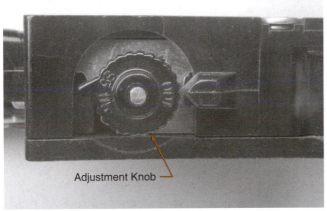

Adjustment Knob

(Source: Delmar/Cengage Learning)

Fig. 3–17 Bimetal strip type overload relays permit the trip rating to be adjusted.

(Source: Delmar/Cengage Learning)

Fig. 3–18 The bimetal strip type overload relay can be set for manual or automatic reset by changing the position of the spring.

this control arrangement, when the overload relay contacts reclose after an overload trip, the motor will restart. Unless the cause of the overload has been removed, the overload relay will trip again. This event will repeat. Soon the motor will burn out because of the accumulated heat from the repeated high inrush and the overload current. (An overload indicating light or alarm can be installed to catch attention before this happens.)

CAUTION: The more important point to consider is the possible danger to personnel. This unexpected restarting of the machine may find the operator or electrician in a hazardous situation as attempts are made to find out why this machine has stopped. *The NEC® prohibits this later installation.*

Most bimetallic relays can be adjusted to trip within a range of 85 to 115 percent of the nominal trip rating of the heater unit. This feature is useful when the recommended heater size may result in unnecessary tripping, while the next larger size will not give adequate protection. Ambient temperatures affect thermally operated overload relays.

This ambient-compensated bimetallic overload relay is recommended for installations when the motor is located in a different ambient temperature from the motor starter. If the controller is located in a changing temperature, the overload relay can be adjusted to compensate for these temperature changes. This thermal overload relay is always affected by the surrounding temperature. If a standard thermal overload relay were used, it would not trip consistently at the same level of motor current whenever the controller temperature changed.

A bimetal strip type overload relay is shown in Figure 3–19. The heater element is located

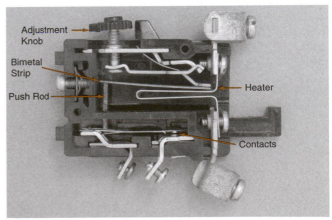

Adjustment Knob

Bimetal Strip

Push Rod

Heater

Contacts

(Source: Delmar/Cengage Learning)

Fig. 3–19 Bimetal strip type overload relay.

closed to the bimetal strip. The temperature of the heater element is depends on the amount of motor current. If the heat becomes great enough, the bimetal strip will warp and cause a push-rod to open a set of normally closed contacts. After the relay has tripped, it is necessary to wait some period of time before the contacts can be reclosed. The cool down time is necessary to permit the bimetal strip to return to its normal position.

When it is necessary to provide overload protection for motors that draw large amounts of current, it is not practical to insert a large heater element in series with the motor leads. Consider, for example, a motor that has a full-load running current of 460 amperes. Overload relays are not designed to handle this amount of current. When overload protection must be provided for a large motor of this type, current transformers are used to reduce the motor current to a lower value, Figure 3–20. In this example, three current transformers with a ratio of 600:5 are used to reduce the motor current. These transformers are designed so that when 600 amperes of current flow through the primary winding, a current of 5 amperes will flow through the secondary winding in a shorted condition. The primary winding is the line supplying power to the motor. The secondary winding is connected to the heater of an overload relay. A simple ratio can be used to determine the secondary current that should flow when 460 amperes of current flow to the motor.

$$\frac{600}{460} = \frac{5}{X}$$

$$600X = 2300 \ (460 \times 5)$$
$$X = 3.833 \text{ amperes } (2300 \div 600)$$

(Source: Delmar/Cengage Learning)

Fig. 3–21 Size 5 starter with current transformers that monitor the load current of each line.

The overload heaters would be sized for a motor with a full-load current draw of 3.833 amperes.

A size 5 starter with current transformers is shown in Figure 3–21. The transformers monitor the load current of each line and have a ratio of 300:5 amperes.

Magnetic Overload Relays

The magnetic overload relay coil is connected in series directly with the motor or is indirectly connected by current transformers (as in circuits with large motors). As a result, the coil of the magnetic relay must be wound with wire large enough in size to pass the motor current. These overload relays operate by current intensity and not heat.

Magnetic overload relays are used when an electrical contact must be opened or closed as the actuating current rises to a certain value. In some cases, the relay may also be used so that it is actuated when the current falls to a certain value. Magnetic overload relays are used to protect large motor windings against continued overcurrent. Typical applications are to stop a material conveyor when conveyors ahead become overloaded, and to limit torque reflected by the motor current.

All overload relays must allow some period of time before tripping to allow the motor to start. Magnetic devices are very fast acting and can trip due to motor inrush current. Magnetic type overload relays employ a timing device called a *dashpot* permit the motor to start. A dashpot timer is basically a container, a piston, and a shaft, Figure 3–22. The piston is placed inside the container and the container is filled with a special type of oil called *dashpot oil*. Dashpot oil

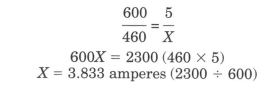

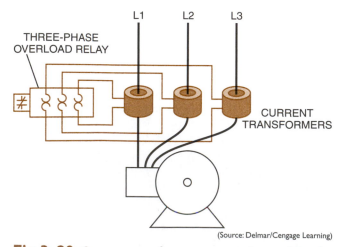

L1 L2 L3

THREE-PHASE OVERLOAD RELAY

CURRENT TRANSFORMERS

(Source: Delmar/Cengage Learning)

Fig. 3–20 Current transformers are used to reduce the current to the overload relay.

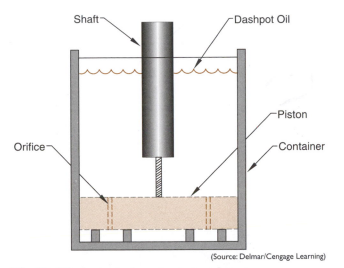

Shaft
Dashpot Oil
Orifice
Piston
Container

(Source: Delmar/Cengage Learning)

Fig. 3–22 Basic design of a dashpot timer.

Orifice

(Source: Delmar/Cengage Learning)

Fig. 3–23 The size of the orifice hole can be adjusted to change the time setting of the dashpot timer.

maintains a constant viscosity over a wide range of temperatures. The type and viscosity of oil used are two of the factors that determine the amount of time delay for the timer. The other factor is setting the opening of the orifice holes in the piston, Figure 3–23. Orifice holes permit the oil to flow through the piston as it rises through the oil. The opening of the orifice holes can be set by adjusting a sliding valve on the piston.

The dashpot overload relay contains a coil that is connected in series with the motor,

Figure 3–24. As current flows through the coil, a magnetic field is developed around the coil. The strength of the magnetic field is proportional to the motor current. This magnetic field draws the shaft of the dashpot timer into the coil. The shaft's movement is retarded by the fact that the piston must displace the oil in the container. If the motor is operating normally, the motor current will drop to a safe level before the shaft is drawn far enough into the coil to open the normally closed contact, Figure 3–25. If the motor is overloaded, however, the magnetic field will be strong enough to continue drawing the shaft into the coil until it opens the overload contact. When power is disconnected from the motor, the magnetic field collapses and the piston returns to the bottom of the container. Check valves permit the piston to return to the bottom of the container almost immediately when motor current ceases.

Dashpot overloads generally provide some method that permits the relay to be adjusted for different full load current values. To make this adjustment, the shaft, is connected to a threaded rod. This permits the shaft to be lengthened or shortened inside the coil. The greater the length of the shaft the less current required to draw the shaft into the coil far enough to open the contacts. A nameplate on the coil lists the different current settings for a particular overload relay. The adjustment is made by moving the shaft until the line on the shaft that represents the desired current is flush with the top of the dashpot container, Figure 3–26. A dashpot overload relay is shown in Figure 3–27.

Instantaneous Trip Current Relays

Instantaneous trip current relays are used to take a motor off the line as soon as a predetermined load condition is reached. For example, when a blockage of material on a woodworking machine causes a sudden high current, an instantaneous trip relay can cut off the motor quickly. After the cause of the blockage is removed, the motor can be restarted immediately because the relay resets itself as soon as the overload is removed. This type of relay is also used on conveyors to stop the motor before mechanical breakage results from a blockage.

The instantaneous trip current relay does not have the inverse time characteristic. Thus, it must not be used in ordinary applications requiring an overload relay. The instantaneous trip current relay should be considered as a special-purpose relay.

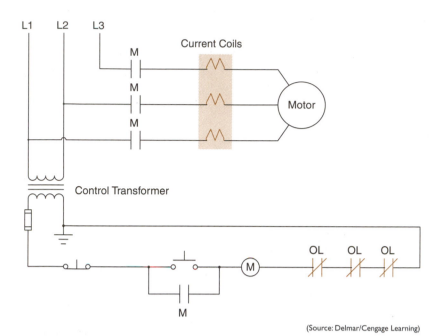

(Source: Delmar/Cengage Learning)

Fig. 3–24 The current coils of dashpot overload relays are connected in series with the motor.

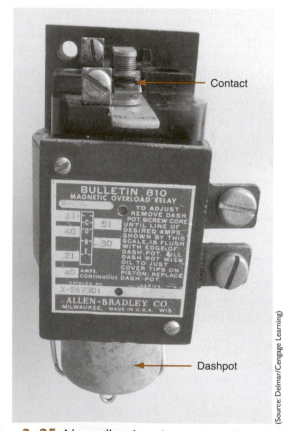

(Source: Delmar/Cengage Learning)

Fig. 3–25 Normally closed contact of a dashpot overload relay.

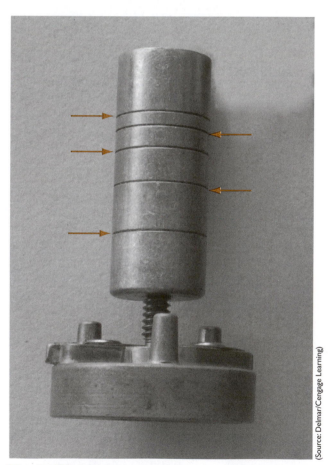

(Source: Delmar/Cengage Learning)

Fig. 3–26 Lines on the shaft are used to adjust the current setting.

The operating mechanism of the trip relay in Figure 3–28 consists of a solenoid coil through which the motor current flows. There is a movable iron core within the coil. Mounted on top of the solenoid frame is a snap-action precision switch that has connections for either a normally open or a normally closed contact. The motor current exerts a magnetic pull upward on the iron core. Normally, however, the pull is not sufficient to lift the core. If an overcurrent condition causes the core to be lifted, the snap-action precision switch is operated to trip the control contact of the relay.

The tripping value of the relay can be set over a wide range of current ratings by moving

the plunger core up and down on the threaded stem. As a result, the position of the core in the solenoid is changed. By lowering the core, the magnetic flux is weakened and a higher current is required to lift the core and trip the relay.

Number of Overload Relays Needed to Protect a Motor

The *NEC*® requires that each phase of a motor be protected by an overload relay. There are two basic ways of protecting a three-phase motor. Some starters use three separate single-phase overload relays. When this is done, the normally closed overload contacts are all connected in series so that any relay can open the circuit to the starter coil in the event of an overload. Other starters employ a three-phase overload relay, Figure 3–29. A three-phase overload relay contains three heaters that are connected in series with the motor, but only one set of contacts. An overload on any phase will open the contacts and de-energize the motor starter.

A balanced supply voltage must be maintained for all polyphase load installations.

A single-phase load on a three-phase circuit can produce serious unbalanced motor currents. A large three-phase motor on the same feeder with a small three-phase motor may not be protected if a single-phase condition occurs, Figure 3–30.

A defective line fuse, an open "leg" through a circuit breaker, a loose or broken wire anywhere in the conduit system or in a motor lead can result in single-phase operation. This will show up as a sluggish, hot-running motor. The motor will not start at all but will produce a distinct magnetic hum when it is energized. The three-phase motor may continue to operate (at reduced

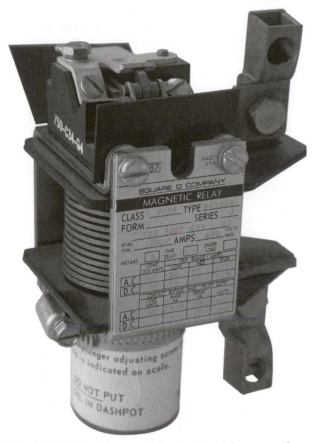

Fig. 3–27 Dashpot overload relay. *(Courtesy of Square D Co.)*

Fig. 3–28 Instantaneous trip current relay. *(Courtesy of Rockwell Automation)*

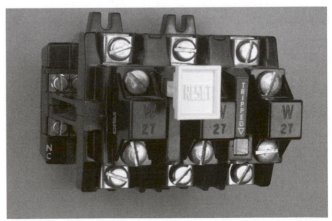

(Source: Delmar/Cengage Learning)

Fig. 3–29 A three-phase overload relay contains three separate heaters, but only one set of contacts.

torque) when single phasing occurs. But once stopped, it will not restart. This is also a sign of a single-phase condition in a three-phase motor.

Unbalanced single-phase loads on three-phase panel boards must be avoided. Problems may occur on distribution systems where one or more large motors may feed back power to smaller motors under open-phase conditions.

OVERLOAD CONTACTS

Although all overload relays contain a set of normally closed contacts, some manufacturers also add a set of normally open contacts as well. These two sets of contacts are either in the form of a single-pole double-throw switch or as two separate contacts. The single-pole double-throw switch arrangement will contain a common terminal (C), a normally closed terminal (NC), and a normally open terminal (NO). There are several reasons for adding the normally open set of contacts. The starter shown if Figure 3–31 uses the normally closed section to disconnect the motor starter in the event of an overload, and the normally open section to turn on a light to indicate that the overload has tripped.

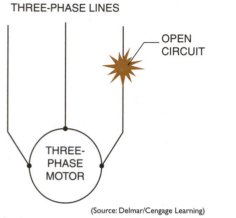

THREE-PHASE LINES

OPEN CIRCUIT

THREE-PHASE MOTOR

(Source: Delmar/Cengage Learning)

Fig. 3–30 A single-phase condition: two high currents, one zero current.

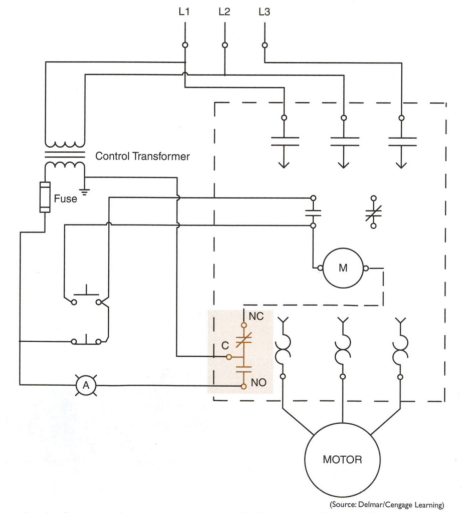

(Source: Delmar/Cengage Learning)

Fig. 3–31 Some overload relays contain a common terminal (C), a normally closed contact terminal (NC), and a normally open contact terminal (NO).

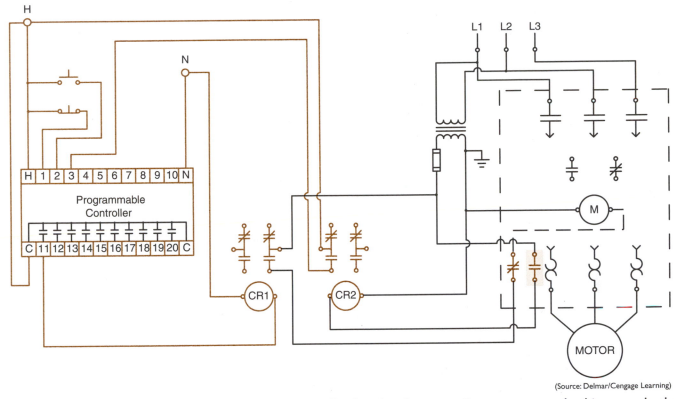

(Source: Delmar/Cengage Learning)

Fig. 3–32 Some overload relays contain both a normally closed and a normally open contact. In this example the normally open contact is employed to signal a programmable logic controller that the motor is tripped.

Another common use for the normally open set of contacts on an overload relay is to provide an input signal to a programmable logic controller (PLC). If the overload trips, the normally closed set of contacts will open and disconnect the starter coil from the line. The normally open set of contacts will close and provide a signal to the input of the PLC, Figure 3–32. Notice that two interposing relays, CR1 and CR2, are used to separate the programmable logic controller and the motor starter. This is often done for safety reasons. The control relays prevent more than one source of power from entering the starter or programmable controller. Note that the starter and programmable controller each have a separate power source. If the power were to be disconnected from the starter, for example, it could cause an injury if the power from the programmable controller were connected to any part of the starter.

AC MAGNETIC STARTERS

There are several manufacturers of magnetic motor starters, and most have a different appearance. Several starters are shown in Figure 3–33. The starter shown in Figure 3–33(A) contains a three-phase overload relay. The reset handle is shown below the overload relay. The starter in Figure 3–33(B) contains three separate single-phase overload relays. The load contacts of each relay must be connected in series. Each overload relay contains a separate reset lever. The starter shown in Figure 3–33(C) is part of a module designed to plug into a motor control center (MCC). The module contains a fused disconnect, control transformer, motor starter, and terminal strips for making connections. Note that the fused disconnect contains only two fuses. A jumper wire replaces the center fuse. This is common on delta connected three-phase systems with a B phase ground. The *NEC®* does not require fuse protection for a grounded phase. The starter shown in Figure 3–33(D) is the same as the starter shown in the module in photograph (C). Note the two normally open and two normally closed auxiliary contacts added to the sides of the starter. Many motor starters have the capability of adding additional auxiliary contacts to be used in the control circuit.

Starter Sizes

Magnetic starters are available in many sizes as shown in Table 3–1. Each size has been assigned a horsepower rating that applies when the motor used with the starter is operated for

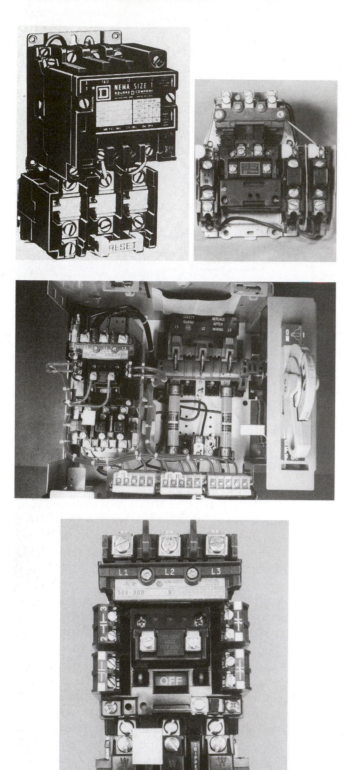

Fig. 3–33 Three-phase AC magnetic motor starters. *(Part A, courtesy of Schneider Electric)*

normal starting duty. All starter ratings comply with the NEMA Standards. The capacity of a starter is determined by the size of its contacts and the wire cross-sectional area. The size of the power contacts is reduced when the voltage is doubled because the current is halved for the same power ($P = 1 \times E$). Power circuit contacts handle the motor load.

Three-pole starters are used with motors operating on three-phase, three-wire AC systems. Two-pole starters are used for single-phase motors.

The number of *poles* refers to the power contacts, or the motor load contacts, and does not include control contacts for control circuit wiring.

AC COMBINATION STARTERS

The circuit breakers and fuses of the motor feeders and branch circuits are normally selected for overcurrent, short-circuit, or ground-fault protection.

With minor exceptions, the *NEC®* and some local codes also require that every motor have a disconnect means. This means may be an attachment cord cap and receptacle, a nonfusible isolation disconnect safety switch, a fusible disconnect motor switch, or a combination starter. A combination starter (Figure 3–34) consists of an across-the-line starter and a disconnect means wired together in a common enclosure. Combination starters may have a blade-type disconnect switch, either fusible or nonfusible, or a thermal-magnetic trip circuit breaker. The starter may be controlled remotely with push buttons or selector switches, or these devices may be installed in the cover of the starter enclosure. The combination starter takes little mounting space and makes compact electrical installation possible.

A combination starter provides safety for the operator because the cover of the enclosure is interlocked with the external, operating handle of the disconnecting means. The door cannot be opened while the disconnecting means is closed. When the disconnecting means is open, all parts of the starter are accessible; however, the hazard is reduced because the readily accessible parts of the starter are not connected to the power line. This safety feature is not available on separately enclosed starters. In addition, the starter enclosure is provided with a means for

Table 3–1 Standard Motor Starter Sizes and Ratings

| *NEMA SIZE (EEMAC SIZE) | LOAD VOLTS | MAXIMUM HORSEPOWER RATING – NONPLUGGING AND NONJOGGING DUTY | | *NEMA SIZE (EEMAC SIZE) | LOAD VOLTS | MAXIMUM HORSEPOWER RATING – NONPLUGGING AND NONJOGGING DUTY | |
		SINGLE PHASE	POLY-PHASE			SINGLE PHASE	POLY-PHASE
00	115	1/3	...	3	115	7 1/2	...
	200 (208)	...	1 1/2		200 (208)	...	25
	230	1	1 1/2		230	15	30
	460	...	2		460	...	50
	575	...	2		575	...	50
0	115	1	...	4	200 (208)	...	40
	200 (208)	...	3		230	...	50
	230	2	3		460	...	100
	460	...	5		575	...	100
	575	...	5				
1	115	2	...	5	200 (208)	...	75
	200 (208)	...	7 1/2		230	...	100
	230	3	7 1/2		460	...	200
	460	...	10		575	...	200
	575	...	10				
2	115	3	...	6	200 (208)	...	150
	200 (208)	...	10		230	...	200
	230	7 1/2	15		460	...	400
	460	...	25		575	...	400
	575	...	25				
				7	230	...	300
					460	...	600
					575	...	600
				8	230	...	450
					460	...	900
					575	...	900

*In Canada, the Electrical and Electronic Manufacturers Association of Canada uses the same sizes, and they are defined as EEMAC size numbers.

padlocking the disconnect in the OFF position. Controller enclosures are available for every purpose and application.

Protective Enclosures

The selection and installation of the correct enclosure can contribute to useful, safe service and freedom from trouble in operating electromagnetic control equipment.

An enclosure is the surrounding controller case, cabinet, or box. Generally, this electrical equipment is enclosed for one or more of the following reasons:

A. To shield and protect workers and other personnel from accidental contact with electrically live parts, thereby preventing electrocution.

B. To prevent other conducting equipment from coming into contact with live electrical parts, thereby preventing unnecessary electrical outages and indirectly protecting per sonnel from electrical contact.

C. To protect the electrical controller from harmful atmospheric or environmental conditions, such as the presence of dust or moisture, to prevent corrosion and interference of operation.

D. To contain the electrical arc of switching within the enclosure, to prevent explosions and fires that may occur with flammable gases or vapors within the area.

Fig. 3–34 Combination starter with circuit breaker, control transformer, control circuit fuse, and motor starter. The door cannot be opened unless the circuit breaker is open. *(Courtesy Square D Co.)*

You may readily understand why some form of enclosure is necessary and required. The most frequent requirement is usually met by a general-purpose, sheet steel cabinet. The conduit is installed with locknuts and bushings. The presence of dust, moisture, or explosive gases often makes it necessary to use a special enclosure to protect the controller from corrosion or the surrounding equipment from possible explosions. Conduit access is through threaded openings, hubs, or flanges. In selecting and installing control apparatus, it is necessary to carefully consider the conditions under which the apparatus must operate. There are many applications where a general-purpose sheet steel enclosure does not give sufficient protection.

Watertight and dusttight enclosures are used for the protection of control apparatus. Dirt, oil, or excessive moisture are destructive to insulation and frequently form current-carrying paths that lead to short circuits or grounded circuits.

Special enclosures for hazardous locations are used for the protection of life and property. Explosive vapors or dusts exist in some departments of many industrial plants as well as in grain elevators, refineries, and chemical plants. The *NEC®* and local codes describe hazardous locations. The Underwriters' Laboratories have defined the requirements for protective enclosures according to the hazardous conditions. The NEMA has standardized enclosures from these requirements. Some examples are as follows.

General-Purpose Enclosures (NEMA 1) These enclosures are constructed of sheet steel, and are designed primarily to prevent accidental contact with live parts. Covers have latches with provisions for padlocking, Figure 3–35. Enclosures are intended for use indoors, in areas where unusual service conditions do not exist. They do provide protection from light splash, dust, and falling debris such as dirt.

Watertight Enclosures (NEMA 4) These enclosures are made of cast construction or of sheet metal of suitable rigidity and are designed to pass a hose test with no leakage of water. Watertight enclosures are suitable for outdoor applications on ship docks, in dairies, in breweries, and other locations where the apparatus is subjected to dripping or splashing liquids, Figure 3–36. Enclosures that meet requirements for more than one NEMA type may be designated by a combination of type numbers, for example, Type 3-4, dusttight and watertight.

Fig. 3–35 General-purpose enclosure (NEMA 1). *(Courtesy of Schneider Electric)*

Fig. 3–36 Watertight enclosure (NEMA 4). *(Courtesy of Schneider Electric)*

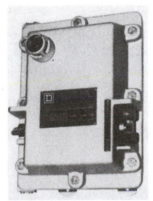

Fig. 3–37 Hazardous location enclosure (NEMA 7). *(Courtesy of Schneider Electric)*

Table 3–2 Motor Controller Enclosure Selection

1. General-purpose, incidental contact with enclosed equipment.
3. Dusttight, raintight
3R. Rainproof, sleet resistant
4. Watertight, dusttight
4X. Watertight, corrosive resistant
6P. Submersible
7. Hazardous locations–Class 1
9. Hazardous locations–Class 2
12. Industrial use, dusttight
13. Oiltight, dusttight

Table 3–2 assists in the proper selection of motor control enclosures. You may note that many enclosures are overlapping in their protection abilities.

NOTE: Applicable and enforced national, state, or local electrical codes and ordinances should be consulted to determine the safe way to make any installation.

VACUUM CONTACTORS

Vacuum contactors enclose their load contacts in a sealed vacuum chamber. A metal bellows connected to the movable contact permits it to move without breaking a seal, Figure 3–38. Sealing contacts inside a vacuum chamber permits them to switch higher voltages with a relatively narrow space between the contacts without establishing an arc. The contactor shown in Figure 3–38 is rated 12 kV and 400 amperes.

Dusttight Enclosures (NEMA 12) These enclosures are constructed of sheet steel and are provided with cover gaskets to exclude dust, lint, dirt, fibers, and flyings. Dusttight enclosures are suitable for use in steel and knitting mills, coke plants, and similar locations where nonhazardous dusts are present. Mounting is by means of outside flanges or mounting feet.

Hazardous Locations (NEMA 7) Class 1 enclosures are designed for use in hazardous locations where atmospheres containing gasoline, petroleum, naphtha, alcohol, acetone, or lacquer solvent vapors are present or may be encountered. Enclosures are heavy, gray iron castings, machined to provide a metal-to-metal seal, Figure 3–37.

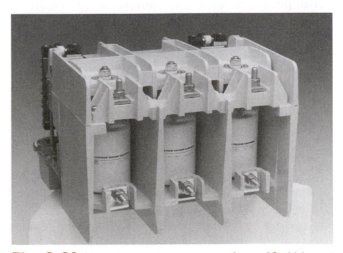

Fig. 3–38 Vacuum contactor rated at 12 kV and 400 amperes. *(Courtesy of GEC Alsthom)*

An electric arc is established when the voltage is high enough to ionize the air molecules between staionary and movable contacts. Medium-voltage contactors are generally large because they must provide enough distance between the contacts to break the arc path. Some medium-voltage contactors use arc suppressors, arc shields, and oil immersion to quench or prevent an arc. Vacuum contactors operate on the principle that if there is no air surrounding the contact, there is no ionization path for the establishment of an arc. Vacuum contactors are generally much smaller in size than other types of medium-voltage contactors.

NEMA AND IEC

NEMA is an acronym for National Electrical Manufacturers Association. Likewise, IEC is an acronym for International Electrotechnical Commission. The IEC establishes standards and ratings for different types of equipment just as NEMA does. The IEC, however, is more widely used throughout Europe than in the United States. Many equipment manufacturers are now beginning to specify IEC standards for their products in the United States also. The main reason is that much of the equipment produced in the United States is also marketed in Europe. Many European companies will not purchase equipment that is not designed with IEC standard equipment.

Although the IEC uses some of the same ratings as NEMA-rated equipment, there is often a vast difference in the physical characteristics of the two. Two sets of load contacts are shown in Figure 3–39. The load contacts on the left are used in a NEMA-rated 00 motor starter, whereas the ones on the right are used in an IEC-rated 00 motor starter. Notice that the surface area of the NEMA-rated contacts is much larger than the IEC-rated contacts. This permits the NEMA-rated starter to control a much higher current than the IEC starter. In fact, the IEC-rated 00 starter contacts are smaller than the contacts of a small eight-pin control relay, Figure 3–40.

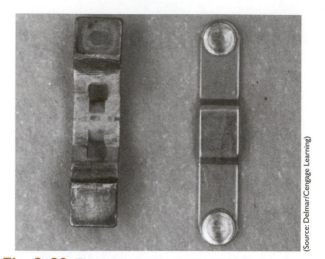

Fig. 3–39 The load contacts on the left are from a NEMA-rated 00 motor starter. The load contacts on the right are from an IEC-rated 00 motor starter.

(Source: Delmar/Cengage Learning)

Fig. 3–40 The control contact of an eight-pin control relay is shown on the left. The load contacts of an IEC-rated 00 starter are shown on the right.

(Source: Delmar/Cengage Learning)

Due to the size difference in contacts between NEMA- and IEC-rated starters, many engineers and designers of control systems specify a one to two larger size for IEC-rated equipment than would be necessary for NEMA-rated equipment. A table showing the ratings for IEC starters is shown in Figure 3–41.

IEC Motor Starters (60 Hz)

Size	Max Amps	Motor Voltage	Maximum Horsepower	
			Single Phase	Three Phase
A	7	115	1/4	
		200		1 1/2
		230	1/2	1 1/2
		460		3
		575		5
B	10	115	1/2	
		200		2
		230	1	2
		460		5
		575		7 1/2
C	12	115	1/2	
		200		3
		230	2	3
		460		7 1/2
		575		10
D	18	115	1	
		200		5
		230	3	5
		460		10
		575		15
E	25	115	2	
		200		5
		230	3	7 1/2
		460		15
		575		20
F	32	115	2	
		200		7 1/2
		230	5	10
		460		20
		575		25
G	37	115	3	
		200		7 1/2
		230	5	10
		460		25
		575		30
H	44	115	3	
		200		10
		230	7 1/2	15
		460		30
		575		40
J	60	115	5	
		200		15
		230	10	20
		460		40
		575		40
K	73	115	5	
		200		20
		230	10	25
		460		50
		575		50
L	85	115	7 1/2	
		200		25
		230	10	30
		460		60
		575		75
M	105	115	10	
		200		30
		230	10	40
		460		75
		575		100
N	140	115	10	
		200		40
		230	10	50
		460		100
		575		125
P	170	115		
		200		50
		230		60
		460		125
		575		125
R	200	115		
		200		60
		230		75
		460		150
		575		150
S	300	115		
		200		75
		230		100
		460		200
		575		200
T	420	115		
		200		125
		230		125
		460		250
		575		250
U	520	115		
		200		150
		230		150
		460		350
		575		250
V	550	115		
		200		150
		230		200
		460		400
		575		400
W	700	115		
		200		200
		230		250
		460		500
		575		500
X	810	115		
		200		250
		230		300
		460		600
		575		600
Z	1215	115		
		200		450
		230		450
		460		900
		575		900

Fig. 3–41 IEC motor starters rated by size, horsepower, and voltage for 60-Hz circuits.

Study Questions

When answering questions related to troubleshooting, you may want to refer to Unit 59—
"Motor Startup and Troubleshooting Basics."

1. What is a magnetic line voltage motor starter?

2. How many poles are required on motor starters for the following motors:
   (a) 240-volt, single-phase induction motor, (b) 440-volt, three-phase induction motor?

3. If a motor starter is installed according to directions but will not start, what is a common cause for the failure to start?

4. Using the time limit overload or the dashpot overload relay, how are the following achieved: time-delay characteristics; tripping current adjustments?

5. What causes the AC hum or chatter in AC electromagnetic devices?

6. What is the phase relationship between the flux in the main pole of a magnet and the flux in the shaded portion of the pole?

7. In what AC and DC devices is the principle of the shaded pole used?

8. What does the electrician look for to remedy the following conditions:
   a. loud or noisy hum?
   b. chatter?

9. What type of protective enclosure is most commonly used and what is its NEMA number?

10. What advantage is there in using combination starters?

11. What safety feature does the combination starter provide that individual motor starting assemblies do not?

12. List the probable causes if the armature does not release after the magnetic starter is de-energized.

13. How is the size of the overload heaters selected for a particular installation?

14. What type of motor starter enclosure (include NEMA number and class) is recommended for an installation requiring safe electrical operation around an outside, inflammable paint–filling pump?

    Select the *best* answer for each of the following.

15. The magnetic starter is held closed
    a. mechanically
    b. by 15 percent undervoltage
    c. by 15 percent overvoltage
    d. electrical magnetically

16. When a motor starter coil is de-energized,
    a. the contacts stay closed
    b. it is held closed mechanically
    c. gravity and spring tension open contacts
    d. it must cool for a restart

17. An AC magnet may hum excessively due to
    a. improper alignment
    b. foreign matter between contact surfaces
    c. loose laminations
    d. all of these

18. AC magnets are made of *laminated* iron
    a. for better induction
    b. to reduce heating effect
    c. for AC and DC use
    d. to prevent chattering

19. The purpose of overload protection on a motor is to protect the
    a. motor from sustained overcurrents
    b. wire from high currents
    c. motor from sustained overvoltage
    d. motor from short circuits

20. The number of magnetic starter poles refers to
    a. the number of power, motor, or load contacts
    b. the number of control contacts
    c. the number of north and south poles
    d. all of these

21. Motors may burn out because of
    a. overloads
    b. high ambient temperatures
    c. poor ventilation
    d. all of the above

22. The purpose of a shading coil on an AC electromagnetic pole tip is to
    a. prevent overheating of the coil
    b. limit the tripping current
    c. limit the closing current
    d. prevent chattering

23. The current drawn by a motor is
    a. low on starting
    b. an accurate measurement of motor load
    c. an inaccurate measurement of motor load
    d. none of these

24. Thermal overload relays react to
    a. high ambient temperatures and excessive heating due to overload currents
    b. heavy mechanical loads
    c. heavy electrical loads
    d. rising starting currents

25. The thermal relay heating element is selected
    a. 15 percent under voltage
    b. 10 percent over voltage
    c. from the motor's FLA and a manufacturer's selection chart
    d. by ambient temperature

26. When the reset button does not reestablish the control circuit after an overload, the probable cause is
    a. the overload heater is too small
    b. the overload trip has not cooled sufficiently
    c. the auxiliary contacts are defective
    d. the overload heater is burned out

27. If an operator pushes a start button on a three-phase induction motor and the motor starts to hum, but not run, the probable trouble is
    a. one fuse is blown and the motor is single phasing
    b. the overload trip needs resetting
    c. the auxiliary contact is shorted
    d. one phase is grounded

28. A combination starter provides
    a. disconnecting means
    b. overload protection
    c. short-circuit protection
    d. all of these

29. Vacuum contactors
    a. are quieter in operation than standard contactors
    b. can control higher voltages in a smaller space
    c. are more economical to purchase than conventional contactors
    d. require the connection of a vacuum pump when in operation

30. What does IEC stand for?
    a. Intensity of current, EMF, and Capacitance
    b. Internal Electrical Connection
    c. International European Committee
    d. International Electrotechnical Commission

Circuit Layout, Connections, and Symbols

UNIT 4

Symbols

Objectives

After studying this unit, the student should be able to

- Identify common electrical symbols used in motor control diagrams.
- Use electrical symbols in drawing schematic and wiring diagrams.
- Draw and define supplementary contact symbols.

If the directions for wiring electrical equipment are written without the use of diagrams, or if diagrams are used but must show each device as it actually appears, then the work and time involved both in preparing the directions and in installing the equipment will be very expensive. Therefore, symbols—a form of pictorial shorthand—are used rather than pictures of electrical equipment to show how separate pieces of electrical equipment are connected in a circuit. Because the symbols may not be similar to the physical appearance of the devices they represent, many symbols must be memorized.

Unfortunately, there is no actual standard used for motor control symbols. Different manufacturers and companies often use their own set of symbols for their in-house schematics. Also, schematics drawn in other countries may use an entirely different set of symbols to represent different control components. Although symbols can vary from one manufacturer to another, or from one country to another, once you have learned to interpret circuit logic, it is generally possible to determine what the different symbols represent by the way they are used in the schematic. The most standardized set of

symbols in the United States is provided by the NEMA. It is these symbols that will be discussed in this unit.

Push Buttons

One of the most used symbols in control schematics is the push button. Push buttons can be shown as normally open or normally closed, Figure 4–1. Most are momentary contact devices in that they make or break connection only as long as pressure is applied to them. When the pressure is removed, they return to their normal position. Push buttons contain both movable and stationary contacts. The stationary contacts are connected to the terminal screws. The normally open push button is characterized by drawing the movable contact above and not touching the stationary contacts. Because the movable contact does not touch the stationary contacts, there is an open circuit and current cannot flow from one stationary contact to the other. The way the symbol is drawn assumes that pressure will be applied to the movable contact. When the button is pressed, the movable contact moves downward and bridges the two stationary contacts to complete a circuit.

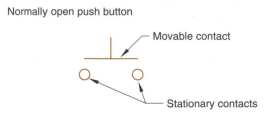

Normally open push button

Movable contact

Stationary contacts

Normally open push buttons are drawn with the movable contact above and not touching the stationary contacts.

Fig. 4–1 NEMA standard push-button symbols.

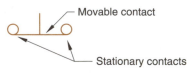

Normally closed push button

Movable contact

Stationary contacts

Normally closed push buttons are drawn with the movable contact below and touching the stationary contacts.

(Source: Delmar/Cengage Learning)

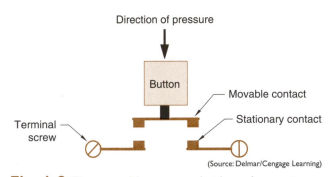

Direction of pressure

Button

Movable contact

Stationary contact

Terminal screw

(Source: Delmar/Cengage Learning)

Fig. 4–2 The movable contact bridges the stationary contacts when the button is pressed.

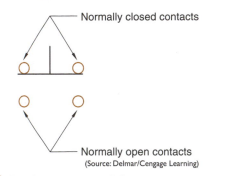

Normally closed contacts

Normally open contacts

(Source: Delmar/Cengage Learning)

Fig. 4–3 Double-acting push button.

(Source: Delmar/Cengage Learning)

Fig. 4–4 The double-acting push button has four screw terminals.

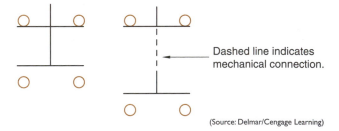

Dashed line indicates mechanical connection.

(Source: Delmar/Cengage Learning)

Fig. 4–5 Other symbols used to represent double-acting push buttons.

Figure 4–2. When pressure is removed from the button, a spring returns the movable contact to its original position.

The normally closed push button symbol is characterized by drawing the movable contact below and touching the two stationary contacts. Because the movable contact touches the two stationary contacts, a complete circuit exists and current can flow from one stationary contact to the other. If pressure is applied to the button, the movable contact will move away from the two stationary contacts and open the circuit.

Another very common push button found throughout industry is the double-acting push button, Figure 4–3. Double-acting push buttons contain both normally open and normally closed contacts. When connecting these push buttons in a circuit, you must make certain to connect the wires to the correct set of contacts.

A typical double-acting push button is shown in Figure 4–4. Note that the double-acting push button has four terminal screws. The symbol for a double-acting push button can be drawn in different ways, Figure 4–5. The symbol on the left is drawn with two movable contacts connected by one common shaft. When the button is pressed, the top movable contact breaks away from the top two stationary contacts and the bottom movable contact bridges the bottom two stationary contacts to complete the circuit. The symbol on the right is very similar in that it also shows two movable contacts. The right-hand symbol, however, connects the two push-button symbols with a dashed line. When components

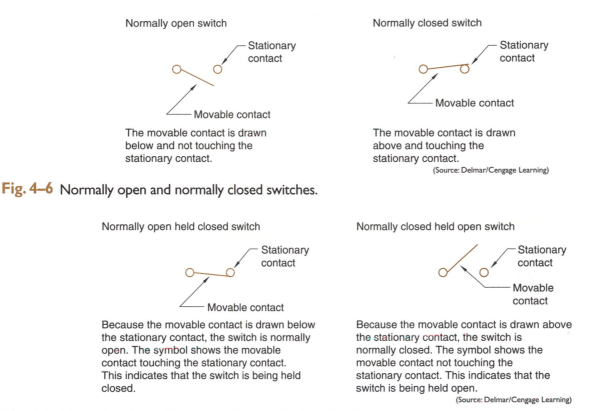

Normally open switch
Stationary contact
Movable contact

The movable contact is drawn below and not touching the stationary contact.

Normally closed switch
Stationary contact
Movable contact

The movable contact is drawn above and touching the stationary contact.

(Source: Delmar/Cengage Learning)

Fig. 4–6 Normally open and normally closed switches.

Normally open held closed switch
Stationary contact
Movable contact

Because the movable contact is drawn below the stationary contact, the switch is normally open. The symbol shows the movable contact touching the stationary contact. This indicates that the switch is being held closed.

Normally closed held open switch
Stationary contact
Movable contact

Because the movable contact is drawn above the stationary contact, the switch is normally closed. The symbol shows the movable contact not touching the stationary contact. This indicates that the switch is being held open.

(Source: Delmar/Cengage Learning)

Fig. 4–7 Normally open held closed and normally closed held open switches.

are shown connected by a dashed line in a schematic diagram, it indicates that the components are mechanically connected. If one is pressed, all that are connected by the dashed line are pressed. This is a very common method of showing several sets of push-button contacts that are actually controlled by one button.

Switch Symbols

Switch symbols are employed to represent many common control-sensing devices. There are four basic symbols: normally open (NO), normally closed (NC), normally open held closed (NOHC), and normally closed held open (NCHO). To understand how these switches are drawn, it is necessary to begin with how normally open and normally closed switches are drawn, Figure 4–6. Normally open switches are drawn with the movable contact **below and not touching** the stationary contact. Normally closed switches are drawn with the movable contact **above and touching** the stationary contact.

The normally open held closed and normally closed held open switches are shown in Figure 4–7. Note that the movable contact of the normally open held closed switch is drawn below the stationary contact. The fact that the movable contact is drawn **below** the stationary contact indicates that the switch is normally

open. Because the movable contact is touching the stationary contact, however, a complete circuit does exist because something is holding the contact closed. A very good example of this type switch is the low-pressure switch found in many air-conditioning circuits, Figure 4–8. The low-pressure switch is being held closed by the refrigerant in the sealed system. If the refrigerant should leak out, the pressure will drop low enough to permit the contact to return to its normal open position. This would open the circuit and de-energize coil C, causing both C contacts to open and disconnect the compressor from the power line. Although the schematic indicates that the switch is closed during normal operation, it would have to be connected as an open switch when it is wired into the circuit.

The normally closed held open switch is shown open in Figure 4–7. Although the switch is shown open, it is actually a normally closed switch because the movable contact is drawn **above** the stationary contact, indicating that something is holding the switch open. A good example of how this type switch can be used is shown in Figure 4–9. This circuit is a low-water warning circuit for a steam boiler. The float switch is held open by the water in the boiler. If the water level should drop sufficiently, the contacts will close and energize a buzzer and warning light.

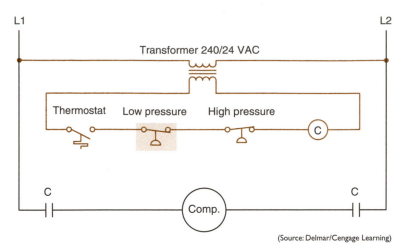

(Source: Delmar/Cengage Learning)

Fig. 4—8 If system pressure should drop below a certain value, the normally open held closed low-pressure switch will open and de-energize coil C.

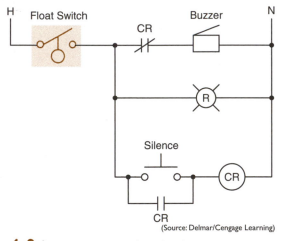

(Source: Delmar/Cengage Learning)

Fig. 4—9 Low-water warning circuit.

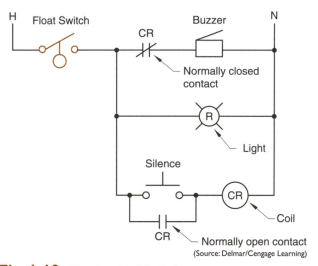

(Source: Delmar/Cengage Learning)

Fig. 4—10 Circuit with labeled components.

Basic Schematics

To understand the operation of the circuit shown in Figure 4–9, you must understand some basic rules concerning schematic or ladder diagrams.

1. Schematics show components in their electrical sequence without regard for physical location. The schematic in Figure 4–9 has been redrawn in Figure 4–10. Labels have been added, and the schematic shows a coil labeled CR and one normally open and one normally closed contact labeled CR. All of these components are physically located on control relay CR.
2. Schematics are always drawn to show components in their de-energized or off state.
3. Any contact that has the same label or number as a coil is controlled by that coil. In this example, both CR contacts are controlled by the CR coil.
4. When a coil energizes, all contacts controlled by it change position. Any normally

open contacts will close and normally closed contacts will open. When the coil is de-energized, the contacts will return to their normal state.

Referring to Figure 4–10, if the water level should drop far enough, the float switch will close and complete a circuit through the normally closed contact to the buzzer and to the warning light connected in parallel with the buzzer. At this time both the buzzer and warning light are turned on. If the silence push button is pressed, the CR coil will energize and both CR contacts will change position. The normally closed contact will open and turn off the buzzer. The warning light, however, will remain on as long as the low water level exists. The normally open CR contact connected in parallel with the silence push button will close. This contact is generally referred to as a holding, sealing, or maintaining contact. Its function is to maintain a current path to the coil when the push button returns to its normal open position. The circuit will remain in this state until the water level becomes high enough to reopen the float switch. When the float switch opens, the warning light and CR coil will turn off. The circuit is now back in it original de-energized state.

SENSING DEVICES

Motor control circuits depend on sensing devices to determine what conditions are occurring. They act very much like the senses of the body. The brain is the control center of the body. It depends on input information such as sight, touch, smell, and hearing to determine what is happening around it. Control systems are very similar in that they depend on such devices as temperature switches, float switches, limit switches, flow switches, and so on to know the conditions that exist in the circuit. These sensing devices will be covered in greater detail later in the text. The four basic types of switches are used in conjunction with other symbols to represent some of these different kinds of sensing switches.

Limit Switches

Limit switches are drawn by adding a wedge to one of the four basic switches, Figure 4–11. The wedge represents the bumper arm. Common industrial limit switches are shown in Figure 4–12.

Float, Pressure, Flow, and Temperature Switches

The symbol for a float switch illustrates a ball float. It is drawn by adding a circle to a line, Figure 4–13. The flag symbol of the flow switch represents the paddle that senses movement. The flow switch symbol is used for both liquid- and air-flow switches. The symbol for a pressure switch is a half circle connected to a line. The flat part of the semicircle represents a diaphragm. The symbol for a temperature switch represents a bimetal helix. The helix will contract and expand with a change of temperature. It

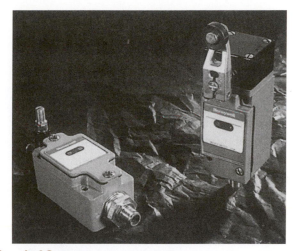

Fig. 4–12 Typical industrial limit switches. (*Courtesy of Micro Switch, a Honeywell division*)

Float Switches		Flow Switches	
NO	NC	NO	NC
Pressure Switches		Temperature Switches	
NO	NC	NO	NC

(Source: Delmar/Cengage Learning)

Fig. 4–13 Schematic symbols for sensing switches.

Proximity Switch X4

(Source: Delmar/Cengage Learning)

Fig. 4–14 Special symbols are often used for sensing devices that do not have a standard symbol.

should be noted that any of these symbols can be used with any of the four basic switches.

There are many other types of sensing switches that do not have a standard symbol. Some of these are photo switches, proximity switches, sonic switches, Hall Effect switches, and others. Some manufacturers will employ some special type of symbol and label the symbol to indicate the type of switch. An example of this is shown in Figure 4–14.

Coils

The most common coil symbol used in schematic diagrams is the circle. The reason for this is that letters and/or numbers are written in the circle to identify the coil. Contacts controlled by the coil are given the same number. Several standard coil symbols are shown in Figure 4–15.

Normally closed limit switch

Normally closed held open limit switch

Normally open limit switch

Normally open held closed limit switch

(Source: Delmar/Cengage Learning)

Fig. 4–11 Limit switches.

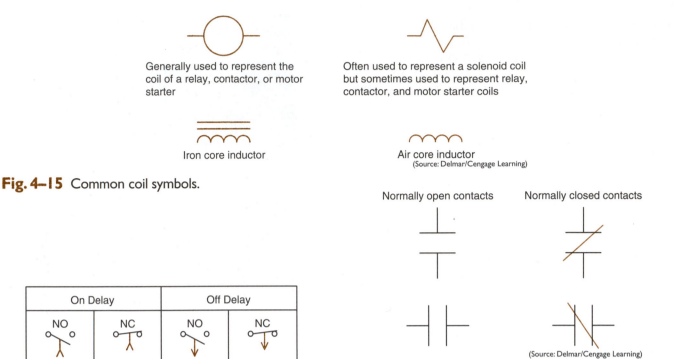

Generally used to represent the coil of a relay, contactor, or motor starter

Often used to represent a solenoid coil but sometimes used to represent relay, contactor, and motor starter coils

Iron core inductor

Air core inductor
(Source: Delmar/Cengage Learning)

Fig. 4–15 Common coil symbols.

Normally open contacts   Normally closed contacts

(Source: Delmar/Cengage Learning)

Fig. 4–17 Normally open and normally closed contact symbols.

On Delay		Off Delay	
NO	NC	NO	NC

(Source: Delmar/Cengage Learning)

Fig. 4–16 Timed contact symbols.

Timed Contacts

Timed contacts are either normally open or normally closed. They are not drawn as normally open held closed or normally closed held open. There are two basic types of timers, on delay and off delay. Timed contact symbols use an arrow to point in the direction that the contact will move at the end of the time cycle. Timers will be discussed in detail in a later unit. Standard timed contact symbols are shown in Figure 4–16.

Contact Symbols

Another very common symbol used on control schematics is the contact symbol. The symbol is two parallel lines connected by wires, Figure 4–17. The normally open contacts are drawn to represent an open connection. The normally closed contact symbol is the same as the normally open symbol with the addition of a diagonal line drawn through the contacts. The diagonal line indicates that a complete current path exists.

Other Symbols

Not only are there NEMA standard symbols for coils and contacts, but there are also symbols for transformers, motors, capacitors, and special types of switches. A chart of both common control and electrical symbols is shown in Figure 4–18.

SWITCHES								
DISCONNECT	CIRCUIT INTERRUPTER	CIRCUIT BREAKER W/THERMAL O.L.	CIRCUIT BREAKER W/MAGNETIC O.L.	CIRCUIT BREAKER W/THERMAL AND MAGNETIC O.L.	LIMIT SWITCHES		FOOT SWITCHES	
					NORMALLY OPEN	NORMALLY CLOSED	N.O.	N.C.
					HELD CLOSED	HELD OPEN		
PRESSURE & VACUUM SWITCHES		LIQUID LEVEL SWITCH		TEMPERATURE ACTUATED SWITCH		FLOW SWITCH (AIR, WATER, ETC.)		
N.O.	N.C.	N.O.	N.C.	N.O.	N.C.	N.O.	N.C.	

Fig. 4–18 Standard wiring diagram symbols.

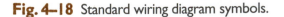

FUSE	STANDARD DUTY SELECTOR		HEAVY DUTY SELECTOR			
POWER OR CONTROL	2 POSITION	2 POSITION	3 POSITION	2 POS SEL PUSH BUTTON		

2 POSITION (Standard Duty)

3 POSITION (Standard Duty)

2 POSITION (Heavy Duty)

J K

	J	K
A1	1	
A2		1

A1
A2

1–CONTACT CLOSED

3 POSITION (Heavy Duty)

J K L

	J	K	L
A1	1		
A2			1

A1
A2

1–CONTACT CLOSED

2 POS SEL PUSH BUTTON

CONTACTS A B	SELECTOR POSITION			
	A		B	
1○ ○2	BUTTON		BUTTON	
3○ ○4	FREE	DEPRES'D	FREE	DEPRES'D
1–2	1			
3–4		1	1	1

1–CONTACT CLOSED

PUSH BUTTONS							PILOT LIGHTS	
MOMENTARY CONTACT						MAINTAINED CONTACT	INDICATE COLOR BY LETTER	
SINGLE CIRCUIT		DOUBLE CIRCUIT	MUSHROOM HEAD	WOBBLE STICK	ILLUMINATED	TWO SINGLE CKT.	ONE DOUBLE CKT.	NON PUSH-TO-TEST	PUSH-TO-TEST
N.O.	N.C.	N.O. & N.C.							

CONTACTS								COILS		OVERLOAD RELAYS		INDUCTORS
INSTANT OPERATING				TIMED CONTACTS–CONTACT ACTION RETARDED WHEN COIL IS				SHUNT	SERIES	THERMAL	MAGNETIC	IRON CORE
WITH BLOWOUT		WITHOUT BLOWOUT		ENERGIZED		DE-ENERGIZED						
N.O.	N.C.	N.O.	N.C.	N.O.	N.C.	N.O.	N.C.					AIR CORE

TRANSFORMERS					MOTORS						
AUTO	IRON CORE	AIR CORE	CURRENT	DUAL VOLTAGE	SINGLE PHASE	3 PHASE SQUIRREL CAGE	WOUND ROTOR	ARMATURE	SHUNT FIELD	SERIES FIELD	COMM OR COMPENS FIELD
									(SHOW 4 LOOPS)	(SHOW 3 LOOPS)	(SHOW 2 LOOPS)

WIRING				CONNECTIONS	RESISTORS			CAPACITORS	
NOT CONNECTED	CONNECTED	POWER	CONTROL	WIRING TERMINAL	MECHANICAL	FIXED	ADJ BY FIXED TAPS	RHEOSTAT. POT OR ADJ TAP	FIXED	ADJ*

MECHANICAL

GROUND

MECHANICAL INTERLOCK

RES

H

HEATING ELEMENT

RES

RH

SPEED PLUGGING		ANTI-PLUGGING	BARREL	BUZZER	HORN	METER	SHUNT METER	HALF-WAVE RECTIFIER	FULL-WAVE RECTIFIER	BATTERY
						INDICATE TYPE BY LETTER				

F ... R

VM

AM

+ DC AC
DC AC

DIODE	SILICONE CONTROLLED RECTIFIER (SCR)	

(Source: Delmar/Cengage Learning)

Fig. 4–18 continued

EXPLANATION OF COMMON SYMBOLS

The control circuit line diagram of Figure 4–19 shows the symbols of each device used in the circuit. The diagram indicates the function of each device. The push-button station wiring diagram on the right of Figure 4–19 represents the physical control station and shows the relative position of each device, the internal wiring, and the connections with the motor starter.

Refer to Figure 4–18. The disconnect switch shown is used to disconnect live equipment from the line voltage. This action is usually done with the equipment unloaded. The load may be interrupted by using contactors, motor starters, or circuit interrupters such as circuit breakers.

Machine operators often find that a foot switch is provided. The use of the foot switch frees the operator's hands. The switch mechanism is usually a heavy-duty, double-break, normally open or closed switch in a rugged enclosure with a broad mushroom-like metallic

pedal or foot operator. This type of switch may be connected by a portable cord or wired in solid, in a stationary position near the machine to be controlled.

The wobble stick push button is controlled by a stem that can be operated from any direction. It is so constructed for fast and easy access by a production machine operator.

The maintained contact push buttons are mechanically held in the selected position.

The push to test pilot light is used to ensure that the lamp and circuit are OK. In the push position it is placed across the control voltage. Otherwise, it may be connected so that it will indicate when a machine or pump motor is on or off.

The timed contacts are described in Unit 9, "Timing Relays." Note that there is a time delay in the contact action when the relay coil is energized (on delay). In a different contact arrangement, there is also a time delay when the coil is de-energized (off delay).

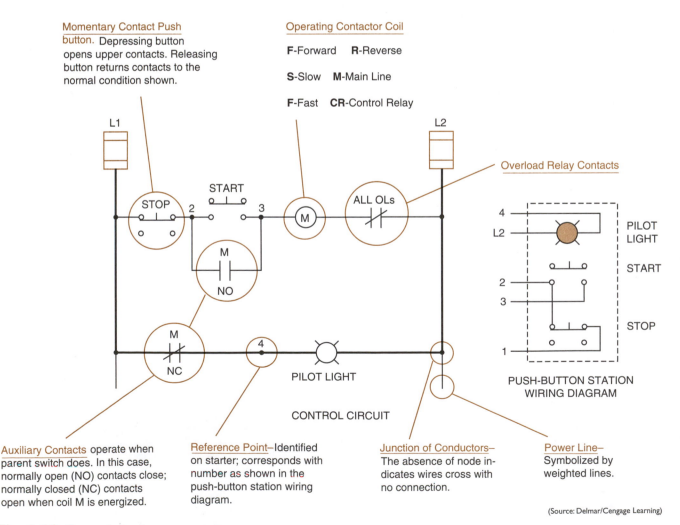

Momentary Contact Push button. Depressing button opens upper contacts. Releasing button returns contacts to the normal condition shown.

Operating Contactor Coil

F-Forward   **R**-Reverse

S-Slow   **M**-Main Line

F-Fast   **CR**-Control Relay

Overload Relay Contacts

Auxiliary Contacts operate when parent switch does. In this case, normally open (NO) contacts close; normally closed (NC) contacts open when coil M is energized.

Reference Point–Identified on starter; corresponds with number as shown in the push-button station wiring diagram.

Junction of Conductors– The absence of node indicates wires cross with no connection.

Power Line– Symbolized by weighted lines.

(Source: Delmar/Cengage Learning)

Fig. 4–19 Control circuit components.

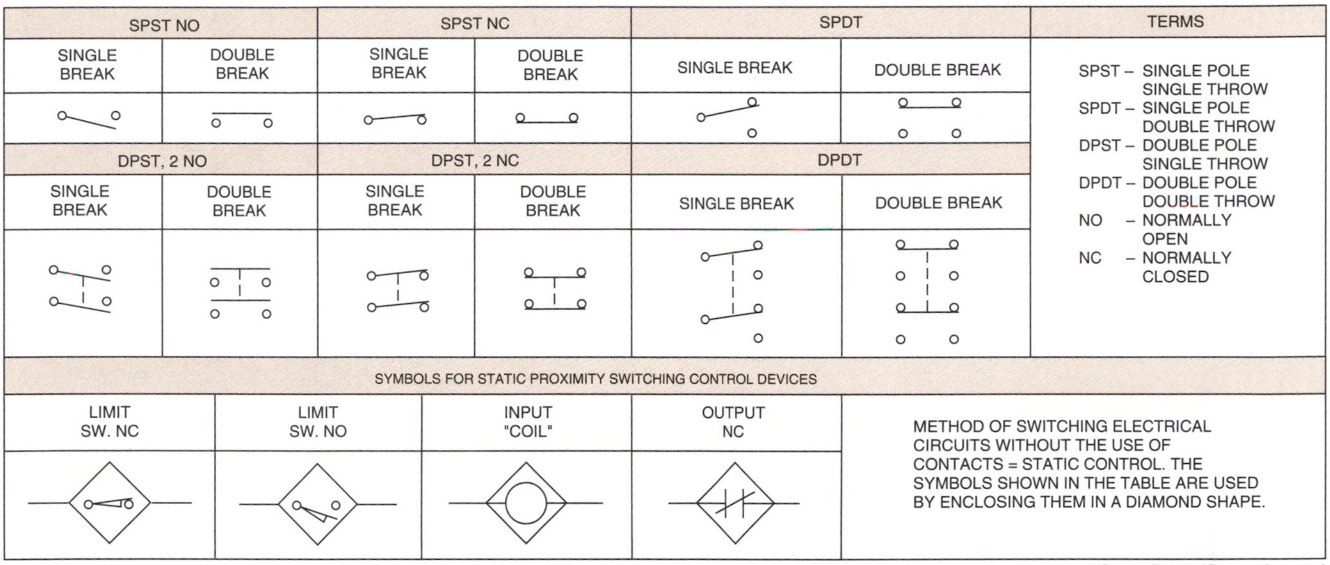

SPST NO		SPST NC		SPDT		TERMS
SINGLE BREAK	DOUBLE BREAK	SINGLE BREAK	DOUBLE BREAK	SINGLE BREAK	DOUBLE BREAK	SPST – SINGLE POLE SINGLE THROW SPDT – SINGLE POLE DOUBLE THROW DPST – DOUBLE POLE SINGLE THROW DPDT – DOUBLE POLE DOUBLE THROW NO – NORMALLY OPEN NC – NORMALLY CLOSED
DPST, 2 NO		DPST, 2 NC		DPDT		
SINGLE BREAK	DOUBLE BREAK	SINGLE BREAK	DOUBLE BREAK	SINGLE BREAK	DOUBLE BREAK	
SYMBOLS FOR STATIC PROXIMITY SWITCHING CONTROL DEVICES						
LIMIT SW. NC		LIMIT SW. NO	INPUT "COIL"	OUTPUT NC		METHOD OF SWITCHING ELECTRICAL CIRCUITS WITHOUT THE USE OF CONTACTS = STATIC CONTROL. THE SYMBOLS SHOWN IN THE TABLE ARE USED BY ENCLOSING THEM IN A DIAMOND SHAPE.

(Source: Delmar/Cengage Learning)

Fig. 4–20 Supplementary contact symbols.

A shunt coil is connected with the full voltage applied. A series coil is connected in series with a load or the full current.

In drawing wiring diagrams, it is not an accepted practice to "jump" a wire. Wires are shown as crossing each other and, unless specified otherwise, are not connected. If a wire is to be connected, a quick method is to show it with a connection node or dot.

Mechanical connections are shown with straight broken lines (they are insulated, non-current carrying parts). One example is when two contacts are operated at the same time on a push button.

The speed plugging and antiplugging switch symbol will be shown in use later in the text.

Figure 4–20 defines abbreviations commonly used in wiring prints and specifications. The symbols further illustrate descriptions of the abbreviation.

Study Questions

1. Identify the following symbols.

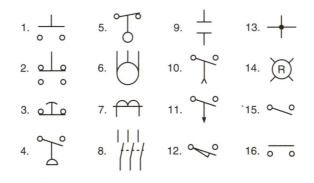

2. Electrical symbols usually conform to which standard?

3. What do the following abbreviations stand for?
   a. SPST   d. DPDT
   b. SPDT   e. NO
   c. DPST   f. NC

4. Single-pole and double-pole switch symbols are shown as
   a. single break
   b. double break
   c. NO and NC
   d. all of the above

5. The symbol shown is
   a. polarized capacitor
   b. normally closed switch
   c. normally open held closed switch
   d. normally open contact

6. The symbol shown is
   a. normally closed float switch
   b. normally open held closed float switch
   c. normally open float switch
   d. normally closed held open float switch

7. The symbol shown is
   a. iron core transformer
   b. auto transformer
   c. current transformer
   d. air core transformer

8. The symbol shown is
   a. normally open pressure switch
   b. normally open flow switch
   c. normally open float switch
   d. normally open temperature switch

H O A

9. The symbol shown is
   a. double-acting push button
   b. two-position selector switch
   c. three-position selector switch
   d. maintained contact pushbutton

10. If you were installing the circuit in Figure 4–9, what type of push button would you use for the silence button?
    a. Normally closed
    b. Normally open

11. Referring to the circuit in Figure 4–9, should the float switch be connected as a normally open or normally closed switch?

12. Referring to the circuit in Figure 4–9, what circuit component controls the actions of the two CR contacts?

13. Why is a circle most often used to represent a coil in a motor control schematic?

14. When reading a schematic diagram, are the control components shown as they should be when the machine is turned off or de-energized, or are they shown as they should be when the machine is in operation?

UNIT 5

Interpretation and Application of Simple Wiring and Elementary Diagrams

Objectives

After studying this unit, the student should be able to

- Draw an elementary line diagram using a wiring diagram as a reference.

- Describe the different advantages of two- and three-wire controls.

- Connect motor starters with two- and three-wire controls.

- Identify common terminal markings.

- Read and use target tables.

- Connect drum reversing controllers.

- Interpret motor nameplate information.

- Connect dual-voltage motors.

LINE DIAGRAMS AND WIRING DIAGRAMS

Most electrical circuits can be represented by two types of diagrams: a wiring diagram and a line diagram. *Wiring diagrams* include all of the devices in the system and show the physical relationships between the devices. All poles, terminals, contacts, and coils are shown on each device. These diagrams are particularly useful in wiring circuits because the connections can be made exactly as they appear on the diagram, a wire for every line. A wiring diagram gives the necessary information for the actual wiring of a circuit and provides a means of physically tracing the wires for troubleshooting purposes or when normal preventative maintenance is necessary. In other words, the actual physical installation and the wiring diagram coincide as far as the locations of the devices and wiring are concerned.

The wiring diagram is quite often such a maze of lines that it may be nearly impossible to determine the electrical sequence of operation of the circuits shown. For this reason, a rearrangement of the circuit elements is made to form what is called a *line diagram*. Line diagrams, also called elementary, schematic, and ladder diagrams, are widely used in industry. It is a decided advantage to an electrician to learn to use these diagrams.

The line diagram also represents the electrical circuit but does so in the simplest manner possible. No attempt is made to show the various devices in their actual physical positions. All control devices are shown located between vertical lines that represent the source of control power. Circuits are shown connected as directly as possible from one of the source lines horizontally—through contacts and current consuming devices—to the other source line. All connections are made so that the functions and sequence of operations of the various devices and circuits can be traced easily. The schematic diagram is invaluable in troubleshooting because it shows the effect that opening or closing various contacts have on other devices in the circuit. The schematic diagram attempts to convey as much information as possible with the least amount of confusion. It shows you in the simplest manner how an electrical control system should work.

Any single operation of a piece of electrical equipment is usually not complicated. However, when a sequence of operations to be completed by electrical means depends on the present, previous, or subsequent operations of several other pieces of equipment, the number of operations forms a complex system that may be difficult to understand. The line diagram reduces the necessary information to the simplest possible form to help the electrician understand the operation of the total control system.

Figure 5–1 illustrates the simplicity of an elementary or schematic diagram. Note how easy it is to trace through the circuit: When the disconnect switch is closed, current flows through the only complete path—L1 down through the normally closed contact CR, through the red pilot lamp to L2. When the normally open push button is closed, CR coil is energized through the complete circuit to L2 from L1. When the control relay is energized, it instantly closes the normally open CR contact to energize the green indicating lamp. At the same time, it opens normally closed contact CR, interrupting the circuit to the red lamp. When the push button is released, the red lamp is again lit and the green lamp goes out because the coil is now de-energized. The normally open CR contact opens and normally closed CR contact closes.

Study Figure 5–2 to become familiar with a motor starter as it is represented by a *wiring diagram*. The principal parts of the starter are labeled on the diagram so that a comparison can be made with the actual starter. This study should help you to visualize the starter when viewing a wiring diagram and will help in making correct connections when the

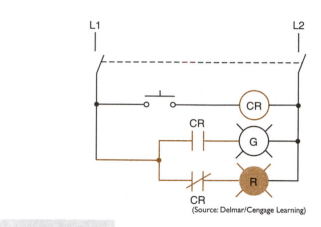
(Source: Delmar/Cengage Learning)

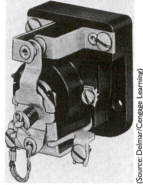
(Source: Delmar/Cengage Learning)

Fig. 5–1 A single-pole, double-throw (SPDT) control relay and its schematic diagram.

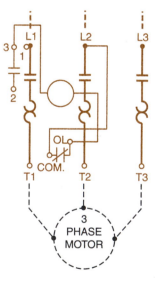

Fig. 5–2 AC magnetic motor starter. (*Courtesy of Schneider Electric*)

starter is actually wired. Note that the wiring diagram shows as many parts as possible in their proper relative positions. It is not necessary to show the armature and crossbar or the overload reset mechanism of the starter in the wiring diagram because these parts are not involved in the actual wiring. The *control circuit* consists of the coil, the contacts of the overload relays, and the coil-holding contact for the control circuit. The *power circuit* consists of the main contacts and the overload heaters to the motor. Note the heavier lines used for the power circuit. The broken lines show where the power lines and motor leads connect to the starter.

The wiring diagram in Figure 5–2 represents a size 1 motor starter. Other starter sizes have similar diagrams and are wired in a similar manner. Some of the connections, however, may not have the same physical locations on the starters.

It should be noted that the control circuit in Figure 5–2 shows the motor starter coil connected directly across the incoming power line. This type of circuit is known as a *line-voltage control* circuit. The coil of the motor starter must have the same voltage rating as the power line connected to the motor. Common industrial voltages are 240, 480, and 560 volts. Although line-voltage control is still employed, most industrial control circuits operate on 120 or 24 volts. Control transformers are used to reduce the line voltage to the control circuit value. Control circuits employing control transformers will be shown throughout the book.

Figure 5–3 is an elementary or line diagram of the starter shown in Figure 5–2. In many cases, the line diagram does not show the power circuit to the motor because of its simplicity. Note that the line 1 (L1) circuit is completed to the motor (when the coil is energized) closing main contact (M) and the overload heater to terminal 1 (T1) of the motor. All of the M contacts close simultaneously when the M coil is energized. The motor then starts with the applied line voltage.

The control wiring of the starter is shown in solid lines, Figure 5–3(B). This completes the wiring of the actual starter. The external wiring completed by the electrician in the field is shown by broken lines. Note how the control wiring is rearranged to form one horizontal path from line 1 (L1) to line 2 (L2). This figure shows how the control wiring diagram is converted in appearance to a line diagram. Although the drawings differ, they are the same electrically. In general, it is easier to trace the operation of the starter by viewing the line drawing than it is to use the wiring diagram of the same starter. Compare the two diagrams carefully. You may be required to convert a complex wiring diagram to an elementary diagram to determine how a machine operates.

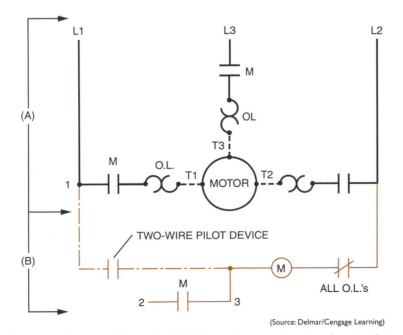

(Source: Delmar/Cengage Learning)

Fig. 5–3 Line diagram of starter shown in Figure 5–2. (A) Line diagram showing the simplicity of the power circuit for the line voltage starter; (B) completed control wiring elementary style. Broken lines and motor are external field wiring that completes the circuit.

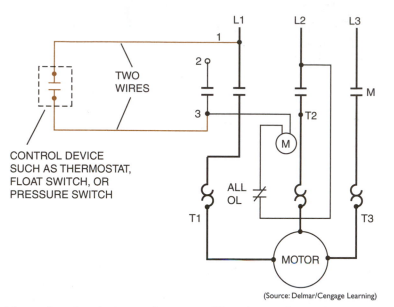

(Source: Delmar/Cengage Learning)

Fig. 5–4 Two wires lead from the pilot device to the starter. The phrases "no-voltage release" and "two-wire control" should indicate to you that an automatic pilot device, such as a limit switch or float switch, opens and closes the control circuit through a single contact.

TWO-WIRE CONTROL

Two-wire control provides no-voltage release or low-voltage release. Two-wire control of a starter means that the starter drops out when there is a voltage failure and picks up as soon as the voltage returns. In Figure 5–4, the pilot device is unaffected by the loss of voltage. Its contact remains closed, ready to carry current as soon as line voltage returns to normal. This form of control is used frequently to operate supply and exhaust blowers and fans.

Two-wire control is also shown in schematic form in Figure 5–3(B). Two-wire control means that an operator does not have to be present to restart a machine following a voltage failure. *CAUTION:* This type of restart can be a safety hazard to personnel and machinery. For example, materials in process may be damaged due to the sudden restart of machines on restoration of power. (Remember, power may return without warning.)

THREE-WIRE CONTROL

Three-wire control involves the use of a *maintaining circuit*. This method eliminates the need of the operator to press continuously on the push button to keep the coil energized.

The Maintaining Circuit

Recall from Figure 5–1 that when pressure is released on the push button, it returns to its normally open position. The circuit to the coil is interrupted and it returns the relay contacts to their normal position. Normally, in starting a motor, it is desirable to energize the motor starter coil by pressing a button and then to keep the coil energized after the pressure is released from the button. Refer to the elementary control circuit diagram in Figure 5–5. When the start button is pressed, coil M is energized across L1 and L2. This closes contact M to place a shunt circuit around 2-3, the start button. A parallel circuit is formed, with one circuit through push-button contacts 2-3 and one circuit through contact M. As a result, current will flow through the M coil. If pressure is removed from the start button, contacts 2-3 open. The other circuit through contact M will remain closed, supplying current to coil M and maintaining a starter closed position. Such a circuit is called a maintaining circuit, a sealing circuit, or a holding circuit.

No-Voltage Protection

Three-wire control providing no-voltage protection or low-voltage protection is also a basic and commonly used control circuit. In this type of circuit, three wires are used to control the contactor or starter.

Three-wire control of a starter means that the starter will drop out when there is a voltage failure but will not pick up automatically when voltage returns. The control circuit in Figure 5–5 is completed through the stop button and also through a holding contact (2-3) on the starter. When the starter drops out, the holding contact

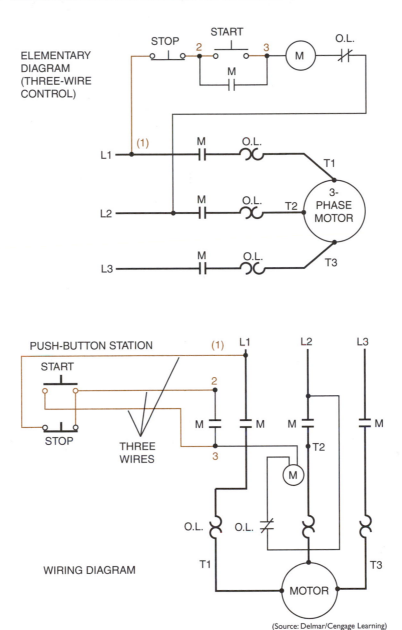

(Source: Delmar/Cengage Learning)

Fig. 5–5 Three wires lead from the pilot device to the starter. The phrases "no-voltage protection" and "three-wire control" should indicate to the electrician that the most common means of providing this type of control is a start-stop push-button station.

opens and breaks the control circuit until the start button is pressed to restart the motor.

The main distinction between the two types of control is that in no-voltage release (two-wire control), the coil circuit is maintained through the pilot switch contacts; in no-voltage protection (three-wire control), the circuit is maintained through a stop contact on the push-button station and an auxiliary (maintaining) contact on the starter. The designations "two-wire" and "three-wire" are used only because they describe the simplest applications of the two types of control. Actually, for more complicated circuits, there may be more wires leading from the pilot device to the starter. However,

the principle of two-wire or three-wire control will still apply in these situations.

Three-wire control prevents the restarting of machinery when power is restored. In other words, an operator must press the start buttons to resume production when it is safe to do so.

In Figure 5–3(B) contact 2-3 was not included in the two-wire circuit. This contact is the holding or maintaining contact, which is connected in a three-wire control circuit as shown in Figure 5–5. It is sometimes shown smaller than power contacts.

Reviewing Figure 5–5, it can be seen that when the start button is pressed, coil M is energized from line 1 (L1) through the normally

Table 5–1 Control and Power Connections—600 Volts or Less—Across-the-Line Starters

	1 PHASE	3 PHASE	DIRECT CURRENT
Line Markings	L1, L2	L1, L2, L3	L1, L2
Ground When Used	L1 Always Ungrounded	L2	L1 Always Ungrounded
O/L Relay Heaters in	L1	T1, T2, T3	L1
Control Circuit Connected to	L1	L1	L1
Control Circuit Switching Connected to	L1, L2	L1, L2	L1, L2
Contactor Coil Connected to	L2	L2	L2
For Reversing Interchange Lines	—	L1, L3	—
Overload Relay Contacts in	L2	L2	L2

closed stop push button, the depressed start button, the coil, and the overload control contacts to line 2 (L2). When coil M is energized, it closes all M contacts. The maintaining contact (2-3) holds the coil circuit when the start button is released. When the stop push button is pressed, the stop contact opens and the coil is de-energized. As a result, all M contacts are opened and the motor stops. In the event of a motor overload, excessive current is drawn from the line, causing the thermal heaters to overheat. When this happens, the normally closed overload relay control contacts are opened and the motor stops.

CONTROL AND POWER CONNECTIONS

The correct connections and component locations for line and wiring diagrams are indicated in Table 5–1. Compare the information given in the table with actual line diagrams to develop an ability to interpret the table quickly and use it correctly. For example, refer to Figure 5–5 and the three-phase column of Table 5–1. Note that the control circuit switching is connected to line 1 (L1) and the contactor coil is connected to line 2 (L2).

TARGET TABLES

The phrase *target table,* or *sequence chart,* refers to a chart that lists the sequence of operation of identified contacts in a circuit. A target table is useful in interpreting circuit drawings and for simplifying troubleshooting.

One particular advantage of using a target table is shown in Figure 5–6. Here, contacts on selector switches need not be drawn close together on the elementary diagram but can be spaced over the diagram if necessary. Broken lines should be drawn on the diagram to connect the various contacts of one switch; they represent an insulated mechanical connection, not to be confused with a wire connection.

Selector Switch Diagramming

Figure 5–6 shows a typical, three-position selector switch, its line diagram symbol, and a target table indicating the condition of each contact for each position of the selector switch. The 1 (one) represents closed contacts. A small target table, such as that for a single two-circuit, double-contact switch, can be shown above the switch contact on the elementary diagram.

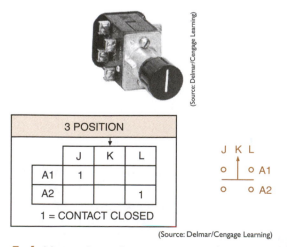

(Source: Delmar/Cengage Learning)

Fig. 5–6 Heavy-duty three-position selector switch, diagram symbol, and target table.

Larger target tables should be shown to one side or below the line diagram.

The target table for the three-position selector switch shows that in the OFF position (arrow), no contacts are closed. In the J position, contact A1 is closed and A2 is open. In the L position, A2 is closed and A1 is open.

Drum Switch Diagramming

A drum switch consists of a set of moving contacts mounted on and insulated from a rotating shaft. The switch also has a set of stationary contacts that make and break contact with the moving contacts as the shaft is rotated, Figure 5–7. In the single-phase sample connection diagram, one motor rotation is shown with solid lines. The other direction is shown with broken lines, reversing one motor winding in relation to the other. A jumper should be added when this switch is used for single phase. The jumper is installed between 3 and 5.

The term *drum switch* is applied to cam switches as well. Drum switches provide a means for starting, stopping, and reversing AC single-phase, polyphase, or DC motors without overload protection, or where overload protection is provided by some other means. Drum switches may be used to carry the starting current of a motor directly or they may be used to handle only the current for the pilot devices that control the main motor current.

In Figure 5–8, the terminal markings (boxed by broken lines) indicate how the line and load are connected to the drum switch. The target table on the right shows which contacts are closed to reverse the three-phase motor and which contacts are closed to cause the motor to run forward.

Target tables are often used in complicated circuits to simplify the interpretation of the operation of large multicircuit drum switches.

MOTOR NAMEPLATE DATA AND WIRING INTERPRETATION

Figure 5–9 shows a typical motor nameplate for a standard three-phase, nine-lead motor.

Serial Number and Product Number

All manufacturers must identify a specific motor somehow. The serial number is a common way to identify a specific motor. This number is

(Source: Delmar/Cengage Learning)

(Source: Delmar/Cengage Learning)

Fig. 5–7 Drum switch with cover removed; diagrams show switch connections to single-phase and three-phase motors.

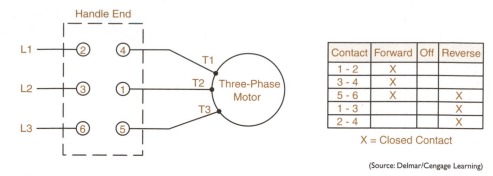

Contact	Forward	Off	Reverse
1 - 2	X		
3 - 4	X		
5 - 6	X		X
1 - 3			X
2 - 4			X

X = Closed Contact

(Source: Delmar/Cengage Learning)

Fig. 5–8 A target table is used to assist in the interpretation of drum switch operation where a three-phase or a single-phase motor is reversed by interchanging two motor leads.

Fig. 5–9 Standard three-phase motor nameplate. *(Courtesy of Reuland Electric Co.)*

unique to a specific motor and is the most common way to identify special equipment made to a specific customer's order. However, there are manufacturers building very large lots of motors for sale through warehouse distribution. Often a computerized product number or catalog model number is used for identification. This type of number will identify a bill of materials for construction, but not a specific motor such as the serial number.

Type and Style

These various characters are unique to a manufacturer. In general, each manufacturer has an individual set of designations. For example, a type of motor may be specified for a particular application, have normal starting torque, normal starting current, and normal slip. A mechanical characteristic is that the motor is drip proof.

Max. Amb.

This is the maximum temperature the motor should operate at safely under full load. The nameplate shown in Figure 5–9 gives the motor frame number 215T, which refers to the frame dimension given on the manufacturer's detailed dimension sheets.

HP and Code

The horsepower (hp) rating given on the nameplate refers to the output of the motor, as measured by mechanical means. The code letter on the nameplate indicates the locked rotor kVA input per horsepower as shown in NEMA standards. This code letter shall be used for determining the branch-circuit short-circuit protection and ground-fault protection from applicable tables in the *NEC*®.

Time Rating

This is sometimes referred to as DUTY. Usually a time reference is used, such as 5, 15, 30, 60 minute duty, or CONT for continuous operation. Occasionally an application name may be used, such as "extractor duty" for a laundry extractor. RPM on the nameplate refers to the speed of the motor in revolutions per minute. The current in amperes drawn by the motor is listed under AMPS and varies according to the voltage system to which the motor is connected. Motors draw half as much current on high voltages as they do on low voltages. However, the same amount of power is used for either high- or low-voltage connections ($P = E \times I$). Low-current connections mean that smaller wire sizes, conduits, feeders, and switch gear can be used. If available, connection to high voltages will mean a financial savings in wiring materials and less wasted electrical energy due to decreased voltage drop in the circuits. The motor starter voltage must be equal to the supply voltage (except for separate control, Unit 17), and the overload heaters must be rated for the proper current value.

In Figure 5–9, the volts indication refers to motor terminal connections 230 volts; the HIGH volts, 460.

Phase and Hertz

Ph refers to phase, such as three-phase voltage. HZ means the voltage hertz, or voltage frequency. The most common electrical service in North America is 60 Hz.

Insulation System (INS SYS)

This is an indication of the materials used for insulating the motor winding. It will give an indication of the temperature capabilities and environmental protection of the wire and varnish used.

Additional Data

There are often additional plates on a motor. The connection plate is used to identify the various reconnections of the motor leads of dual-voltage and multispeed motors. Besides motor winding connections, there may be a space heater, temperature sensor, lubrication, bearings, lifting data, and safety approval plates. Motor space heaters are an optional item used to keep motors dry in humid climates. See Figure 5–10. Motor connection plates, or data, are sometimes found inside the motor terminal connection box cover. Such information is shown in Figure 5–11.

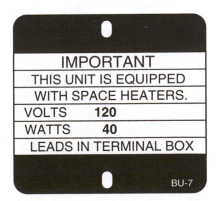

Fig. 5–10 Motor space heater nameplate. *(Courtesy of Reuland Electric Co.)*

Fig. 5–11 Typical nameplate of motor connections. *(Courtesy of Reuland Electric Co.)*

For a high-voltage connection, the nameplate shows that motor lead terminals 4 and 7 are connected, as are 5 and 8, and 6 and 9. The connections are insulated from each other in the motor terminal connection box. The three remaining leads (1, 2, and 3) are connected to three-phase supply lines L1, L2, and L3. Actually, this connection is to the load side of the starter terminals (T1, T2, and T3).

The nameplates for some motors include a service factor reference. The service factor of a general-purpose motor is an allowable overload. The amount of allowable overload is indicated by a multiplier. When this multiplier is applied to the normal horsepower rating, the permissible loading is indicated.

In keeping with a greater awareness of energy conservation, many motor manufacturers indicate a percentage efficiency value on the nameplates of selected motors. The nameplate shown in Figure 5–12 contains both an efficiency rating (94.1 percent) and a service factor value (1.15).

MOTOR POWER AND GREATER EFFICIENCY

The only power that the electric motor consumes is used up in losses of about 6 to 7 percent. Major losses are copper losses (I^2R), iron losses (eddy currents and hysteresis), and windage. A common error is that motors use over 60 percent of all electrical power consumed, but they only convert and deliver it to mechanical loads. The much greater amount of energy is used up by the driven load.

Mechanical equipment experiences far greater losses than electric motors. Greater energy efficiency can be achieved by improving motors, using more efficient motors, and by designing and using more efficient mechanical

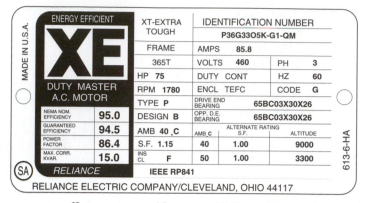

Fig. 5–12 Nameplate for an energy efficient motor. *(Courtesy of Reliance Electric Company)*

equipment. For example, a motor-pump shaft seal with a carbon disc-to-flat metal ring seal, instead of the common shaft compression seal, creates drag. Throttle valves and narrow pipes can waste huge amounts of power; so can inefficient pumps. Proper lubrication to driven machines makes them easier on the motor. Of course a motor that is larger than necessary wastes away still more energy. We not only save energy, we also get better performance with some attention to driven equipment. Using an adjustable speed drive is one of the fastest ways to save energy, but it must be a good one. Adjustable speed drives are sometimes hard to select. Although upgrading motors and the inefficient equipment that they run involves major changes for businesses, the energy savings will make the effort worthwhile.

THREE-PHASE, DUAL-VOLTAGE MOTOR CONNECTIONS

Figure 5–13 shows a recommended method of determining the three-phase motor terminal connections in the event that the motor nameplate is inaccessible, destroyed, or lost when the motor is installed. Test equipment is used to determine the unknown internal wiring connections of the motor, either star or delta. A continuity test will find one group of three common leads. For example, such a test will establish the center of the star high-voltage connection (at T7, T8, and T9) in Figure 5–13, upper left. There should be only one group of three motor leads common to each other for a star-connected motor, and three groups of three leads common to each other for a delta, internally connected motor (T1-T4-T9, T2-T5-T7, T3-T6-T8).

SINGLE-PHASE, DUAL-VOLTAGE MOTOR CONNECTIONS

The connections for dual-voltage, single-phase AC motors are shown in Figure 5–14. For low-voltage operation, the stator coils are connected in parallel; for high-voltage operation, they are connected in series. The instantaneous current direction is indicated by arrows at the motor coils. In general, starting windings are factory connected to the terminal screws for 120 volts.

All external wiring motor leads are located in the motor terminal connection or conduit box.

REVERSING A SINGLE-PHASE MOTOR WITH A DRUM SWITCH

To reverse a single-phase AC motor, either the starting or running windings connections (but not both) must be reversed on the line. In this example it will be assumed that the run winding leads of the motor are labeled T1 and T2, and the start winding leads are labeled T5 and T8. The drum switch previously discussed will be used in this illustration, Figure 5–15.

The circuit is connected as shown in Figure 5–16. A jumper is connected between terminals 3 and 5 of the drum switch. When the switch is in the off position no connection is made across the terminals of the switch. Although terminal T5 of the motor is connected directly to the terminal 2 of the drum switch and the neutral conductor, there is no connection between terminals 5 and 6 of the drum switch. Line 1 is, therefore, not connected to the start winding.

When the drum switch is moved to the forward position, connection is made as shown in Figure 5–17. Terminal T8 of the start winding is connected to Line 1 and T5 is connected to the

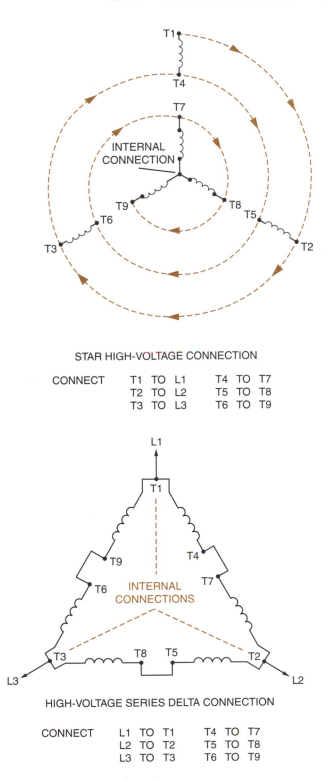

STAR HIGH-VOLTAGE CONNECTION

CONNECT   T1 TO L1   T4 TO T7
            T2 TO L2   T5 TO T8
            T3 TO L3   T6 TO T9

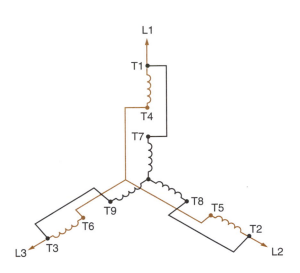

TWO-CIRCUIT, STAR LOW-VOLTAGE CONNECTION

CONNECT   T1 TO L1
            T2 TO L2
            T3 TO L3

T1 TO T7	T2 TO T8
T3 TO T9	T4 TO T5 T6

HIGH-VOLTAGE SERIES DELTA CONNECTION

CONNECT   L1 TO T1   T4 TO T7
            L2 TO T2   T5 TO T8
            L3 TO T3   T6 TO T9

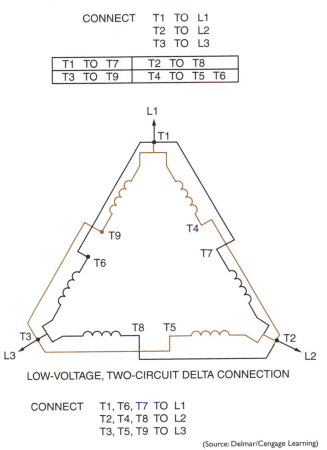

LOW-VOLTAGE, TWO-CIRCUIT DELTA CONNECTION

CONNECT   T1, T6, T7 TO L1
            T2, T4, T8 TO L2
            T3, T5, T9 TO L3

(Source: Delmar/Cengage Learning)

Fig. 5–13 Three-phase motor terminal connections.

neutral conductor. Terminal T1 of the run wind-ing is connected to Line 1 and terminal T2 is connected to neutral.

When the drum switch is moved to the re-verse position, connection is made as shown in Figure 5–18. Terminal T8 of the start wind-ing is still connected to Line 1 and T5 is still connected to the neutral conductor. Terminal T1 of the run winding, however is now con-nected to neutral and T2 is now connected to Line 1. Note that the start winding connec-tions remain the same regardless of the posi-tion of the drum switch, but the run windings change between forward and reverse.

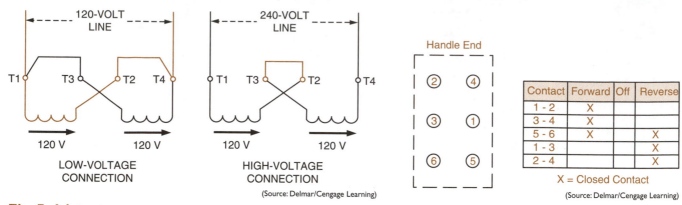

LOW-VOLTAGE
CONNECTION

HIGH-VOLTAGE
CONNECTION

(Source: Delmar/Cengage Learning)

Fig. 5–14 Single-phase, dual-voltage motors.

Handle End

Contact	Forward	Off	Reverse
1 - 2	X		
3 - 4	X		
5 - 6	X		X
1 - 3			X
2 - 4			X

X = Closed Contact

(Source: Delmar/Cengage Learning)

Fig. 5–15 Drum Switch.

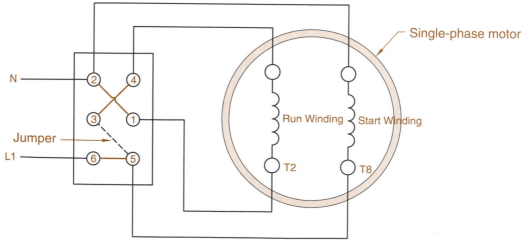

(Source: Delmar/Cengage Learning)

Fig. 5–16 Single-phase motor connected to a drum switch.

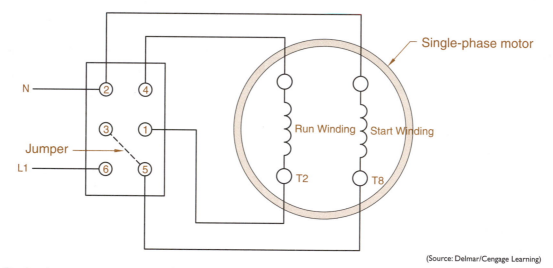

(Source: Delmar/Cengage Learning)

Fig. 5–17 The drum switch is moved to the forward position.

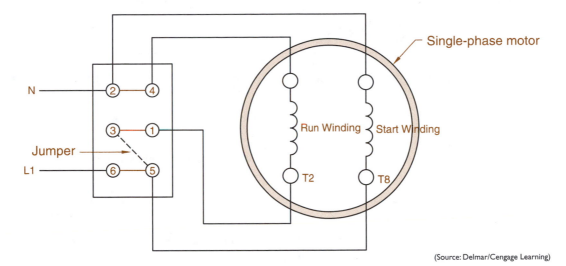

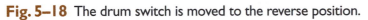

(Source: Delmar/Cengage Learning)

Fig. 5–18 The drum switch is moved to the reverse position.

Study Questions

1. In which diagram do the physical locations of the wiring and the devices match the drawing?

2. Which diagram conveys as much information as possible with the least amount of confusion?

3. Why is it not necessary to show the armature and crossbar or the overload reset mechanism in a wiring diagram?

4. Which control wiring scheme can cause damage to industrial production or processing equipment with the sudden restart of machines on restoration of the power?

5. In developing line or wiring diagrams, in which lines should the overload relays be placed for three-phase motor starters?

6. In which control circuit wire are the overload relay contacts connected?

7. The control circuit is connected to which lines for a three-phase motor starter?

8. Identify the terminal markings for the delta-wound motor shown in Figure 5–19.

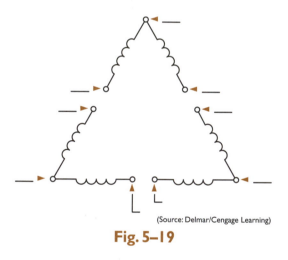

(Source: Delmar/Cengage Learning)

Fig. 5–19

9. The contactor coil is connected to which line of a three-phase contactor?

10. What does the "X" or "1" mean on a target table?

11. A motor nameplate reads 230/460 volts, 30-15 amperes. When connected to a 460-volt supply, how much current will the motor draw fully loaded?

12. To what does the code letter on a motor nameplate refer?

13. Draw an elementary line diagram of the control circuit from the wiring diagram shown in Figure 5–20. Exclude power or motor circuit wiring.

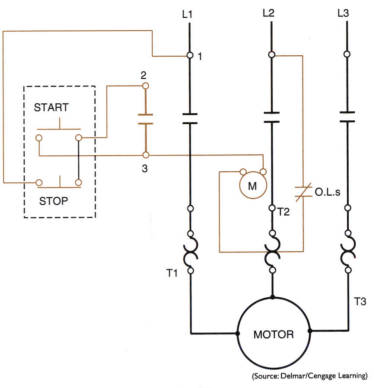

(Source: Delmar/Cengage Learning)

Fig. 5–20

14. Draw a line diagram of the control circuit shown in Figure 5–21. (Exclude power or motor circuit wiring.)

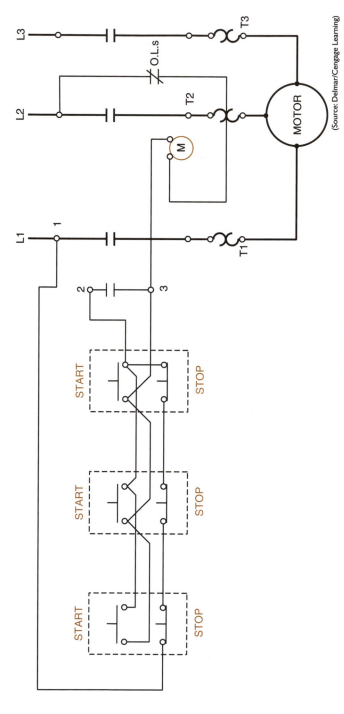

(Source: Delmar/Cengage Learning)

Fig. 5–21

SECTION

3

Control Pilot Devices

Push Buttons and Control Stations

Objectives

After studying this unit, the student should be able to

- Describe the difference between closed and open push buttons.
- Draw the wiring diagram symbols for push buttons and pilot lights.
- Draw simple circuits using normally open and closed push buttons.
- Connect combination push buttons in simple circuits.
- Explain single- and double-break contacts.
- List the types of control operators on control stations.
- Draw simple circuits using a selector switch.
- Wire simple circuits with a selector switch.

PUSH BUTTONS

A push-button station is a device that provides control of a motor through a motor starter by pressing a button that opens or closes contacts. It is possible to control a motor from as many places as there are stations—through the same magnetic controller. This can be done by using more than one push-button station. A single-circuit push-button station is shown in Figure 6–1.

Two sets of momentary contacts are usually provided with push buttons so that when the button is pressed, one set of contacts is opened and the other set is closed. The use of this combination push button is simply illustrated in the wiring diagram of Figure 6–2. When the push button is in its normal position as shown in

Figure 6–2(A), current flows from L1 through the normally closed contacts, through the red pilot light (on top) to L2 to form a complete circuit; this lights the lamp. Because the power push-button circuit is open by the normally open contact, the green pilot lamp does not glow. Note that this situation is reversed when the button is pushed, as shown in Figure 6–2(B). Current now flows from L1 through the pushed closed contact and lights the green pilot lamp when completing the circuit to L2. The red lamp is now out of the circuit and does not glow. However, because the push buttons are momentary contact (spring loaded), we return to the position shown in Figure 6–2(A) when pressure on the button is released. Thus, by connecting to the proper set of contacts, either a normally open or a normally closed situation is

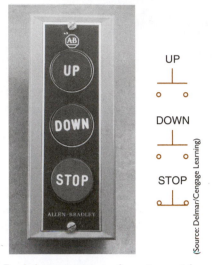

(Source: Delmar/Cengage Learning)

Fig. 6–1 Push-button control station with a general-purpose enclosure. The enclosure is usually made of molded plastic or sheet metal. This station contains three push buttons; two are normally open and one is normally closed.

Fig. 6–3 Open push-button unit with mushroom head. *(Courtesy of Rockwell Automation Co.)*

obtained. *Normally open* and *normally closed* mean that the contacts are in a rest position, held there by spring tension, and are not subject to either mechanical or electrical external forces. (See the Glossary for more detailed definitions.)

Push-button stations are made for two types of service: *standard-duty* stations for normal applications safely passing coil currents of motor starters up to size 4, and *heavy-duty* stations when the push buttons are to be used frequently and subjected to hard or rough usage. Heavy-duty push-button stations have higher contact ratings.

The push-button station enclosure containing the contacts is usually made of molded plastic or sheet metal. Some double-break contacts are made of copper. However, in most push buttons, silver-to-silver contact surfaces are provided for better electrical conductivity and longer life.

Figure 6–3 shows a combination, normally open and normally closed, push-button unit with a mushroom head for fast and easy access such as may be required in a safety circuit. This is also called a *palm-operated push button,* especially when it is used as a safety device for both operator and machine, such as a punch press or emergency stop in a control station. The push-button terminals in Figure 6–2 represent one-half of the

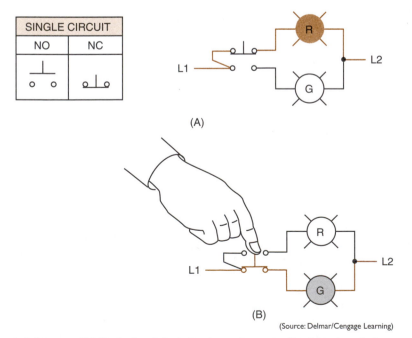

(Source: Delmar/Cengage Learning)

Fig. 6–2 Combination push button. (A) Red pilot light is lit through normally closed push-button contact; (B) green pilot light is lit when the momentary contact button is pushed.

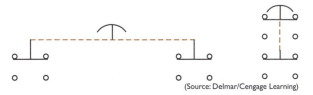

(Source: Delmar/Cengage Learning)

Fig. 6–4 Terminal configurations used by different manufacturers of Figure 6–3 push button. Both configurations represent the same push button.

terminals shown in Figure 6–3 and Figure 6–4, for a double-pole, double-throw push button.

Because control push buttons are subject to high momentary voltages caused by the inductive effect of the coils to which they are connected, good clearance between the contacts and insulation to ground and operator is provided.

The push-button station may be mounted adjacent to the controller or at a distance from it. The amount of current broken by a push button is usually small. As a result, operation of the controller is hardly affected by the length of the wires leading from the controller to a remote push-button station.

Push buttons can be used to control any or all of the many operating conditions of a motor, such as *start, stop, forward, reverse, fast,* and *slow*. Push buttons also may be used as remote stop buttons with manual controllers equipped with potential trip or low-voltage protection.

SELECTOR SWITCHES

Selector switches, as can be seen in Figure 6–5, are usually "maintained" contact position, with three and sometimes two selector positions. Selector switch positions are made by turning the operator knob—not pushing it. Figure 6–5(A) is a single-break contact arrangement, and Figure 6–5(B) is a double-break contact, disconnecting the control circuit at two points. These switches may also have a spring return to give momentary contact operation.

Figure 6–6(A) shows a single-break contact selector switch connected to two lights. The red light may be selected to glow by *turning* the switch to the red position, Figure 6–6(B). Current now flows from line 1 through the selected red position of the switch, through the red lamp to line 2, completing the circuit. Note this switch has an *off* position in the center. There are two position selector switches available with no *off* position. Here, whichever light is selected would burn continuously.

Figure 6–7(A) is an elementary diagram of the heavy-duty, double-break selector switch

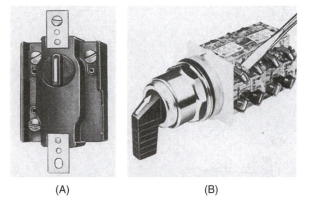

(A) (B)

Fig. 6–5 Open selector switches. [Part (A), *courtesy of Eaton Corporation;* Part (B), *courtesy of Schneider Electric*]

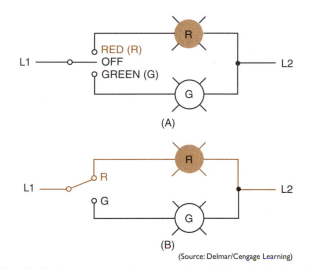

(Source: Delmar/Cengage Learning)

Fig. 6–6 Elementary diagram using (A) a single-break, three-position selector switch and (B) two-position, single-break switch.

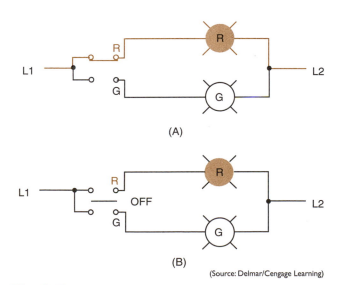

(Source: Delmar/Cengage Learning)

Fig. 6–7 Heavy-duty, double-break selector switch for (A) two-position and (B) three-position switches.

similarly shown in Figure 6–5. Note that the red light will glow, as shown in Part (A), with the two-position switch in this position, the green indicating lamp is de-energized, there is no "off" position. Figure 6–7(B) illustrates a three-position selector switch containing an "off" position. Both lamps may be turned off using this switch, but not in Part (A).

CONTROL STATIONS

A control station may contain push buttons alone, Figure 6–8, or a combination of push buttons, selector switches, and pilot lights (indicating lights), Figure 6–9. Indicating lights may be mounted in the enclosure. These lights are usually red or green and are used for communication and safety purposes. Other common colors available are amber, blue, white, and clear. They indicate when the line is energized, the motor is running, or any other condition is designated.

Control stations may also include switches that are key, coin, or hand-operated wobble sticks. A wobble stick is a stem-operated push button for operation from any direction. There are ball lever push-button operators for a gloved hand or for frequent operations.

Nameplates are installed to designate each control operation. These can be seen in Figure 6–1 and Figure 6–8. These control operators are

Fig. 6–9 Control station with push buttons, selector switches, and indicating lights. *(Courtesy of Schneider Electric)*

Fig. 6–10 Mechanical lockout-on-stop push button. *(Courtesy of Schneider Electric)*

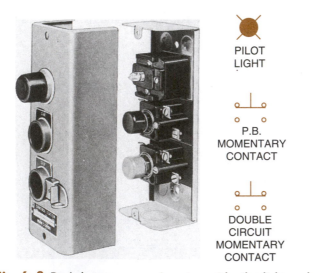

Fig. 6–8 Push-button control station with pilot light and associated wiring symbols for components. *(Courtesy of Schneider Electric)*

commonly used in control circuits of magnetic devices in factory production machinery.

Combination indicating light, nameplate, and push-button units are available. These illuminated push buttons and indicating lights are designed to save space in a wide variety of applications such as control and instrument panels, laboratory instruments, and computers. Miniature buttons are also used for this purpose.

Standard control station enclosures are available for normal general-purpose conditions, whereas special enclosures are used in situations requiring watertight, dust-tight, oil-tight, explosion-proof, or submersible protection. Provisions are often made for *padlocking* stop buttons in the open position (for safety purposes), Figure 6–10. Relays, contactors, and starters cannot be energized while an electrician is working on them with the stop button in this position.

Study Questions

1. What is meant by normally open contacts and normally closed contacts?

2. Why is it that normally open and normally closed contacts cannot be closed simultaneously?

3. How are colored pilot lights indicated in wiring diagrams?

   Select the *best* answer for each of the following.

4. A single-break control is a
   a. heavy-duty selector switch
   b. single-circuit push button
   c. standard-duty selector switch
   d. double-circuit push button

5. Control stations may contain
   a. push buttons
   b. selector switches
   c. indicating lights
   d. all of these

6. A wobble stick is
   a. operated from any direction
   b. knob controlled
   c. palm controlled
   d. glove operated

7. Common selector switches are
   a. one position
   b. two positions
   c. three positions
   d. two and three positions

8. Most push buttons are
   a. momentary contact
   b. single contact
   c. double contact
   d. a combination

9. Control station enclosures are
   a. general purpose
   b. explosion proof
   c. watertight
   d. all of these

10. The diagram that illustrates a single-circuit, normally closed push button is

    a.
    b.

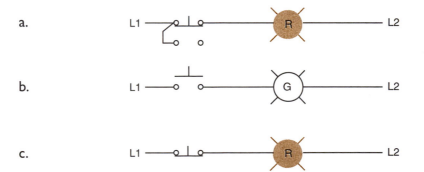

    c.

    d.

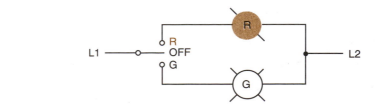

UNIT 7

Relays

Objectives

After studying this unit, the student should be able to

- Tell how relays differ from motor starters.
- List the principal uses of each magnetic relay.
- Identify single- and double-throw contacts; single- and double-break contacts.
- Draw the wiring diagram symbols for relays.
- Describe the operation and use of mechanically held relays.
- Draw elementary diagrams of control and load for mechanically held relays.
- Connect wiring for mechanically held relays.
- Identify dusttight, plug-in, and solid-state relays.
- Identify and connect a thermostat relay.

CONTROL RELAYS

Control magnetic relays are used as auxiliary devices to switch control circuits and large motor starter and contactor coils, and to control small loads such as small motors, solenoids, electric heaters, pilot lights, audible signal devices, and other relays, Figure 7–1 and Figure 7–2.

A magnetically held relay is operated by an electromagnet, which opens or closes electrical contacts when the electromagnet is energized. The position of the contacts changes by spring and gravity action when the electromagnet is de-energized.

Relays are generally used to enlarge or amplify the contact capability, or multiply the switching functions of a pilot device by adding more contacts to the circuit, Figure 7–3 and Figure 7–4.

Most relays are used in control circuits; therefore, their lower ratings (0–15 amperes maximum to 600 volts) show the reduced current levels at which they operate.

Magnetic relays do not provide motor overload protection. This type of relay is ordinarily used in a two-wire control system (any electrical

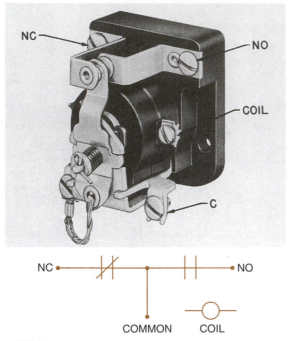

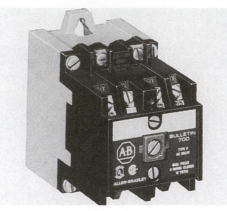

Fig. 7–1 Single-pole, double-throw, single-break AC general-purpose control relay with wiring symbols. *(Courtesy of Schneider Electric)*

Fig. 7–3 Four-pole control relay with four normally open contacts. To change from the normally open to the normally closed reversing contact, invert 180° and reinstall the cartridges. *(Courtesy of Rockwell Automation)*

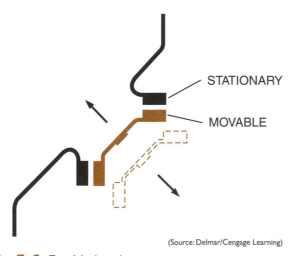

(Source: Delmar/Cengage Learning)

Fig. 7–2 Double-break contacts.

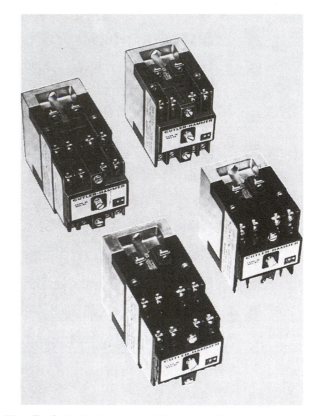

Fig. 7–4 Multiple relays. *(Courtesy of Eaton Corporation)*

contact-making device with two wires). Whenever it is desired to use momentary contact pilot devices, such as push buttons, any available normally open contact can be wired as a holding circuit in a three-wire system (see Unit 5 and Glossary). The contact arrangement and a description of the magnetic structure of relays were presented in Unit 3, as motor starters. Starters, contactors, and relays are similar in construction and operation but are not identical.

Control relays are available in single- or double-throw arrangements with various combinations of normally open and normally closed contact circuits. Although there are some single-break contacts used in industrial relays,

most of the relays used in machine tool control have double-break contacts. The comparison can be seen in Figure 7–1 and Figure 7–2. Looking at the relay contacts in Figure 7–2, note the upper contact being open at two points, making it a double break. One set of normally closed and one set of normally open contacts can be seen with the wiring symbols in Figure 7–1. This is also a double-throw contact because it has a common connection between the normally open and normally closed contacts, such as can be seen in Figure 7–1. The common terminal between the normally closed and normally open contacts make this a "single-pole, double-throw" relay.

It may be of particular interest to an electrician to know about changing contacts that are normally open to normally closed, or the other way around, normally closed to normally open. Most machine tool relays have some means to make this change. It ranges from a simple flip-over contact to removing the contacts and relocating with spring location changes.

Also, by overlapping contacts in this case, one contact can be arranged to operate at a different time relative to another contact on the same relay. For example, the normally open contact closes (makes) before the normally closed contact opens (breaks).

Relays differ in voltage ratings, number of contacts, contact rearrangement, physical size, and in attachments to provide accessory functions such as mechanical latching and timing. (These are discussed later.)

In using a relay for a particular application, one of the first steps should be to determine the control (coil) voltage at which the relay will operate. The necessary contact rating must be made, as well as the number of contacts and other characteristics needed. Because of the variety of styles of relays available, it is possible to select the correct relay for almost any application.

Relays are used more often to open and close control circuits than to operate power circuits. Typical applications include the control of motor starter and contactor coils, the switching of solenoids, and the control of other relays. A relay is a small but vital switching component of many complex control systems. Low-voltage relay systems are used extensively in switching residential and commercial lighting circuits and individual lighting fixtures.

Although control relays from various manufacturers differ in appearance and construction, they are interchangeable in control wiring systems if their specifications are matched to the requirements of the system.

Control relays are available in many shapes and configurations. Figure 7–5 shows a dust-proof, transparent enclosure of a general-purpose eight-pin control relay. These relays are available in both eight- and eleven-pin models. They plug into eight- and eleven-pin tube sockets, Figure 7–6. These relays are reliable, inexpensive, and can be replaced quickly without having to rewire the circuit. Eight-pin relays generally contain two sets of double-acting contacts. A connection diagram of an eight-pin relay is shown in Figure 7–7. Eleven-pin relays contain

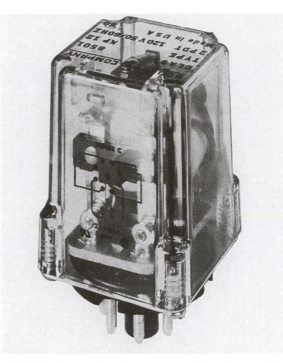

Fig. 7–5 Dust-proof, transparent enclosed control relay. *(Courtesy of Schneider Electric)*

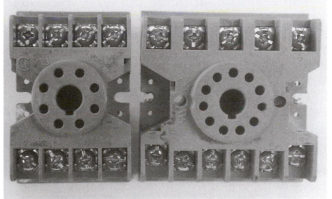

(Source: Delmar/Cengage Learning)

Fig. 7–6 Eight- and eleven-pin tube sockets.

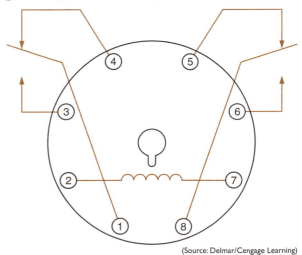

(Source: Delmar/Cengage Learning)

Fig. 7–7 Connection diagram of an eight-pin control relay.

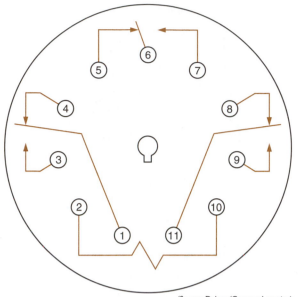

(Source: Delmar/Cengage Learning)

Fig. 7–8 Connection diagram of an eleven-pin control relay.

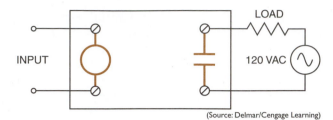

(Source: Delmar/Cengage Learning)

Fig. 7–9 The load side and control side are separated from each other.

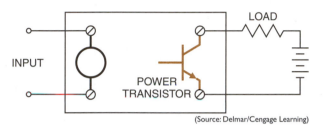

(Source: Delmar/Cengage Learning)

Fig. 7–10 A power transistor is used to connect a DC load to the line.

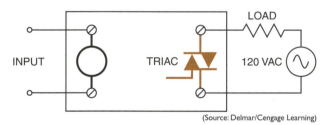

(Source: Delmar/Cengage Learning)

Fig. 7–11 A solid-state relay intended to connect an AC load to the line uses a triac.

three sets of double-acting contacts. A connection diagram for an eleven-pin relay is shown in Figure 7–8.

Another type of control relay contains small reed contacts that are enclosed in glass. A coil causes these contacts to open or close magnetically.

Consulting various industrial motor control catalogs, one will find many different voltages and current values. Basically, the following values are generally found:

Industrial Control Relays
NEMA
    contact voltage ratings—300 V and 600 V
    contact ampere ratings—most 10 amps, some
      20 amps
IEC (International Electrotechnical Control)
    contact voltage rating—600 V
    contact ampere rating—most 10 amps, some
      15 amps
GENERAL-PURPOSE RELAYS
    contact voltage rating—600 V
    contact ampere rating—most 10 amps, some
      to 30 amps
GENERAL PURPOSE—PLUG-IN RELAYS
    contact voltage rating—277 V
    contact ampere rating—10 amps per pole
    not to exceed 20 amps per relay

SOLID-STATE RELAY

Solid-state relays have several advantages over relays with mechanical contacts. Solid-state relays have no moving parts, are resistant to shock and vibration, and are sealed against dirt and moisture. The greatest advantage of the solid-state relay is the fact that the input or control voltage is isolated from the line voltage. This prevents voltage spikes and electrical noise from being transmitted into the control section of the circuit, Figure 7–9.

Solid-state relays can be used to control DC or AC loads. The relay illustrated in Figure 7–9 shows a set of normally open contacts being used to connect the load to the power line. In reality, this set of contacts is actually a solid-state device. If the relay is intended to control a DC load, a power transistor is used to connect the load to the line, Figure 7–10. The transistor is a semiconductor device that has the ability to control the flow of direct current. Although the transistor has the ability to regulate the flow of current in much the same way a water valve can control the flow of water, in this application the transistor is used as a switch. When it is off, it is completely off. When it is on, it is turned completely on.

Solid-state relays intended to control the operation of AC loads use an electronic device called a triac to connect the load to the line, Figure 7–11. The triac is a bidirectional device

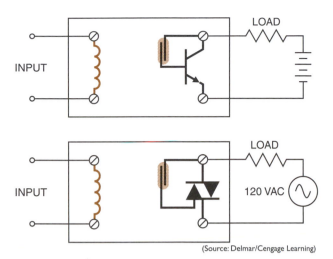

Fig. 7–14 Solid-state relay.

(Source: Delmar/Cengage Learning)

Fig. 7–12 Reed relay used to control the output device.

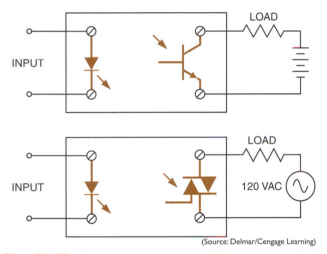

(Source: Delmar/Cengage Learning)

Fig. 7–13 Optoisolation uses a light-emitting diode (LED) to control a phototransistor or phototriac.

that will permit current to flow through in either direction. It has the ability to control alternating current in much the same way that a transistor controls direct current. Like the transistor, the triac is used as a solid-state switch in this circuit.

Solid-state relays generally apply 3 to 32 volts DC to the input to control the operation of the relay. Two basic methods are employed. The first method uses a reed relay. The coil of the relay is connected to the input voltage. The magnetic field produced by the direct current closes a reed contact and connects the base of the power transistor to the line or the gate of the triac to the line, Figure 7–12.

The second method involves the use of a light-emitting diode (LED) and photo-sensitive components. This method is generally referred to as *optoisolation*. The input voltage operates an LED, Figure 7–13. The output device is a phototransistor or phototriac. When the LED

turns on, it has the effect of connecting the base of the transistor to the line or the gate of the triac to the line. Optoisolation is the most widely used method because it prevents voltage spikes and electrical noise from being transmitted from the load side of the circuit to the control side. These types of solid-state relays are commonly used in the output section of programmable controllers.

Solid-state relays can be obtained in a variety of case styles and sizes. Some are housed in four-pin in-line IC circuit cases and are intended to control other solid-state devices. Others may be housed in cases intended to be heat sinked and are capable of controlling several amperes. The solid-state relay shown in Figure 7–14 is rated to control a 10-ampere load at 240 volts. This relay should be mounted on a heat sink.

SURGE PROTECTION

The coils of magnetic devices such as relays often generate large voltage spikes and electrical noise when they are switched on or off. As a general rule, magnetic relays are not affected by the electric noise and the voltage spikes induced in the system, but electronic devices are. Because electronic control is becoming more and more prevalent in industry, it is often necessary to eliminate as much electrical noise and transient voltages as possible. One method of eliminating noise and voltage spikes is to use a transient suppressor, Figure 7–15. The transient suppressor connects directly across the coil of the relay as shown in Figure 7–16.

The transient suppressor is a solid-state device called a metal oxide varistor (M.O.V.). The M.O.V. is basically a voltage-sensitive resistor. As long as the voltage remains below a certain point, the M.O.V. exhibits a very high resistance in the range of several hundred thousand ohms. However, when the voltage reaches a certain point, the M.O.V. suddenly drops to a very low value of resistance, acting almost like a short circuit. This change in resistance occurs rapidly, generally in the range of 3 to 5 nanoseconds, preventing the voltage spike from going any higher, Figure 7–17. The energy in the spike is dissipated as heat by the M.O.V. The M.O.V. will have a voltage rating slightly higher than the circuit voltage. A 120-volt coil would typically have an M.O.V. with a voltage rating of about 140 volts connected across it.

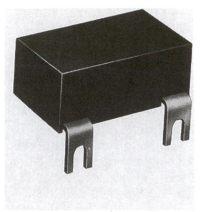

Fig. 7–15 Transient suppressor. *(Courtesy of Schneider Electric)*

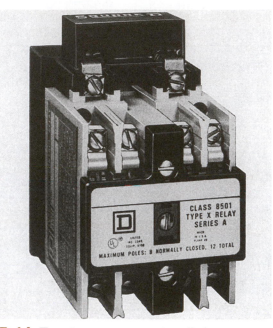

Fig. 7–16 Transient suppressor installed on a magnetic relay. *(Courtesy of Schneider Electric)*

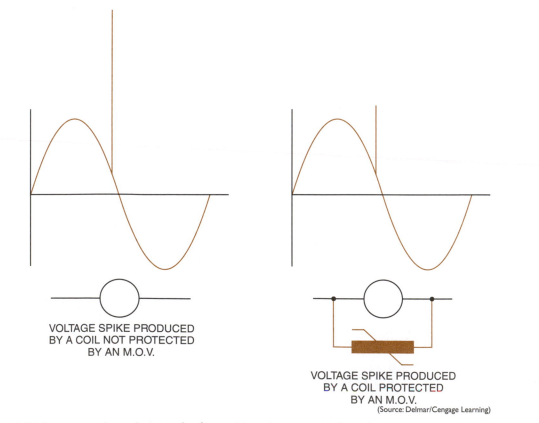

VOLTAGE SPIKE PRODUCED BY A COIL NOT PROTECTED BY AN M.O.V.

VOLTAGE SPIKE PRODUCED BY A COIL PROTECTED BY AN M.O.V.

(Source: Delmar/Cengage Learning)

Fig. 7–17 The M.O.V. prevents the voltage spike from rising above a certain point.

THERMOSTAT RELAY

Thermostat-type relays, Figure 7–18, are used with three-wire, gauge-type thermostat controls or other pilot controls having a slowly moving element, which makes a contact for both the closed and open positions of the relay. The contacts of a thermostat control device usually cannot handle the current to a starter coil; therefore, a thermostat relay must be used between the thermostat control and the starter, Figure 7–19.

When the moving element of the thermostat control touches the closed contact, the relay

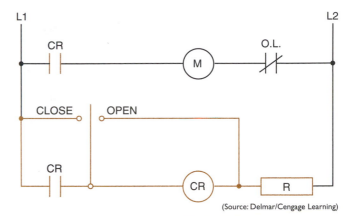

(Source: Delmar/Cengage Learning)

Fig. 7–19 Starter coil (M) is controlled by thermostat relay.

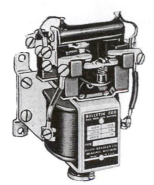

Fig. 7–18 Thermostat relay. *(Courtesy of Rockwell Automation)*

closes and is held in this position by a maintaining contact. When the moving element touches the open contact, the current flow bypasses the operating coil through a small resistor and causes the relay to open. The resistor is usually built into the relay and serves to prevent a short circuit.

The thermostat contacts must not overlap or be adjusted too closely to one another as this may result in the resistance unit being burned out. It is also advisable to compare the inrush current of the relay with the current rating of the thermostat.

Study Questions

1. What are control relays?

2. What are typical uses for control relays?

3. What are the advantages of a mechanically held relay?

4. Why are mechanically held relays energy-saving devices?

5. Why does the coil (CR) in Figure 7–19 drop out when the thermostat touches the open position?

6. Match each item in the alphabetical column with the appropriate item in the numbered columns.
    a. Double-break contact
    b. Industrial control relay
    c. Single pole, double throw
    d. Normally closed contact
    e. Single-break contact
    f. Relay coil
    g. Normally open contact

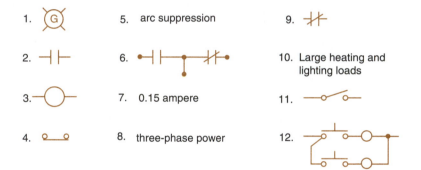

1.	5. arc suppression	9.
2.	6.	10. Large heating and lighting loads
3.	7. 0.15 ampere	11.
4.	8. three-phase power	12.

7. What device is used to connect the load to the line in a solid-state relay intended to control a DC load?

8. What device is used to connect the load to the line in a solid-state relay intended to control an AC load?

9. What is optoisolation?

10. What type of device is used to suppress voltage spikes in relay circuits?

Contactors

Objectives

After studying this unit, the student should be able to

- Tell how contactors differ from relays and motor starters.

- Draw the wiring diagrams and symbols for contactors.

- List the principal uses of contactors.

- Describe the operation of mechanically held contactors.

- Connect wiring for different types of contactors.

- List the advantages of mechanically held contactors.

- Describe the operation of magnetic blowout coils and how they provide arc suppression.

CONTACTORS

Magnetic contactors are electromagnetically operated switches that provide a safe and convenient means for connecting and interrupting branch circuits, Figure 8–1. The principal difference between a contactor and a motor starter is that the contactor does not contain overload relays. Contactors are used in combination with pilot control devices to switch lighting and heating loads and to control AC motors in those cases where overload protection is provided separately. The larger contactor sizes are used to provide remote control of relatively high-current circuits where it is too expensive to run the power leads to the remote controlling location,

Figure 8–2. This flexibility is one of the main advantages of electromagnetic control over manual control. Pilot devices such as push buttons, float switches, pressure switches, limit switches, and thermostats are provided to operate the contactors. Time clocks and photo cells are generally used for lighting loads.

Magnetic Blowout

The contactors shown in Figure 8–3 and Figure 8–4 operate on alternating current. Heavy-duty contact arc-chutes are provided on most of these larger contactors, Figure 8–3. The chutes contain heavy copper coils called blowout coils mounted above the contacts in series with

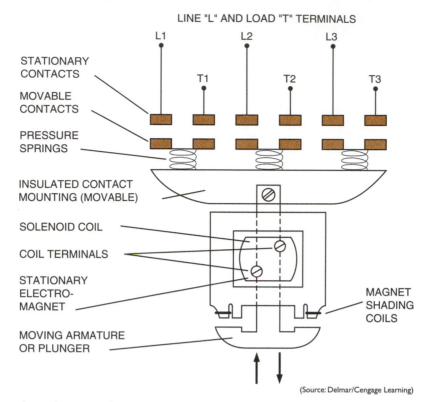

LINE "L" AND LOAD "T" TERMINALS

STATIONARY CONTACTS

MOVABLE CONTACTS

PRESSURE SPRINGS

INSULATED CONTACT MOUNTING (MOVABLE)

SOLENOID COIL

COIL TERMINALS

STATIONARY ELECTRO-MAGNET

MOVING ARMATURE OR PLUNGER

MAGNET SHADING COILS

(Source: Delmar/Cengage Learning)

Fig. 8–1 Three-pole, solenoid-operated magnetic switch contactor.

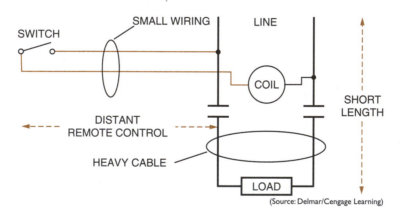

SWITCH SMALL WIRING LINE

COIL

SHORT LENGTH

DISTANT REMOTE CONTROL

HEAVY CABLE

LOAD

(Source: Delmar/Cengage Learning)

Fig. 8–2 An advantage of a remote control load.

Fig. 8–3 Open magnetic contactor "clapper" action, size 6. *(Courtesy of Schneider Electric)*

Fig. 8–4 Contactor, size 1. Contacts are accessible by removing the two front screws. *(Courtesy of Schneider Electric)*

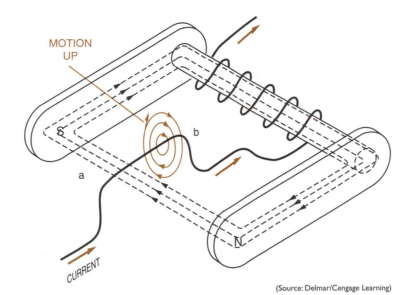

MOTION UP

CURRENT

(Source: Delmar/Cengage Learning)

Fig. 8–5 Illustration of the magnetic blowout principle. Straight conductor simulates arc.

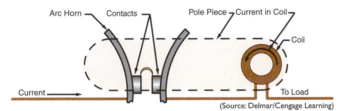

Arc Horn — Contacts — Pole Piece — Current in Coil

Coil

Current — To Load

(Source: Delmar/Cengage Learning)

Fig. 8–6 Section of blowout magnet with straight conductor replaced by a set of contacts. An arc is conducting between the contacts.

the load to provide better arc suppression. These magnetic blowout coils help to extinguish an electric arc at contacts opening under AC and DC loads. These arcs may be similar in intensity to the electric arc welding process. An arc-quenching device is used to ensure longer contact life. Because the hot arc is transferred from the contact tips very rapidly, the contacts remain cool and so last longer.

Contactor and motor starter contacts that frequently break heavy currents are subject to a destructive burning effect if the arc is not quickly extinguished. The arc that is formed when the contacts open can be lengthened and extinguished by motor action if it is in a magnetic field. This magnetic field is provided by

the magnetic blowout coil. Because the coil of the magnet is usually in series with the line, the field strength and extinguishing action are in proportion to the size of the arc.

Figure 8–5 is a sketch of a blowout magnet with a straight conductor (ab) located in the field and in series with the magnet. This figure can represent either DC polarity or instantaneous AC. With AC current, the blowout coil magnetic field and the conductor (arc) magnetic field will reverse simultaneously. According to Fleming's left-hand rule, motor action will tend to force the conductor in an upward direction. The application of the right-hand rule for a single conductor shows that the magnetic field around the conductor aids the main field on the bottom and opposes it on the top, thus producing an upward force on the conductor.

Figure 8–6 shows a section of Figure 8–5 with the wire (ab) replaced by a set of contacts. The contacts have started to open and there is an arc between them.

Figure 8–7 shows what happens because of the magnetic action. Part (A) shows the beginning deflection of the arc because of the effect of the motor action. Part (B) shows that the contacts are separated more than in Part (A) and the arc is

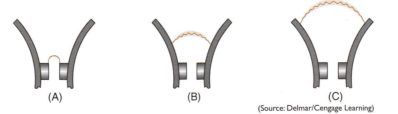

(A)   (B)   (C)

(Source: Delmar/Cengage Learning)

Fig. 8–7 Arc deflection between contacts.

beginning to climb up the horns because of the motor action and the effect of increased temperature. Part (C) of Figure 8–7 shows the arc near the tips of the horns. At this point, the arc is so lengthened that it will be extinguished.

The function of the blowout magnet is to move the arc upward at the same time that the contacts are opening. As a result, the arc is lengthened at a faster rate than will normally occur because of the opening of the contacts alone. It is evident that the shorter the time the arc is allowed to exist, the less damage it will do to the contacts. Most arc-quenching action is based upon this principle.

AC MECHANICALLY HELD CONTACTORS AND RELAYS

A mechanically held relay, or contactor, is operated by electromagnets, but the electromagnets are automatically disconnected by contacts within the relay. Accordingly, these relays are mechanically held in position and no current flows through the operating coils of these electromagnets after switching. It is apparent, therefore, that near continuous operation of multiple units of substantial size will lower the electrical energy requirements. Also, the magnetically held relay, in comparison, will change contact position on loss of voltage to the electromagnet, whereas the mechanically held relay will respond only to the action of the control device.

Sequence of Operation

Referring to Figure 8–8, when the "on" push button is pressed momentarily, current flows from L1 through the *on* push-button contact energizing the M coil through the now closed clearing contact to L2. The relay now closes and latches mechanically. At the same time it closes M contacts (in Figure 8–9), lighting a bank of lamps when the circuit breaker is closed. To unlatch the relay, thereby turning the lamps off, the *off* button is pressed momentarily, unlatching the relay and opening contacts M, turning off the lamps. Most operating coils are not designed for continuous duty. Therefore, they are disconnected automatically by contacts to prevent an accidental coil burnout. These coil clearing contacts change position alternately with a change in contactor latching position.

Figure 8–10 shows a control circuit with two locations: for controlling a latching contactor

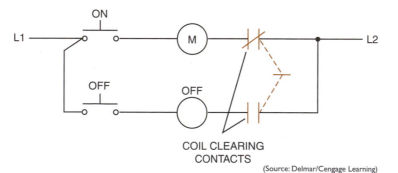

(Source: Delmar/Cengage Learning)

Fig. 8–8 Mechanically held contactor control circuit.

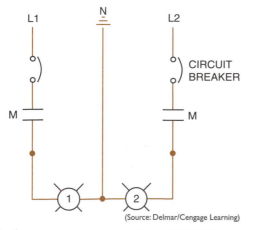

(Source: Delmar/Cengage Learning)

Fig. 8–9 Load connections for a 115/230-volt, three-wire lighting load.

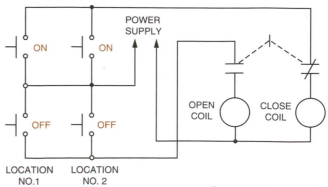

(Source: Delmar/Cengage Learning)

Fig. 8–10 Control circuit for a mechanically held lighting contactor controlled from two locations.

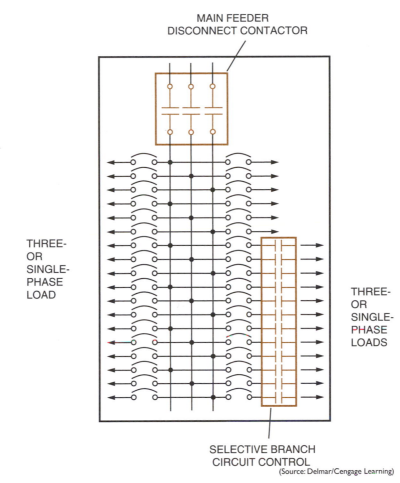

MAIN FEEDER
DISCONNECT CONTACTOR

THREE-
OR
SINGLE-
PHASE
LOAD

THREE-
OR
SINGLE-
PHASE
LOADS

SELECTIVE BRANCH
CIRCUIT CONTROL

(Source: Delmar/Cengage Learning)

Fig. 8–11 Mechanically held contactor loads for three-phase power.

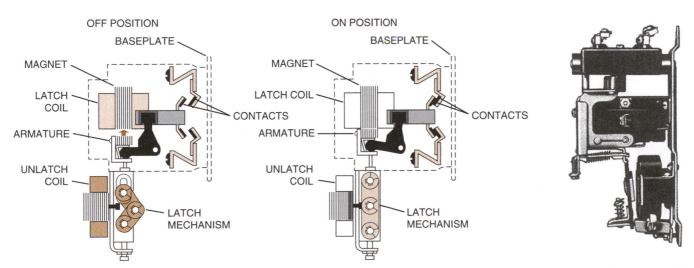

OFF POSITION

BASEPLATE

MAGNET

LATCH
COIL

CONTACTS

ARMATURE

UNLATCH
COIL

LATCH
MECHANISM

ON POSITION

BASEPLATE

MAGNET

LATCH COIL

CONTACTS

ARMATURE

UNLATCH
COIL

LATCH
MECHANISM

Fig. 8–12 Two types of latched-in or mechanically held contactors in service. The upper coil is energized momentarily to close contacts, and the lower coil is energized momentarily to open the contact circuit. The momentary energizing of the coil is an energy-saving feature. *(Courtesy of Schneider Electric)*

and for industrial lighting. All multiple control stations are connected parallel as shown. Figure 8–11 shows a three-phase power load application using one main contactor to disconnect the distribution panel. Selective single- or three-phase, branch circuits may be switched independently by other mechanically held contactors or relays.

These mechanically held contactors and relays are electromechanical devices, Figure 8–12. They provide a safe and convenient means of switching circuits where quiet operation, energy efficiency, and continuity of circuit connection are requirements of the installation. For example, circuit continuity during power failures is often important in automatic processing equipment, where a sequence of operations must continue from the point of interruption after power is resumed—rather than return to the beginning of the sequence. Quiet operation of contactors and relays is required in many control systems used in hospitals, schools, and office buildings. Mechanically held contactors and relays are generally used in locations where the slight hum, characteristic of AC magnetic devices, is objectionable.

In addition, mechanically held relays are often used in machine tool control circuits. These relays can be latched and unlatched through the operation of limit switches, timing relays, starter interlocks, time clocks, photoelectric cells, other control relays, or push buttons. Generally, mechanically held relays are available in 10- and 15-ampere sizes; mechanically held contactors are also available in sizes ranging from 30 amperes up to 1200 amperes. Industrial rated magnetic contactors built to International Electrotechnical Control (IEC), as well as U.S. standards, are generally acceptable throughout the world.

Study Questions

1. Why are contactors described as both control pilot devices and large magnetic switches controlled by pilot devices?

2. What is the principal difference between a contactor and a motor starter?

3. What causes the arc to move upward in a blowout magnet?

4. Why is it desirable to extinguish the arc as quickly as possible?

5. What will happen if the terminals of the blowout coil are reversed?

6. Why will the blowout coil also operate on AC?

7. What does IEC stand for?

8. "ON" control stations used in multiple locations are connected in series or parallel, for latched-in contactors?

UNIT 9

Timing Relays

Objectives

After studying this unit, the student should be able to

- Identify the primary types of timing relays.
- Explain the basic steps in the operation of the common timing relays.
- List the factors that affect the selection of a timing relay for a particular use.
- List applications of several types of timing relays.
- Draw simple circuit diagrams using timing relays.
- Identify *on-* and *off-*delay timing wiring symbols.

INTRODUCTION

A timing relay is similar to a control relay, except that certain of its contacts are designed to operate at a preset time interval, or time lag, after the unit is energized, *or* de-energized.

Many industrial control applications require timing relays that can provide dependable service and are easily adjustable over the timing ranges. The proper selection of timing relays for a particular application can be made after a study of the service requirements and with a knowledge of the operating characteristics inherent in each available device. A number of timing devices are manufactured with features suitable for a wide variety of applications.

APPLICATION

Timing relays are used for many applications. A few examples are

- to control the acceleration of contactors of motor starters
- to time the closing or opening of valves on refrigeration equipment
- for any application where the operating sequence requires a delay

PNEUMATIC TIMERS

The construction and performance features of the pneumatic (air) timer make it suitable for the majority of industrial applications. Pneumatic timers have the following characteristics:

- unaffected by normal variations in ambient temperature or atmospheric pressure
- adjustable over a wide range of timing periods
- good repeat accuracy
- available with a variety of contact and timing arrangements

This type of relay has a pneumatic time-delay unit that is mechanically operated by a magnet structure. The time-delay function depends on the transfer of air through a restricted orifice by the use of a reinforced synthetic rubber bellows or diaphragm. The timing range is adjusted by positioning a needle valve to vary the amount of orifice or vent restriction.

The process of energizing or de-energizing pneumatic timing relays can be controlled by pilot devices such as push buttons, limit switches, or thermostatic relays. Because the power drawn by a timing relay coil is small, sensitive control devices may be used to control the operating sequence.

Pneumatic timing relays are used for motor acceleration and in automatic control circuits. Automatic control is necessary in applications where repetitive accuracy is required, such as controls for machine tools and control of sequence operations, industrial process operation, and conveyor lines.

Pneumatic timers provide time delay through two arrangements. The first, *on delay,* means that the relay provides time delay when it is *energized;* the second arrangement, *off delay,* means the relay is *de-energized* when it provides time delay. Figure 9–1 is a typical pneumatic timer attachment. It is an "add-on" attachment for a type of relay. The lower screw is the timing adjustment that allows the limited passage of air to escape. The timer here has a flat rubber diaphragm. Figure 9–2 shows a rubber bellows type of timing action.

The speed with which the bellows rises is set by the position of the needle valve at the bottom. The setting of the needle valve determines the time interval that must elapse between the solenoid closing and the rise of the bellows to

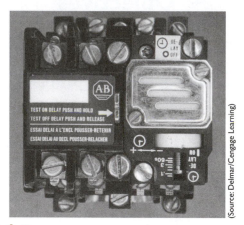

(Source: Delmar/Cengage Learning)

Fig. 9–1 Pneumatic timer attachment.

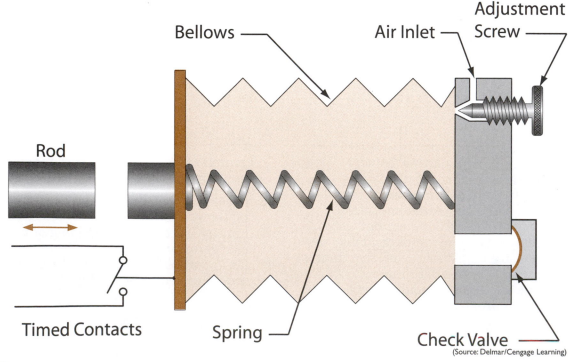

(Source: Delmar/Cengage Learning)

Fig. 9–2 Bellows-type timing unit.

operate the switch. If the needle valve is almost closed, an appreciable length of time is required for air to pass the valve and cause the bellows to rise.

When the solenoid is de-energized, the plunger drops by force of gravity and by the action of the reset spring. The downward movement of the plunger forces down, thus resetting the timer almost instantaneously.

Both types of pneumatic timing controls do the same thing, that is, to control the timing period, or cycle, of the timing period to open or close contacts.

TIMED CONTACTS

Table 9–1 illustrates standard diagram symbols for timed contacts. These symbols are used by manufacturers and others in wiring diagrams. The first symbol, from left to right, represents a normally open contact (when the relay is de-energized). When the relay is energized, there is a time delay in closing (on delay). Thus, it is a normally open, timed closing (NOTC) contact. Similarly, the second symbol is for a normally closed contact. After the relay coil is energized, there is a time delay before this contact opens (NCTO). The third symbol is for a normally open contact, at rest. When the relay coil is energized, this contact closes instantly. It will stay closed as long as the coil is energized. When the coil is de-energized, there is a time delay before the contact opens (NOTO). This mode is called off delay. The last symbol represents a normally closed timing contact. When the timer is energized, this contact opens rapidly and remains open. When current to the coil is disconnected, there is a time delay before the contact closes to its normal position.

Table 9–1 Standard Elementary Diagram Symbols for Timed Contacts

TIMED CONTACTS—CONTACT ACTION RETARDED AFTER COIL IS:			
Energized		De-energized	
NOTC	NCTO	NOTO	NCTC
On Delay		Off Delay	

INSTANTANEOUS CONTACTS

Most pneumatic timers may have nontimed contacts in addition to timing contacts. These nontimed contacts are controlled directly by the timer coil, as in a general-purpose control relay. These auxiliary contacts are often used on pneumatic timers to combine the functions of a standard or conventional control relay and a timing relay. The regular, or instantaneous contacts, are easily installed, or added, to the timer on the job site.

MOTOR-DRIVEN TIMERS

When a process has a definite on and off operation, or a sequence of successive operations, a motor-driven timer is generally used, Figure 9–3 and Figure 9–4. A typical application of a motor-driven timer is to control laundry washers where the loaded motor is run for a given period in one direction, reversed, and then run in the opposite direction. Motor-driven timers are also used where infrequent starting of large motors is required.

Generally, this type of timer consists of a small, synchronous motor driving a cam-dial assembly on a common shaft. A motor-driven timer successively closes and opens switch contacts, which are wired in circuits to energize control relays or contactors to achieve desired operations.

Min. Time Delay: 0.05 second

Max. Time Delay: 3 minutes

Minimum Reset Time: .075 second

Accuracy: ±10 percent of setting

Contact Ratings:

AC
6.0 A, 115 V
3.0 A, 230 V
1.5 A, 460 V
1.2 A, 550 V

DC
1.0 A, 115 V
0.25 A, 230 V

Operating Coils: Coils can be supplied for voltages and frequencies up to 600 volts, 60 hertz AC and 250 volts DC.

Types of Contacts: One normally open and one normally closed. Cadmium silver alloy contacts.

Fig. 9–3 Typical specifications.

Fig. 9–4 Motor-driven process timer in a general-purpose enclosure. *(Courtesy of Rockwell Automation)*

DC SERIES RELAY

Generally, the name of a relay is descriptive of its major purpose, construction, or principle of operation. A common application of DC series relays is to time the acceleration of DC motors. For example, the coil of the DC series relay, Figures 9–5(A) and 9–5(B), is connected in series with the starting resistance so that the starting cur rent of the motor passes through it. The contacts of the series relay are connected to an auxiliary circuit.

The relay contacts shown in Figure 9–5(A) are usually connected to control the coils of a magnetic contactor. The armature is light and constructed so that it is very fast in operation. As the starting current passes through the coil, the armature is pulled down (overcoming the resistance of a spring), causing small contacts to open. When the current in the coil has decreased to a predetermined value, the spring pulls the armature back and the contacts close. The value of current at which the coil loses control of the armature is determined by the spring setting. Crane control is another common application.

CAPACITOR TIME LIMIT RELAY

Assume that a capacitor is charged by connecting it momentarily across a DC line and then the capacitor DC is discharged through a relay coil. The current induced in the coil will decay slowly, depending on the relative values of capacitance, inductance, and resistance in the discharge circuit.

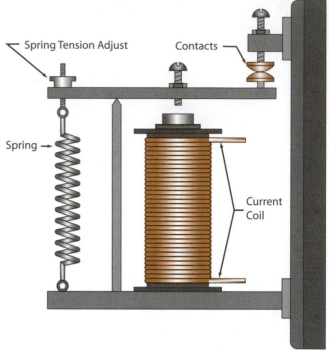

(Source: Delmar/Cengage Learning)

Fig. 9–5 (A) Series type relay; (B) DC series relay. *(Courtesy of General Electric Company)*

If a relay coil and a capacitor are connected in parallel to a DC line, Figure 9–6, the capacitor is charged to the value of the line voltage and a current appears in the coil. If the coil and capacitor combination is now removed from the line, the current in the coil will start to decrease along the curve shown in Figure 9–6.

If the relay is adjusted so that the armature is released at current i_1, a time delay of t_1 is obtained. The time delay can be increased to a value of t_2 by adjusting the relay so that the armature will not be released until the current is reduced to a value of i_2.

A potentiometer is used as an adjustable resistor to vary the time. This resistance-capacitance (RC) theory is used in industrial electronic and

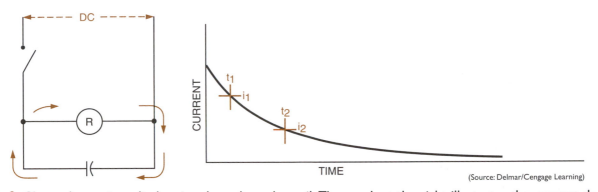

Fig. 9–6 Charged capacitor discharging through a relay coil. The graph at the right illustrates the current decrease in the coil.

solid-state controls also. This timer is highly accurate and is used in motor acceleration control and in many industrial processes.

ELECTRONIC TIMERS

Electronic timers have become increasingly popular in industrial applications for several reasons. They generally have better repeat accuracy, are less expensive, can be set for a greater range of time delays, and many perform multiple timing functions. Electronic timers use some type of electronic circuit to accomplish a time delay. Some use a basic RC time constant and others contain quartz clocks and integrated circuit timers. Many of these timers can be set for time delays with an accuracy of 0.01 second.

Electronic timers are available in various case styles. Many are designed to plug into eight- or eleven-pin tube sockets. These have the advantage of being easy to replace in the event they fail. Timers intended for use only as on-delay timers are generally designed to plug into an eight-pin socket, Figure 9–7. Timers that can be set for multiple functions or as off-delay timers generally plug into eleven-pin tube sockets, Figure 9–8.

Although electronic timers may be similar in function, they may connect differently. Different manufacturers stipulate different connection methods. Always refer to the manufacturer's specifications before trying to connect a timer into a circuit. The connection diagrams for two different eleven-pin timers are shown in Figure 9–9. Another type of electronic timer is shown in Figure 9–10. This timer has the case style of a standard control relay.

It is sometimes necessary to connect electronic timers into the circuits in a different manner than other types of timers. Mechanical timers, such as pneumatic timers, depend on

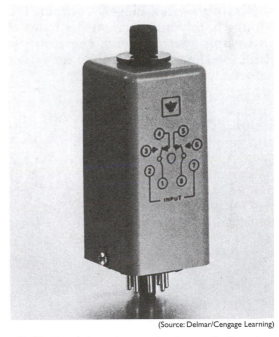

(Source: Delmar/Cengage Learning)

Fig. 9–7 On-delay electronic timer designed to plug into an eight-pin tube socket.

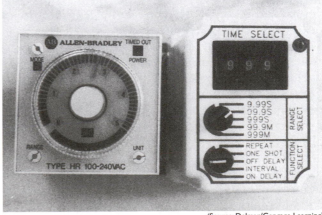

(Source: Delmar/Cengage Learning)

Fig. 9–8 Electronic timers that perform multiple functions or are intended for use as off-delay timers plug into eleven-pin sockets.

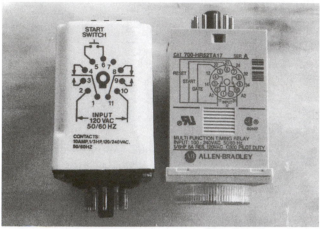

(Source: Delmar/Cengage Learning)

Fig. 9–9 Connection diagram for two eleven-pin timers.

Fig. 9–10 Solid-state timing relay. *(Courtesy of Eagle Signal)*

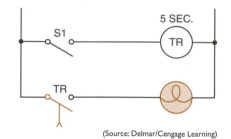

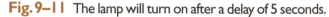

(Source: Delmar/Cengage Learning)

Fig. 9–11 The lamp will turn on after a delay of 5 seconds.

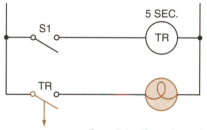

(Source: Delmar/Cengage Learning)

Fig. 9–12 The lamp will turn on immediately when switch S1 closes. The lamp will remain on for 5 seconds when switch S1 opens.

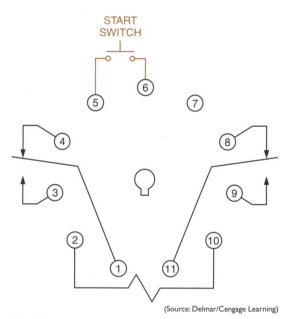

(Source: Delmar/Cengage Learning)

Fig. 9–13 Connection diagram for an eleven-pin timer.

power being applied to or removed from the coil to initiate the action of the timer. In the circuit shown in Figure 9–11, the action of the timer starts when switch S1 is closed. After a delay of 5 seconds the lamp will turn on. The lamp in Figure 9–12 will turn on immediately when switch S1 is closed. It will remain on for a period of 5 seconds when switch S1 is opened. The timers in Figure 9–11 and Figure 9–12 are pneumatic timers; their time-delay mechanism is accomplished by air refilling a bellows or diaphragm.

Electronic timers must have power connected to them in order to provide a time delay. Electronic on-delay timers are not problematic because their time delay starts when power is applied. Off-delay timers, however, must have power applied at all times to permit the timing circuit to operate. These timers use a separate contact to initiate the action of the timer.

Figure 9–13 illustrates the connection diagram for one of the timers shown in Figure 9–8. The diagram shows a normally open push button labeled START SWITCH connected between pins 5 and 6. This indicates that when the timer is used as an off-delay timer, the action of the timer is initiated by shorting pins 5 and 6. A simple circuit for connecting this timer is shown in Figure 9–14. Notice that pins 2 and 10 are connected directly to the power line. When the initiating switch closes, contact TR will close immediately and turn on the lamp. When

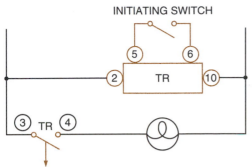

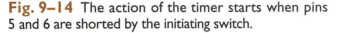

(Source: Delmar/Cengage Learning)

Fig. 9–14 The action of the timer starts when pins 5 and 6 are shorted by the initiating switch.

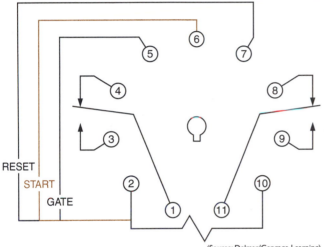

(Source: Delmar/Cengage Learning)

Fig. 9–15 Connection diagram for an Allen Bradley eleven-pin timer.

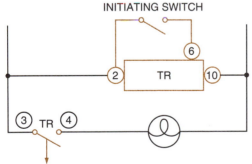

(Source: Delmar/Cengage Learning)

Fig. 9–16 The action of the timer starts when the initiating switch connects pins 2 and 6.

the initiating switch opens, the timer starts counting. At the end of the time delay, contact TR reopens and turns off the lamp.

The connection diagram for an Allen Bradley eleven-pin timer is shown in Figure 9–15. Although this timer is similar to the Dayton timer, the action of the timer is started by connecting pins 2 and 6 together. A schematic diagram for connecting this timer is shown in Figure 9–16.

SELECTING A TIMING RELAY

In selecting a timing relay for a specific application, the following factors should be carefully considered:

- Length of time delay required
- Timing range required
- Allowable error
- Cycle of operation and reset time
- Cost
- Additional requirements

Solid-state timers, like relays, are very reliable provided their cycle of operation is not exceeded. Overworking is harmful in that it creates internal heat, destroying the unit.

Length of Time Delay Required

The length of time delay required is determined by the type of machine or process that the timer will control. This time delay will range from a fraction of a second to as long as several minutes.

Timing Range Required

The phrase "timing range" means the various time intervals over which the timer can be adjusted. Timers are available that can be set for a time delay of 1 second, 100 seconds, or any value of delay between 1 and 100 seconds. When selecting a timer for use with a machine or process, the range should be wide enough to handle the various time-delay periods that may be required by the machine or process.

The exact timing value for any position within the timing range must be found by trial and error. A scale provided with a timer is intended primarily to permit a quick reset of the timer to the timing position previously determined to be correct for a given operation.

Allowable Error

All timers are subject to some error; that is, there may be a plus or minus time variation between successive timing operations for the same setting. The amount of error varies with the type of timer and the operating conditions. The error is usually stated as some percentage of the time setting.

The percentage of error for any timer depends on the type of timer, the ambient temperature (especially low temperatures), coil temperature, line voltage, and the length of time between operations.

Cycle of Operation Required and Reset Time

For one type of timer, the timer becomes operative when an electrical circuit opens or closes. A time delay then occurs before the application process begins. As soon as the particular process action is complete, the timer circuit resets itself. The circuit must be energized or de-energized each time the timing action is desired.

A second type of timer is called a process timer. When connected into a circuit, this timer provides control for a sequence of events, one after another. The cycle is repeated continuously until the circuit is de-energized.

An important consideration in the selection of a timer is the speed at which the timer resets. Reset time is the time required for the relay mechanism to return to its original position. Some industrial processes require that the relay reset instantaneously. Other processes require a slow reset time. The reset time varies with the type of timing relay and the length of the time delay.

Cost

When there are several timers that meet the requirements of a given application, it is advisable to select the timer with the smallest number of operating parts. In other words, select the simplest timer. In most cases, this timer will probably be the lowest in cost.

Additional Requirements

Several additional factors must be considered in selecting electromagnetic and solid-state timers:

- Type of power supply.
- Contact ratings.
- Timer contacts—a choice of normally open or normally closed contacts can usually be made.
- Temperature range—the accuracy of some timers varies with the temperature; however, temperatures below freezing may affect the timer accuracy.
- Dimensions—the amount of space available may have a bearing on the selection of a timer.

Study Questions

1. List several applications for a motor-driven timer.

2. Which timer is the most commonly used in industrial applications?

3. How is the pneumatic timer adjusted?

4. List six factors to be considered when selecting a timing relay for a particular use.

5. List several additional factors that will affect the selection of a timing relay.

6. On a separate sheet of paper, draw an elementary diagram of a fractional horsepower, manual motor starter. When the starter contact is closed, it will energize a pneumatic timing relay coil through the overload heater. After the timer coil is energized, the circuit will show a delayed closing contact that energizes a small motor on 120 volts.

7. Connect these components in Figure 9–17:
   a. The start push button on the timer coil
   b. Load 1 to NOTC
   c. Load 2 to NCTO
   d. Show proper symbols

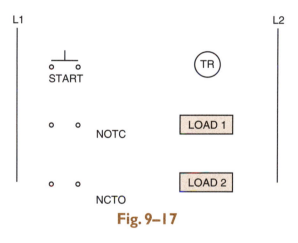

Fig. 9–17

8. Connect the following components in Figure 9–18:
   a. Start push button to the timer coil
   b. Load 1 to NCTC
   c. Load 2 to NOTO
   d. Show proper symbols

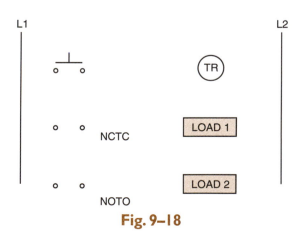

Fig. 9–18

Pressure Switches and Regulators

Objectives

After studying this unit, the student should be able to

- Describe how pressure switches, vacuum switches, and pressure regulators may control motors.

- List the adjustments that can be made to pressure switches.

- Identify wiring symbols used for pressure switches.

Any industrial application that has a pressure-sensing requirement can use a pressure switch, Figure 10–1(A). Figure 10–1(B) shows the internal construction of a typical pressure switch. A large variety of pressure switches are available to cover the wide range of control requirements for pneumatic or hydraulic machines such as welding equipment, machine tools, high-pressure lubricating systems, and motor-driven pumps and air compressors.

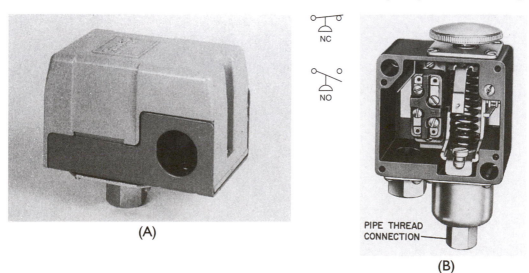

(A)

NC

NO

(B)

PIPE THREAD CONNECTION

Fig. 10–1 (A) Exterior view of a pressure switch. *(Courtesy of Siemens Energy & Automation)*. (B) Industrial pressure switch with the cover removed. Note the operating knob. Also note the wiring diagram symbols for normally closed and normally open contacts. *(Courtesy of Schneider Electric)*

The pressure ranges over which pressure switches can maintain control also vary widely. For example, a diaphragm-actuated switch can be used when a sensitive response is required to small pressure changes at low-pressure ranges, Figure 10–2. A metal bellows-actuated control is used for pressures up to 2000 pounds per square inch (psi). Piston-operated hydraulic switches are suitable for pressures up to 15,000 psi. In all of these pressure-controlled devices, a set of contacts is operated.

The most commonly used pressure switches are single-pole switches. Two-pole switches are also used for some applications. Field adjustments of the range and the differential pressure (or the difference between the cut-in and cut-out pressures) can be made for most pressure switches. The spring pressure determines the pressures at which the switch closes and opens its contacts.

Pressure regulators provide accurate control of pressure or vacuum conditions for systems. When they are used as pilot control devices with magnetic starters, pressure regulators are able to control the operation of liquid pump or air compressor motors in a manner similar to that of pressure switches. Reverse-action regulators can be used on pressure system interlocks to prevent the start of an operation until the pressure in the system has reached the desired level.

Pressure regulators consist of a Bourdon-type pressure gauge and a control relay. Delicate contacts on the gauge energize the relay and cause it to open or close. The relay contacts are used to control a large motor starter in order to avoid damage or burning of the gauge contacts. Standard regulators will open a circuit at high pressure and close it at low pressure. Special reverse-operation regulators will close the circuit at high pressure and open the circuit at low pressure.

Typical Application

Pressure switches are used in many common industrial applications. A circuit that is used to turn off a motor and turn on a pilot warning light is shown in Figure 10–3. In this circuit, a pressure switch is connected to a control relay. If the pressure should become too great, the

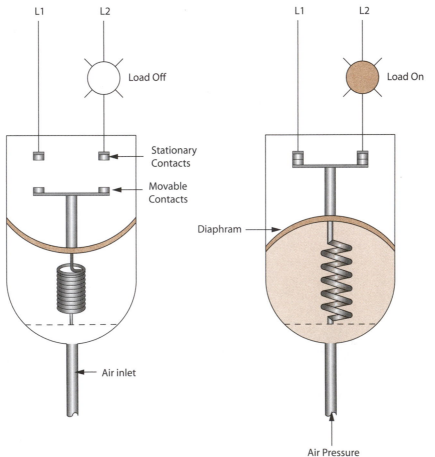

(Source: Delmar/Cengage Learning)

Fig. 10–2 Simplified pressure switch control.

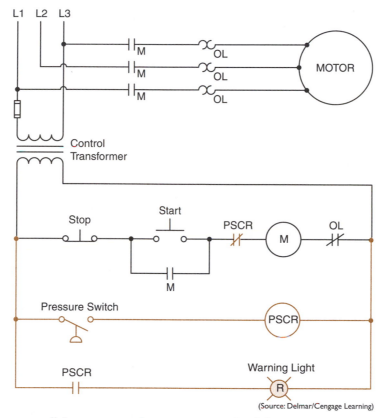

Fig. 10–3 High pressure turns off the motor and turns on a warning light.

(Source: Delmar/Cengage Learning)

control relay will open a normally closed contact connected to a motor starter to stop the motor. A normally open pressure switch control relay (PSCR) contact will close and turn on a pilot light to indicate a high-pressure condition. Notice that in this example circuit, the pressure switch needs both normally open and normally closed contacts. This is not a common contact arrangement for a pressure switch. To solve the problem, the pressure switch controls the action of a control relay. This is a very common practice in industrial control systems.

Study Questions

1. Describe how pressure switches are connected to start and stop (a) small motors and (b) large motors.

2. What type of pressure switch is generally used to sense small changes in low-pressure systems?

3. A pressure switch is set to cut in at a pressure of 375 psi and cut out at 450 psi. What is the pressure differential for this switch?

4. A pressure switch is to be installed on a system with pressures that can range from 1500 psi to 1800 psi. What type of pressure switch should be used?

5. A pressure switch is to be installed in a circuit that requires it to have three normally open contacts and one normally closed contact. The switch actually has one normally open contact. What must be done to permit this pressure switch to operate in this circuit?

6. Refer to the circuit shown in Figure 10–3. If the pressure should become high enough for the pressure switch to close and stop the motor, is it possible to restart the motor before the pressure drops to a safe level?

7. Refer to the circuit shown in Figure 10–3. Assume that the motor is running and an overload occurs and causes the OL contact to open and disconnect coil M to stop the motor. What effect does the opening of the overload contact have on the pressure switch circuit?

UNIT 11

Float Switches

Objectives

After studying this unit, the student should be able to

- Describe the operation of float switches.

- List the sequence of operation for sump pumping or tank filling.

- Draw wiring symbols for float switches.

A float switch is used when a pump motor must be started and stopped according to changes in the water (or other liquid) level in a tank or sump. Float switches are designed to provide automatic control of AC and DC pump motor magnetic starters and automatic direct control of light motor loads.

The operation of a float switch is controlled by the upward or downward movement of a float placed in a water tank. The float movement causes a rod-operated, Figure 11–1, or chain and counterweight, Figure 11–2, assembly to open or close electrical contacts. The float switch contacts

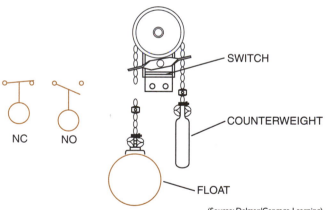

(Source: Delmar/Cengage Learning)

Fig. 11–2 Chain-operated float switch with normally closed (NC) and normally open (NO) wiring symbols.

may be either normally open or normally closed and may not be submerged. Float switches may be connected to a pump motor for tank or sump pumping operations or tank filling, depending on the contact arrangement. See Figure 11–3 for the elementary wiring circuits.

Electronic methods of control may be used in place of the basic float switch for liquid level control. These methods include electrode sensing, electronic wave transmission and detection, and other combinations.

(Source: Delmar/Cengage Learning)

Fig. 11–1 Rod-operated float switch.

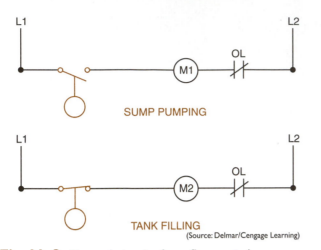

SUMP PUMPING

TANK FILLING

(Source: Delmar/Cengage Learning)

Fig. 11–3 Control circuits for a float switch.

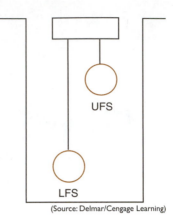

(Source: Delmar/Cengage Learning)

Fig. 11–4 Two sump float switches.

Figure 11–4 shows a sump pump installation with two float switches. Figure 11–5 is a control circuit of the installed switches. The lower float switch closes due to rising liquid. The sump continues to fill until the upper float causes its switch to close. When it closes, the pump motor starts and closes maintaining contact "M," sealing its circuit. The motor will pump out until the lower float opens the circuit after emptying. This arrangement extends the range of pumping.

An electrode, or stainless steel probe system (Figure 11–6), is similar to Figure 11–5 but

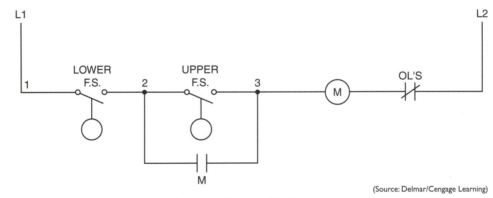

(Source: Delmar/Cengage Learning)

Fig. 11–5 Control circuit for two float switches as in Figure 11–4.

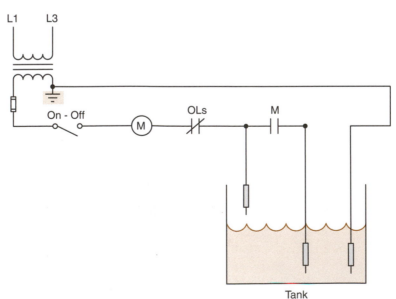

Tank

(Source: Delmar/Cengage Learning)

Fig. 11–6 The electrode or stainless steel probe system.

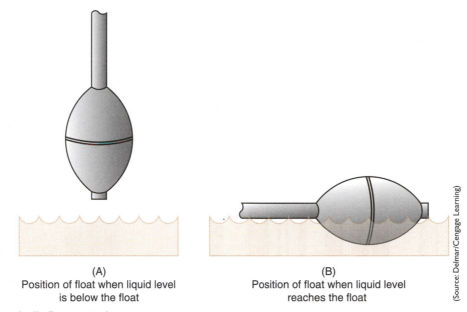

(A)
Position of float when liquid level
is below the float

(B)
Position of float when liquid level
reaches the float

(Source: Delmar/Cengage Learning)

Fig. 11–7 Mercury-bulb float switches.

conducts a circuit in the liquid instead of using floats. For satisfactory operation, the system must be well grounded.

MERCURY BULB FLOAT SWITCH

Another float switch that has become increasing popular is the mercury bulb-type of float switch. This type of float switch does not depend on a float rod or chain to operate. The mercury bulb switch appears to be a rubber bulb connected to a conductor. A set of mercury contacts is located inside the bulb. When the liquid level is below the position of the bulb, it is suspended in a vertical position, Figure 11–7(A). When the liquid level rises to the position of the bulb, it changes to a horizontal position, Figure 11–7(B). This change of position changes the state of the contacts in the mercury switch. Because the mercury bulb float switch does not have a differential setting as does the rod- or chain-type of float switch, it is necessary to use more than one mercury bulb float switch to control a pump motor. The differential level of the liquid is determined by suspending mercury bulb switches at different heights in the tank. Figure 11–8 illustrates the use of four mercury bulb-type switches used to operate two pump motors and provide a high-liquid level alarm. The control circuit is shown in Figure 11–9. Float switch FS1 detects the lowest point of liquid level in the tank and is used to turn both pump motors off. Float switch FS2 starts the

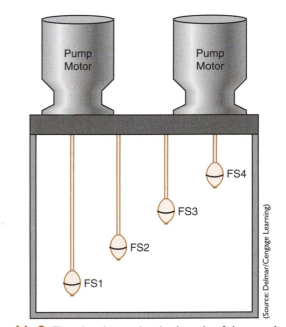

(Source: Delmar/Cengage Learning)

Fig. 11–8 Float level is set by the length of the conductor.

first pump when the liquid level reaches that height. If pump 1 is unable to control the level of the tank, float switch FS3 will start pump motor 2 if the liquid level should rise to that height. Float switch FS4 operates a warning light and buzzer to warn that the tank is about to overflow. A reset button can be used to turn off the buzzer, but the warning light will remain on until the water level drops below the level of float switch FS4.

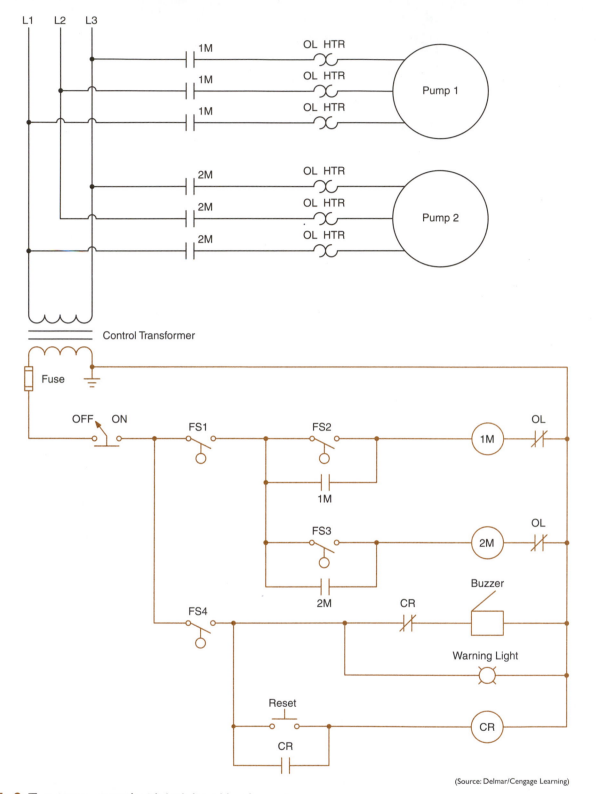

(Source: Delmar/Cengage Learning)

Fig. 11–9 Two-pump control with high liquid level warning.

Study Questions

1. Describe the sequence of operations required to (a) pump sumps and (b) fill tanks.

2. In Figure 11–5, why does the sump pump motor not start when the lower float switch closes?

3. In Figure 11–5, why does the pump not stop while emptying when the upper float switch opens its circuit?

4. Refer to the circuit in Figure 11–9. What is the purpose of control relay CR in this circuit?

5. In Figure 11–9, the reset button is
    a. single-acting normally closed
    b. single-acting normally open
    c. double-acting normally closed
    d. double-acting normally open

6. In Figure 11–9, assume that both motors are running. What would be the action of the circuit if the overload contact connected in series with coil 1M should open?
    a. Both motors will continue to run.
    b. Both motors will stop running.
    c. Pump motor 1 will continue to operate and pump motor 2 will stop running.
    d. Pump motor 1 will stop running and pump motor 2 will continue to operate.

7. In Figure 11–9, what is the purpose of float switch FS4?

UNIT 12

Flow Switches

Objectives

After studying this unit, the student should be able to

- Describe the purpose and functions of flow switches.
- Connect a flow switch to other electrical devices.
- Draw and read wiring diagrams of systems using flow switches.

A flow switch is a device that can be inserted in a pipe so that when liquid or air flows against a part of the device called a paddle, a switch is activated (Figure 12–1). This switch either closes or opens a set of electrical contacts. The contacts may be connected to energize motor starter coils, relays, or indicating lights. In general, a flow switch contains both normally open and normally closed electrical contacts, Figure 12–2.

Figure 12–3 shows a flow switch installed in a pipeline tee. Half couplings are welded into larger pipes for flow switch installations.

Typical applications of flow switches are shown in Figure 12–4 through Figure 12–8. These applications are commonly found in the chemical and petroleum industries. Vaporproof electrical connections must be used with vaporproof switches.

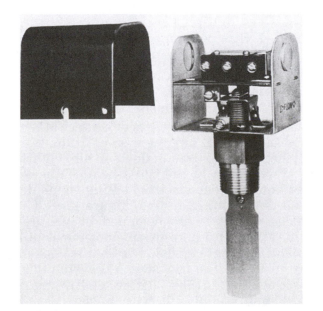

Fig. 12–1 Flow switch. *(Courtesy of ITT McDonnell & Miller)*

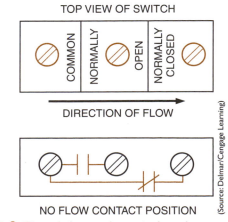

TOP VIEW OF SWITCH

DIRECTION OF FLOW

NO FLOW CONTACT POSITION

(Source: Delmar/Cengage Learning)

Fig. 12–2 Typical electrical terminals and contact arrangement of a flow switch.

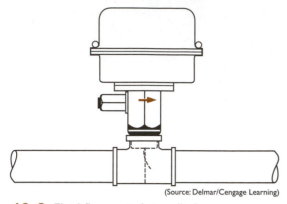

(Source: Delmar/Cengage Learning)

Fig. 12–3 Fluid flow switch installed.

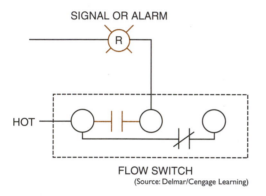

(Source: Delmar/Cengage Learning)

Fig. 12–4 Flow switch used to sound alarm or light signal when flow occurs using normally open contacts.

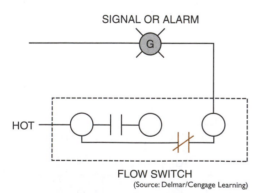

(Source: Delmar/Cengage Learning)

Fig. 12–5 Flow switch used to sound alarm or light signal when there is no flow using normally closed contacts.

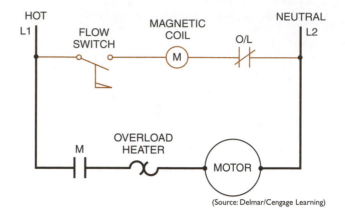

(Source: Delmar/Cengage Learning)

Fig. 12–6 Flow switch used with single-phase circuit starts the motor when flow occurs and stops the motor when there is no flow.

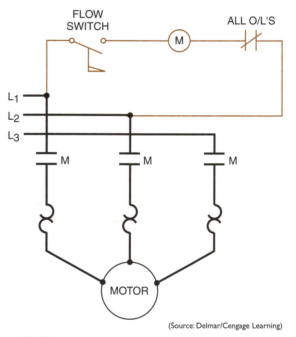

(Source: Delmar/Cengage Learning)

Fig. 12–7 Flow switch used with three-phase circuit starts the motor when flow occurs and stops the motor when there is no flow.

The insulation of the wire leading to the switches must be adequate to withstand the high temperature of the liquid inside the pipe. (Consult the *NEC©* for insulation temperature ratings.)

The elementary diagram in Figure 12–6 shows the electrical connections for a flow switch. When the flow switch contact closes with sufficient fluid or air movement, the M coil is energized. This closes larger contact M in the motor circuit, starting the motor. Airflow or sail switches are also used in ducts in air conditioning systems. An M coil could be a contactor to close and energize electric heating elements in an air duct heating system, Figure 12–8. These heating elements are designed to operate at their rating with a sufficient movement of air. Without adequate airflow, the heating elements may burn out. The flow switch prevents this by detecting the airflow before energizing the heater and stopping it with a flow loss. The action is similar in cooling systems. Iced-over

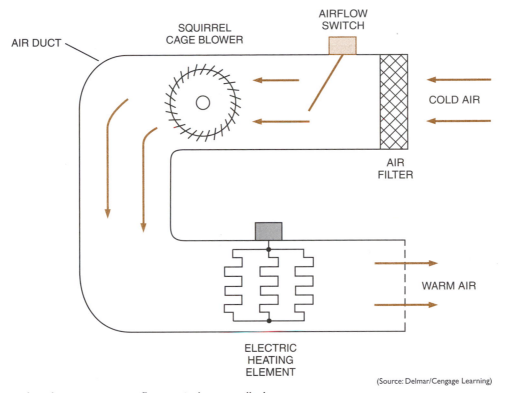

AIRFLOW
SWITCH

SQUIRREL
CAGE BLOWER

AIR DUCT

COLD AIR

AIR
FILTER

WARM AIR

ELECTRIC
HEATING
ELEMENT

(Source: Delmar/Cengage Learning)

Fig. 12–8 An air duct heating system, flow switch controlled.

refrigeration coils would restrict airflow. The flow switch is used to detect insufficient air movement over the refrigeration coils, stopping the refrigeration compressor and starting a defrost cycle. Although the construction of airflow switches differs from that of liquid flow switches, the electrical connections are similar.

An airflow switch is shown in Figure 12–9.

(Source: Delmar/Cengage Learning)

Fig. 12–9 Airflow switch.

Study Questions

1. What are typical uses of flow switches?

2. Draw a line diagram to show that a green light will glow when liquid flow occurs.

3. Draw a one-line diagram showing a bell that will ring in the absence of flow. Include a switch to turn off the bell manually.

Limit Switches and Proximity Control

Objectives

After studying this unit, the student should be able to

- Explain the use of limit switches in the automatic operation of machines and machine tools.

- Wire a simple two-wire circuit using a limit switch.

- Read and draw normally open (NO) and normally closed (NC) wiring symbols in a simple circuit.

- Explain the use and operation of solid-state proximity switches.

LIMIT SWITCHES

The automatic operation of machinery requires the use of switches that can be activated by the motion of the machinery. The limit switch, Figure 13–1, is used to convert this mechanical motion into an electrical signal to switch circuits. The repeat accuracy of the switches must be reliable and the response virtually instantaneous.

The size, operating force, stroke, and manner of mounting are all critical factors in the installation of limit switches due to mechanical limitations in the machinery. The electrical ratings of the switches must be carefully matched to the loads to be controlled.

The limit switch contacts should not be overloaded. Contacts must be selected according to the proper load, that is, the proper current and voltage, according to the manufacturer's specifications. Figure 13–2 shows the switch used to its capacity, and how it is used to control a load above its capacity by controlling a contactor to switch a heavier load.

In general, the operation of a limit switch begins when the moving machine or moving part of a machine strikes an operating lever, which actuates the switch, Figure 13–1. The

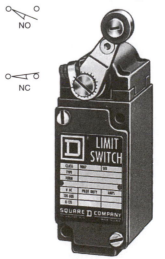

Fig. 13–1 Limit switch shown with wiring symbols. (*Courtesy of Schneider Electric*)

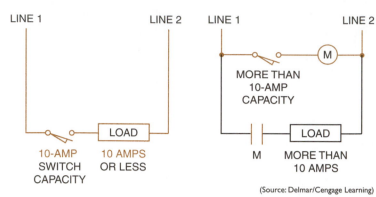

(Source: Delmar/Cengage Learning)

Fig. 13–2 Proper use of limit switches.

limit switch, in turn, affects the electrical circuit controlling the machine and its movement.

Limit switches are used as pilot devices in the control circuits of magnetic starters to start, stop, speed up, slow down, or reverse electric motors. Limit switches may be used either as control devices for regular operation or as emergency switches to prevent the improper functioning of machinery. They may be momentary contact (spring return) or maintained contact types. Most limit switches, as shown in Figure 13–1, have normally open and normally closed contacts that are mechanically connected, or interlocked.

MICRO-LIMIT SWITCHES

Another type of limit switch often used in different types of control circuits is the micro-limit switch or *microswitch*. Microswitches are much smaller in size than the limit switch shown in Figure 13–1. This permits them to be used in small spaces that would not be accessible to the larger device. Another characteristic of the microswitch is that the actuating plunger requires only a small amount of travel to cause the contacts to change position. Most microswitches require a movement of approximately 0.015 inch or 0.38 mm to change the contact position. Switching the contacts with this small amount of movement is accomplished by spring loading the contacts as shown in Figure 13–2. A small amount of movement against the spring will cause the movable contact to snap from one position to another, Figure 13–3.

Electrical ratings for the contacts of the basic microswitch are generally in the range of 250 volts AC and 10 to 15 amperes, depending on the type of switch. The basic microswitch can be obtained in different case styles and with different operating levers and mechanisms, Figure 13–4.

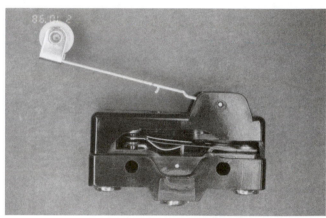

(Source: Delmar/Cengage Learning)

Fig. 13–3 Spring-loaded contact of a micro-limit switch.

(Source: Delmar/Cengage Learning)

Fig. 13–4 Microswitches in different case styles.

SOLID-STATE PROXIMITY SWITCHES

Solid-state proximity switches use various methods to sense objects. They can be switched by a nearby or passing object. No physical contact is necessary to operate a proximity switch as it is with a limit switch. These self-contained,

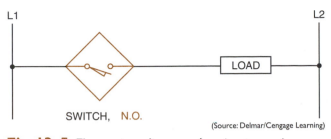

Fig. 13–5 Elementary diagram showing two-wire construction (AC or DC).

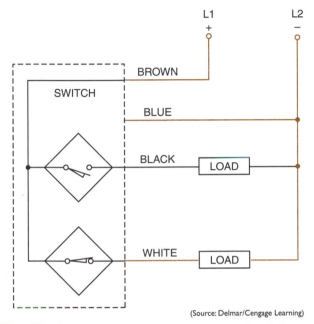

Fig. 13–6 Typical four-wire construction for DC, normally open and normally closed.

Fig. 13–7 Capacitive proximity sensor with a glass-reinforced plastic housing. *(Courtesy of Eaton Corporation)*

Fig. 13–8 Compact inductive proximity sensor with pin base connector. *(Courtesy of Eaton Corporation)*

two-wire devices are available in normally open or normally closed switching positions, both AC and DC. Figure 13–5 shows a simple, normally open proximity switch in a circuit. The switch can be connected in series or parallel. Figure 13–6 shows normally open and normally closed switch wiring for DC, four-wire operation. Manufacturers color code or adequately mark their terminals for wiring. Always keep the manufacturer's installation instructions for future reference.

Capacitive Proximity Sensor (Switches)

Capacitive proximity sensors are designed to detect both metallic and nonmetallic targets. They are ideally suited for liquid level control (such as float switch control) and for sensing powdered or granulated materials. For best operation, they should be used in an environment with relatively constant temperature and humidity. Figure 13–7 features glass-reinforced plastic housing capacitive sensors. Different shapes and mountings are available.

Inductive Proximity Sensors

Description. Compact inductive proximity sensors are also complete, self-contained switches. They are designed to detect the presence of all metals without making contact. Some models are quite elaborate and include LEDs to indicate "power on," "switch start," a "shorted condition," and when a "target is present." Figure 13–8 shows an inductive proximity sensor.

The head of the sensor can be rotated in the field to any one of five sensing positions: top, or any one of four side positions, Figure 13–9.

Pancake-type (low-profile) solid-state sensors are designed to detect targets at distances up to 100 mm (4.0 in.). The sensors can be used for applications where the distance between the sensor and target varies greatly or where long-distance sensing is required. The sensors are two-wire devices that can be corrected in the same manner as a mechanical limit switch, Figure 13–10.

Fig. 13–9 Proximity sensor with rotating sensor head. *(Courtesy of Pepperl and Fuchs, Inc.)*

Fig. 13–10 Low-profile, pancake-type inductive proximity sensor. *(Courtesy of Eaton Corporation)*

Fig. 13–11 Solid-state cylindrical proximity sensor. *[Part (A), courtesy of Eaton Corporation; Part (B), courtesy of Pepperl and Fuchs, Inc.]*

A solid-state cylindrical-shaped proximity switch can be used in small or hard-to-reach places. This type of switch can be wired in series or parallel, Figure 13–11.

The sensing head is mounted where the object is to be sensed. It is then wired to a chassis (amplifier/power supply). The chassis usually contains the electronic circuitry and a screwdriver slot for a sensitivity adjustment.

Operation. Inductive proximity switches operate on the eddy current principle. When a metal object moves into the electromagnetic field of the sensing head, eddy currents are induced in the object. This causes a change in the loading of the oscillator (electronics producing the field), which then operates the output device.

Application. Proximity switches are typically used on production assembly lines in the following ways: to detect metal objects through non-metal barriers, where metal objects must be differentiated from nonmetal objects; where touching of the objects must be avoided, as with freshly painted or highly polished parts; as void and jam detectors; for counting; in special and extreme environments; and in many other types of applications.

Photodetectors

Photodetectors are very popular throughout the industry because of their reliability and flexibility. Photodetectors basically consist of a light source and a light detector. The light source is generally a long-life incandescent lamp or LED and the light detector can be a photodiode or a photoconductive device called a cad cell. Photodiodes are used if high-speed operation is an important factor. Photodetectors used in burglar alarm systems commonly use infrared LEDs as the light source because it is invisible to the human eye.

Some photodetectors use a separate transmitter and receiver, Figure 13–12. In this example, the photodetector is used to count cans on a

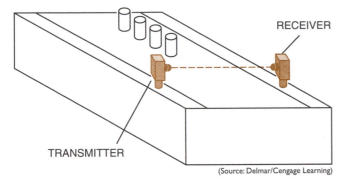

Fig. 13–12 Photodetector with separate transmitter and receiver.

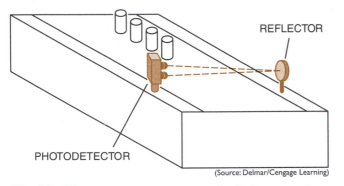

Fig. 13–13 Photodetector contains both the transmitter and receiver.

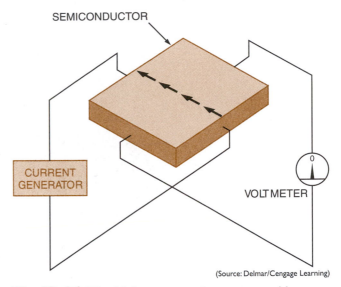

Fig. 13–14 The Hall generator is constructed by passing a constant current through a semiconductor.

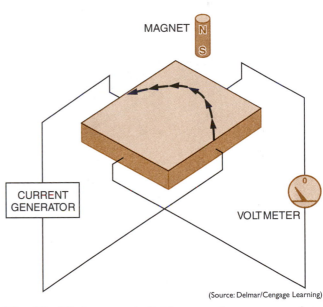

Fig. 13–15 A magnetic field causes the current path to bend.

conveyor line. Each time a can breaks the light beam a counter counts up once. Other types of photodetectors house both the transmitter and receiver in the same case. This type of unit uses a reflector, Figure 13–13. Because both the transmitter and receiver are mounted at the same position, the advantage of this unit is that control wiring is connected at only one location. This permits the reflector to be mounted in places where it would be difficult to run control wires. Some reflector units have a range of 20 feet or more.

Hall Effect Sensors

Hall effect sensors have become increasingly popular for many applications. Hall effect sensors detect the presence of a magnetic field but do not work on the principle of induction. Hall generators operate by passing a constant current through a piece of semiconductor material, Figure 13–14. A zero center voltmeter is connected across the opposite sides of the semiconductor. As long as the current flows through the center of the semiconductor, there is no voltage drop across it. However, if a magnet is brought near the semiconductor the current path will deflect to one side, producing a voltage across the semiconductor, Figure 13–15. As long as the

magnetic field is present, a voltage will be produced across the semiconductor.

Hall effect sensors are fast acting and are often used to measure the speed of rotating objects, Figure 13–16. In this example, a magnet with salient (projecting) poles is mounted on a motor shaft. The Hall sensor can determine the speed of the motor by measuring the number of pulses per second.

Another use for the Hall sensor in industrial applications is the bumperless limit switch, Figure 13–17. In this example, a piece of moving equipment has a permanent magnet attached to it. A Hall sensor is held stationary at the position where a limit switch is needed. When the

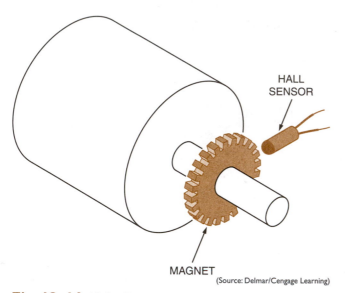

Fig. 13–16 Hall effect sensor used to measure motor speed.

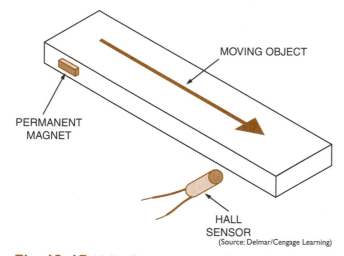

Fig. 13–17 Hall effect sensor used as a bumperless limit switch.

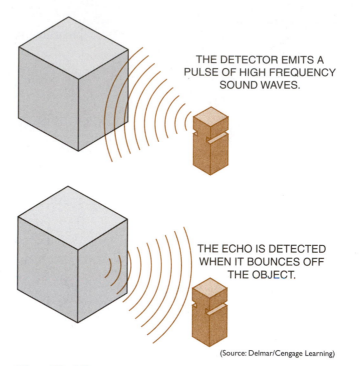

THE DETECTOR EMITS A PULSE OF HIGH FREQUENCY SOUND WAVES.

THE ECHO IS DETECTED WHEN IT BOUNCES OFF THE OBJECT.

(Source: Delmar/Cengage Learning)

Fig. 13–18 Ultrasonic detectors operate by emitting high-frequency sound waves.

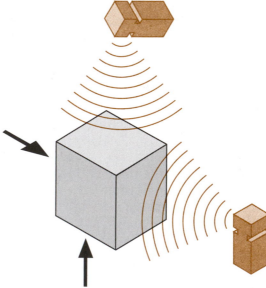

(Source: Delmar/Cengage Learning)

Fig. 13–19 Ultrasonic detectors can be used to position objects.

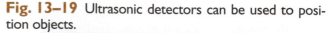

Fig. 13–20 Ultrasonic proximity detector. (*Courtesy of TURCK, Inc.*)

magnet moves in line with the sensor, it has the same effect as hitting the arm of a limit switch. Because the magnet does not make physical contact with the sensor, however, there is no wear on the sensor and no mechanical arm to break or get out of adjustment.

Ultrasonic Proximity Detectors

Another type of proximity detector is the ultrasonic detector. This detector emits a high-frequency sound wave and then detects the echo when it bounces off the object, Figure 13–18. Ultrasonic detectors can accurately calculate the distance to an object by measuring the time delay between pulse emission and pulse return. For this reason ultrasonic detectors are often used to position objects, Figure 13–19. An ultrasonic proximity detector is shown in Figure 13–20.

Study Questions

1. Draw a simple circuit showing how a red pilot light is energized when a limit switch is operated by a moving object.

2. On what principle does an inductive proximity sensing unit work?

3. What is the advantage of a photodiode in a photodetector?

4. What type of photodetector is often used in burglar alarm systems? Why?

5. What do Hall effect sensors detect?

6. Hall effect sensors do not depend on electromagnetic induction to operate. Explain how they can sense the presence of a magnetic field if induction is not involved.

7. How do ultrasonic proximity detectors operate?

8. Explain how an ultrasonic detector can measure the distance to an object.

9. What is the average amount of movement necessary to cause the contacts of a micro-limit switch to change position?

10. How is it possible for the miscroswitch to change contact position with such a small amount of movement of the activating arm?

Phase Failure Relays

Objectives

After studying this unit, the student should be able to

- Explain the purpose of phase failure relays.
- List the hazards of phase failure and phase reversal.
- Connect solid-state phase monitors.

If two phases of the supply to a three-phase induction motor are interchanged, the motor will reverse its direction of rotation. This action is called phase reversal. In the operation of elevators and in many industrial applications, phase reversal may result in serious damage to the equipment and injury to people using the equipment. In other situations, if a fuse blows or a wire to a motor breaks while the motor is running, the motor will continue to operate on single phase but will experience serious overheating. To protect motors against these conditions, phase failure and reversal relays are used.

One type of phase failure relay, Figure 14–1, uses coils connected to two lines of the three-phase supply. The currents in these coils set up a rotating magnetic field that tends to turn a copper disc clockwise. This clockwise torque actually is the result of two torques. One polyphase torque tends to turn the disc clockwise, and one single-phase torque tends to turn the disc counterclockwise.

The disc is kept from turning in the clockwise direction by a projection resting against a stop. However, if the disc begins to rotate in the counterclockwise direction, the projecting arm will move a toggle mechanism to open the line contactors and remove the motor from the line. In other words, if one line is opened, the polyphase

torque disappears and the remaining single-phase torque rotates the disc counterclockwise. As a result, the motor is removed from the line. In case of phase reversal, the polyphase torque helps the single-phase torque turn the disc counterclockwise, and again, the motor is disconnected from the line.

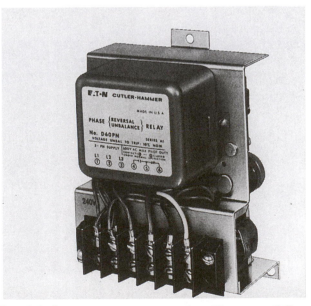

Fig. 14–1 Phase monitoring relay. (*Courtesy of Eaton Corporation*)

135

Other designs of phase failure and phase reversal relays are available to protect motors, machines, and personnel from the hazards of open phase or reverse phase conditions. For example, one type of relay consists of a static, current-sensitive network connected in series with the line and a switching relay connected in the coil circuit of the starter. The sensing network continuously monitors the motor line currents. If one phase opens, the sensing network immediately detects it and causes the relay to open the starter coil circuit to disconnect the motor from the line. A built-in delay of five cycles prevents nuisance dropouts caused by fluctuating line voltages.

A solid-state phase monitoring relay is shown in Figure 14–2. This relay provides protection in the event of a voltage unbalance or a phase reversal. The unit automatically resets after the correct voltage conditions return. Indicating lights show when the relay is activated. See the electrical connections in Figure 14–3.

LARGE MOTOR CURRENT MONITOR

Figure 14–3 shows the elementary wiring diagram for large motor current monitoring. Current transformers are used for the input to the motor protector module. These modules, or relays, feature ambient temperature–compensated, solid-state overload protection, phase loss, phase reversal, locked motor rotor, and instantaneous fault protection. Terminal connections are clearly marked for proper installation.

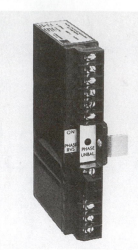

Fig. 14–2 Solid-state phase monitor relay. *(Courtesy of Eaton Corporation)*

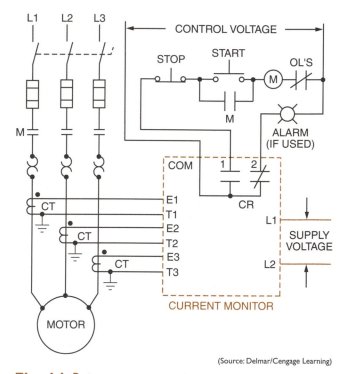

(Source: Delmar/Cengage Learning)

Fig. 14–3 Large motor solid-state current monitor connections.

Study Questions

1. What is the purpose of phase failure relays?

2. What are the hazards of phase failure and phase reversal?

UNIT 15

Solenoid Valves

Objectives

After studying this unit, the student should be able to

- Describe the purpose and operation of two-way solenoid valves.
- Describe the purpose and operation of four-way solenoid valves.
- Connect and troubleshoot solenoid valves.
- Read and draw wiring symbols for solenoid valves.

Valves are mechanical devices designed to control the flow of fluids such as oil, water, air, and other gases. Many valves are manually operated, but electrically operated valves are most often used in industry because they can be placed close to the devices they operate, thus minimizing the amount of piping required. Remote control is accomplished by running a single pair of control wires between the valve and a control device such as a manually operated switch or an automatic device.

A solenoid valve is a combination of two basic units: an assembly of the solenoid (the electromagnet) and plunger (the core), and a valve containing an opening in which a disc or plug is positioned to regulate the flow. The valve is opened or closed by the movement of the magnetic plunger. When the coil is energized, the plunger (core) is drawn into the solenoid (electromagnet). The valve operates when current is applied to the solenoid. The valve returns automatically to its original position when the current ceases.

Most control pilot devices operate a single-pole switch, contact, or solenoid coil. The wiring

diagrams of these devices are not difficult to understand, and the actual devices can be connected easily into systems. It is recommended that the electrician know the *purpose* of and understand the *action* of the *total* industrial system for which various electrical control elements are to be used. In this way, the electrician will find it easier to design or assist in designing the electrical control system. It will also be easier for the electrician to install and maintain the control system.

TWO-WAY SOLENOID VALVES

Two-way (in and out) solenoid valves, Figure 15–1, are magnetically operated valves that are used to control the flow of Freon, methyl chloride, sulfur dioxide, and other liquids in refrigeration and air-conditioning systems. These valves can also be used to control water, oil, and airflow.

Standard applications of solenoid valves generally require that the valve be mounted directly in line in the piping with the inlet and

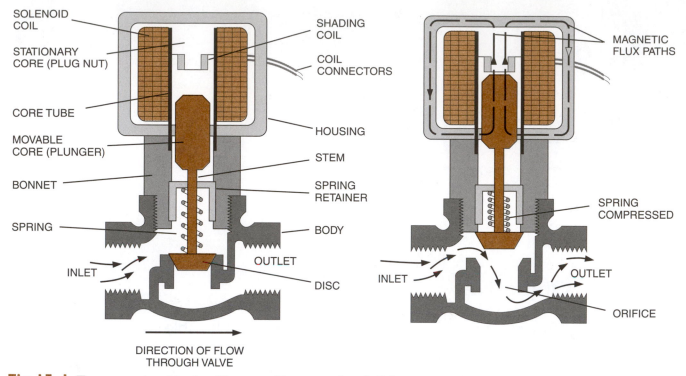

Fig. 15–1 Two-way solenoid valve. *(Courtesy of Automatic Switch Co.)*

outlet connections directly opposite each other. Simplified valve mounting is possible with the use of a bottom outlet, which eliminates elbows and bends. In the bottom outlet arrangement the normal side outlet is closed with a standard pipe plug.

The valve body is usually a special brass forging that is carefully checked and tested to ensure that there will be no seepage due to porosities. The armature, or plunger, is made from a high-grade stainless steel. The effects of residual magnetism are eliminated by the use of a kickoff pin and spring, which prevent the armature from sticking. A shading coil ensures that the armature will make a complete seal with the flat surface above it to eliminate noise and vibration.

It is possible to obtain DC coils with a special winding that will prevent the damage that normally results from an instantaneous voltage surge when the circuit is broken. Surge capacitors are not required with this type of coil.

To ensure that the valve will always seat properly, it is recommended that strainers be used to prevent grit or dirt from lodging in the orifice or valve seat. Dirt in these locations will cause leakage. The inlet and outlet connections of the valve must not be reversed. The tightness of the valve depends to a degree on pressure acting downward on the sealing disc. This pressure is possible only when the inlet is connected to the proper point as indicated on the valve.

FOUR-WAY SOLENOID VALVES

Electrically operated, four-port, four-way air valves are used to control a double-acting cylinder, Figure 15–2.

When the coil is de-energized, one side of the piston is at atmospheric pressure, and the other side is acted on by the line pressure compressed air. When the valve magnet coil is energized, the valve exhausts the high-pressure side of the piston to atmospheric pressure. As a result, the piston and its associated load reciprocate in response to the valve movement.

Four-way valves are used extensively in industry to control the operation of the pneumatic cylinders used on spot welders, press clutches, machine and assembly jig clamps, tools, and lifts.

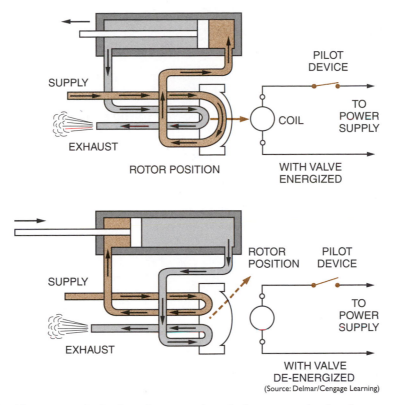

Fig. 15–2 Control of double-acting cylinder by a four-way, electrically operated valve shown with elementary diagrams.

Study Questions

1. Why is it important to understand the purpose and action of the total operational system when working with controls?

2. If an electrically controlled, two-way solenoid valve is leaking, what is the probable cause?

3. What is the difference between a two-way solenoid valve and a flow switch?

Temperature Switches

Objectives

After studying this unit, the student should be able to

- Describe the operation of temperature switches.
- List several applications of temperature switches.
- Read and draw wiring symbols and circuits for temperature switches.
- Describe a solid-state control system.

Temperature switches are designed to provide automatic control of temperature regulating equipment, Figure 16–1. Industrial temperature controllers are used for applications where the temperature to be controlled is higher than the normal or ambient temperature. In general, the applications of temperature control are more concerned with the temperature regulation of liquids than of gases. This is a result of the relatively greater conductivity between the bulb and a liquid, as compared with the conductivity between the bulb and a gas (such as air). Thus, if air or gas temperatures are to be controlled, the sensitivity of the sensor decreases and the difference between the on and off points widens.

To operate a temperature switch, the gas vapor or liquid pressure expands a metal bellows against the force of a spring. The expanding bellows moves an operating pin. The pin snaps a precision switch to its operating position when a preset point is reached. (This action is similar to the operation of industrial pressure switches.) The pressure that operates the switch is proportional to the temperature of a liquid in a closed bulb. The precision switch

snaps back to its normal position when the pressure in the bellows drops enough to allow the main spring to compress the bellows.

There are many types of thermostats that can be used to provide automatic control of space heating and cooling equipment. A typical thermostat is a temperature-actuated, two-wire (or more) control pilot device (switch). Temperature-actuated switches are used to control circuits in order to operate heaters, blowers, fans, solenoid valves, pumps, and other devices. Figure 16–2 illustrates a two-wire temperature switch controlling a cooling fan motor. When a high temperature closes the temperature switch, it energizes coil F. Coil F closes F contacts, starting the motor.

SOLID-STATE TEMPERATURE CONTROL SYSTEMS

There are many different types of temperature switches, both mechanical and electronic. Figure 16–3 shows a panel mounting, solid-state temperature control system. This system

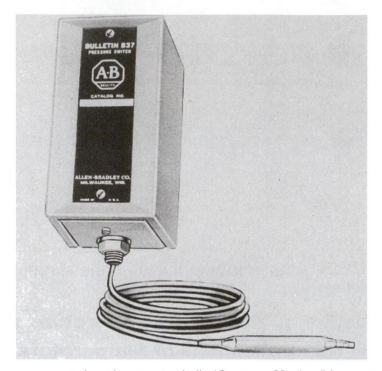

Fig. 16–1 Industrial temperature switch with extension bulb. *(Courtesy of Rockwell Automation)*

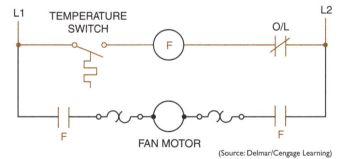

(Source: Delmar/Cengage Learning)

Fig. 16–2 Temperature switch controlling a cooling fan.

Fig. 16–3 Solid-state temperature control system using thermistor-type temperature sensors. *(Courtesy of Schneider Electric)*

is designed with three high-accuracy thermistor temperature sensors. These sensors can be used to protect three-phase transformer coils from overheating. The sensors transmit the internal coil temperatures to a microprocessor (minicomputer), which is programmed to provide a digital display in degrees on the front panel of the control module. Electrical hookup terminals are provided on the back for alarms and to energize fan motors.

Study Questions

1. What are some applications of temperature-actuated switches?

2. How is the average electrician involved with electronic temperature controls?

SECTION

4

Basic Control Circuits

UNIT 17

Two-Wire Controls

Objectives

After studying this unit, the student should be able to

- List the advantages of two-wire controls.
- Connect two-wire devices to motor starters.
- Read and draw simple diagrams for two-wire controls.

A two-wire control may be a toggle switch, pressure switch, float switch, limit switch, thermostat, or any other type of switch having definite *on* and *off* positions. As indicated in Unit 16, devices of this type are generally designed to handle small currents. Two-wire control devices will not carry sufficient current to operate large motors. In addition, 230-volt motors and three-phase motors require more contacts than the one contact usually provided on two-wire devices.

Two-wire controls may be connected to operating coils of magnetic switches, as shown in Figure 17–1. When the switch is closed, the control circuit is completed through the coil (M). When the coil is energized, it closes the contacts at M and runs the motor. When the switch is opened, the coil is de-energized and the contacts open to stop the motor. In the case of an overload, the thermal heaters open the overload contacts in the control circuit and de-energize the coil, thus stopping the motor. Two-wire control provides no-voltage (or low-voltage) release. When the starter is wired, as shown in Figure 17–1, it will operate automatically in response to the control device. A human operator is not required. The control maintaining

contact 2-3 (shown in the wiring diagram) is furnished with the starter. However, this contact is not used in two-wire control. For simplicity, this contact is omitted from the two-wire elementary diagram. The motor starter in Figure 17–1(A) is a line voltage, or across-the-line, starter (described in Unit 3).

The circuit shown in Figure 17–1 employs the use of line voltage controls. This simply means that the control components must be rated to operate on the voltage of the line supplying power to the motor. If the power line is 480 volts, the contactor coil must be rated at 480 volts and the contacts of the two-wire control device and the overload relay must be capable of interrupting this voltage. Two-wire controls often use a control transformer to reduce the control voltage to a lower value, Figure 17–2. Typical control voltages are 120 and 24 volts. Control systems that operate on 24 volts are often used in hazardous areas. The *NEC*® permits the use of intrinsically safe systems in areas that contain hazardous vapors. Intrinsically safe systems cannot provide enough energy to ignite the surrounding atmosphere. These systems are much more cost effective than installing explosion proof fixtures.

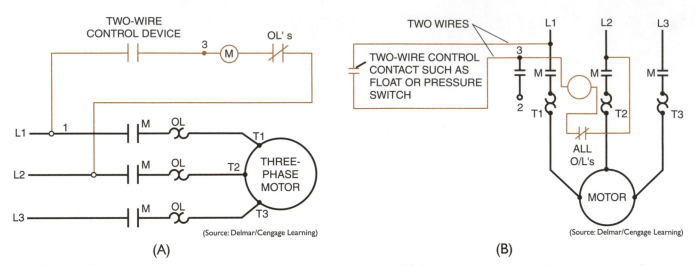

(A)   (B)

Fig. 17–1 (A) Basic two-wire control circuit—elementary diagram. (B) Basic two-wire control circuit—wiring diagram.

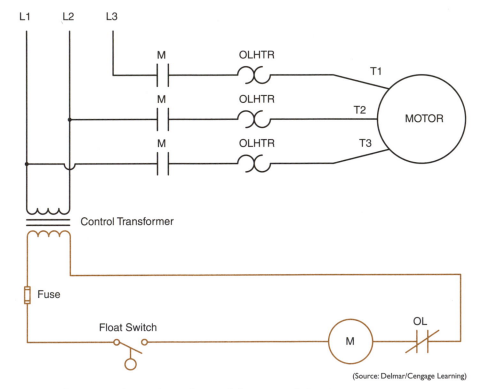

(Source: Delmar/Cengage Learning)

Fig. 17–2 A control transformer reduces the voltage of the control circuit.

Study Questions

1. What are some advantages of the use of two-wire control?

2. How many wires are required for a two-wire control device?

3. In the conduit diagram shown in Figure 17–3, determine how many wires are in each conduit between each piece of equipment. Identify each wire terminal connection and the quantity of wires, as shown in the example, above the disconnect switch.

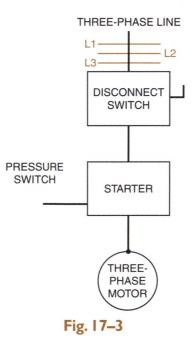

Fig. 17–3

4. Draw an elementary diagram of a pressure switch controlling a motor starter only.

5. What is an intrinsically safe system?

6. What is the most common control voltage used in an intrinsically safe system?

Three-Wire and Separate Controls

Objectives

After studying this unit, the student should be able to

- Describe the basic sequence of operations for a three-wire control circuit.

- Describe the basic sequence of operations for a separate control circuit.

- List the advantages of each type of circuit.

- Connect three-wire and separate control circuits.

- Connect pilot indicating lights.

- Connect an alarm silencing circuit.

- Read and draw three-wire and separate control circuits.

THREE-WIRE CONTROLS

A three-wire control circuit uses momentary contact, start-stop stations, and a holding circuit interlock connected in parallel with the start button to maintain the circuit.

In general, three-wire devices are connected, as shown in Figure 18–1. Although the arrangement of the various parts may vary from one manufacturer's switch to another, the basic circuit remains the same.

The sequence of operation for this circuit is as follows: When the start button is pushed, the circuit is completed through the coil (shown as M) and the contacts at M close. The power circuit contacts to the motor also close (not shown). When the start button is released, the holding

contact on M keeps this auxiliary contact on. When the starter is used in this manner, it is said to be "maintaining" or "sealing." With the holding contact closed, the circuit is still complete through the coil. If the stop button is pushed, the circuit is broken, the coil loses its energy, and the contacts at M open. When the stop button is released, the circuit remains open because both the holding contact and the start button are open. The start button must be pushed again to complete the circuit. The operation of the overload protection opens the control circuit, resulting in the same effect.

If the supply voltage fails, the circuit is de-energized. When the supply voltage returns, the circuit remains open until the start button is pushed again. This arrangement is called

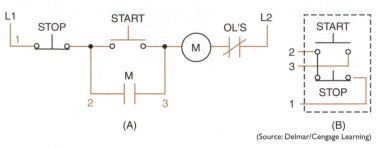

Fig. 18–1 Basic three-wire control circuit.

no-voltage protection and protects both the operator and the equipment.

The push-button station wiring diagram, Figure 18–1(B), represents the physical station. It shows the relative positions of the units, the internal wiring, and the connections to the starter. The wires to the terminals are labeled 1, 2, and 3 (giving rise to the name "three-wire control"). For similar diagrams, review Figure 4–2 and Figure 5–5. A pilot light or alarm can be added to this circuit to indicate when the motor is running (Figure 18–2). For this case, the pilot light is connected between control terminal 3 and line 2. Normally open auxiliary contacts are used to switch the pilot light on and off. When the motor is running, these contacts are closed; when the motor is stopped, the contacts are open and the pilot light is off. Except for this

modification, the circuit is a basic three-wire, push-button control circuit.

PUSH-TO-TEST PILOT LIGHTS

It is necessary to restart a motor after it has been stopped by a three-wire control circuit with *low-voltage protection*. An indicating lamp is often used to signal this when a motor stops so that it can be restarted after the problem causing the stoppage is cleared. Because pilot lights are an important component in such cases, they are tested frequently to ensure operation. Push-to-test pilot lights show immediately if a circuit is off or if the lamp is burned out. Part of Figure 18–2 shows the schematic wiring for such a circuit. The three-wire motor starter control circuit is wired as usual. Note that the pilot lamp is energized from terminal 3 down to C, through a normally closed push button to L2. To test, the lamp is pushed, opening the circuit at C and closing it across L1. With a push-button arrangement, the lamp is tested directly across lines 1 and 2.

ALARM SILENCING CIRCUIT

Sirens, horns, and loud buzzers are also used in production systems to call attention to malfunctions. The problem is acknowledged by attempts to silence this "noise pollution." A typical circuit is shown in Figure 18–3. Assume a high pressure in an industrial system is dangerous to continue. Such a condition will close a pressure switch. When this switch closes, the alarm is sounded through normally closed contact S. In addition, the red indicating lamp lights. When alerted, maintenance personnel can silence the alarm by depressing the *off* push button. The red light continues to silently announce the problem until it is cleared. After the pressure switch opens, the alarm system can be reactivated by pressing the *on* button. There are virtually unlimited control circuits using the three-wire system.

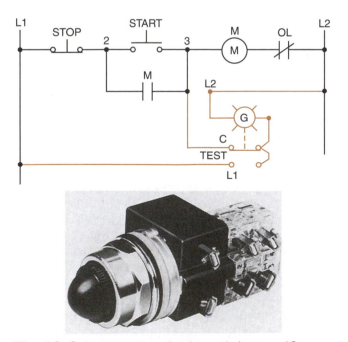

Fig. 18–2 Push-to-test oil-tight push button. *(Courtesy of Schneider Electric)*

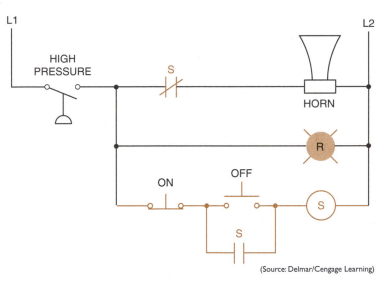

Fig. 18–3 Alarm silencing circuit.

SEPARATE CONTROL

It is sometimes desirable to operate push buttons or other control devices at some voltage lower than the motor voltage. In the control system for such a case, a separate source—such as an isolating transformer or an independent voltage supply—provides the power to the control circuit. This independent voltage is separate from the main power supply for the motor.

One form of separate control is shown in Figure 18–4. This is an elementary diagram of a cooling circuit for a commercial air-conditioning installation. When the thermostat calls for cooling, the compressor motor starter coil (shown as M) is energized through the stepdown isolating transformer. When coil M is energized, power contacts in the 240-volt circuit close to start the refrigeration compressor motor. Because the control circuit is separated from the power circuit by the isolating control transformer, there

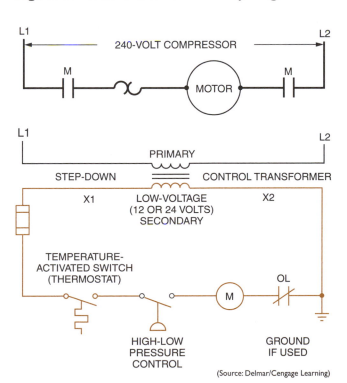

Fig. 18–4 Separate control used in air-conditioning cooling circuit.

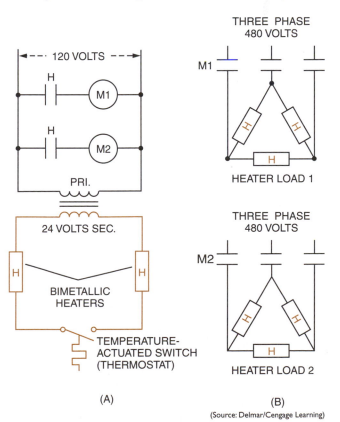

Fig. 18–5 Separate control used in air-conditioning heating circuit.

is no electrical connection between the two circuits. For this reason, the wire jumper attached to L2 on a starter should be removed for different voltages. However, the overload relay control contact must be included in the separate control wiring. The maintenance technician must also ensure that the control transformer voltages match the voltages used and that the proper connections are made.

Another example of separate control is the thermal relay shown in Figure 18–5. This control consists of a line to a low-voltage transformer, two low-voltage heaters (H), and two normally open line voltage contacts. When heat is required, the thermostat closes the 24-volt circuit to the bimetallic heaters. In seconds, the action of the heaters causes the normally open contacts (H) to close on the 120-volt side. As a result, contactor coils M1 and M2 are energized. When the M1 and M2 contactors close, they energize the three-phase heaters on 480 volts, Figure 18–5(B). Circulating blowers (not shown) will start when the heaters are energized. Three separate voltages are shown in Figure 18–5. Both two-wire and three-wire controls are used with separate controls.

Study Questions

1. Draw a control circuit with no-voltage protection. Describe how this method of wiring protects the machine operator.

2. In the conduit layout shown in Figure 18–6, determine how many wires are contained in each conduit between each piece of equipment. Identify each wire with a proper terminal marking, as shown in the example, above the disconnect switch.

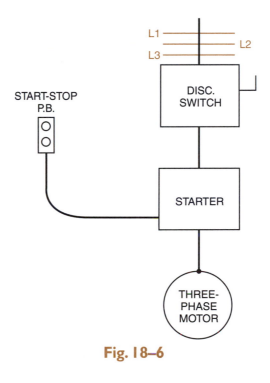

Fig. 18–6

UNIT 19

Hand-Off Automatic Controls

Objectives

After studying this unit, the student should be able to

- State the purpose of hand-off automatic controls.
- Connect hand-off automatic controls.
- Read and draw diagrams using hand-off automatic controls.

Hand-off automatic switches are used to select the function of a motor controller either manually or automatically. This selector switch may be a separate unit or built into the starter enclosure cover. A typical control circuit using a single-break selector switch is shown in Figure 19–1.

With the switch turned to the HAND (manual) position, coil (M) is energized all the time and the motor runs continuously. In the OFF position, the motor does not run at all. In the AUTO position, the motor runs whenever the two-wire control device is closed. An operator does not need to be present. The control device may be a pressure switch, limit switch, thermostat, or other two-wire control pilot device.

The heavy-duty, three-position double-break selector switch shown in Figure 19–2 is also used for manual and automatic control. When the switch is turned to HAND, the coil is energized, bypassing the automatic control device in the AUTO position.

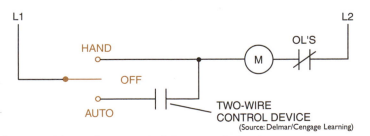

Fig. 19–1 Standard-duty, three-position selector switch in control circuit.

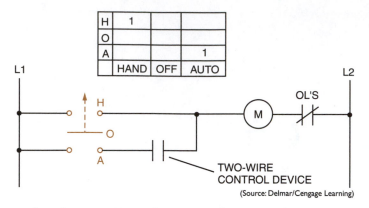

	HAND	OFF	AUTO
H	1		
O			
A			1

Fig. 19–2 Diagram of a heavy-duty, three-position selector switch in control circuit.

(Source: Delmar/Cengage Learning)

Study Questions

1. A selector switch and two-wire pilot device cannot be used to control a large motor directly but rather must be connected to a magnetic starter. Explain why this is true.

2. Determine the minimum number of wires in each conduit shown in Figure 19–3.

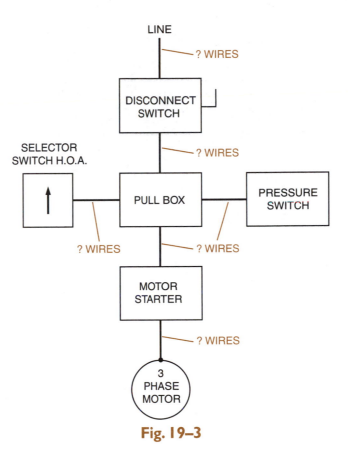

Fig. 19–3

3. The circuit shown in Figure 19–4(A) is used to control the operation of a motor. The motor can be operated manually or it can be operated automatically by a pressure switch. The circuit components necessary to construct this circuit are shown in Figure 19–4(B). In Figure 19–4(B), dashed lines on the starter indicate factory connections. It is also assumed that the output side of the load contacts are connected to the input side of the overload heaters. Using the schematic in Figure 19–4(A), draw lines to connect the components in Figure 19–4(B).

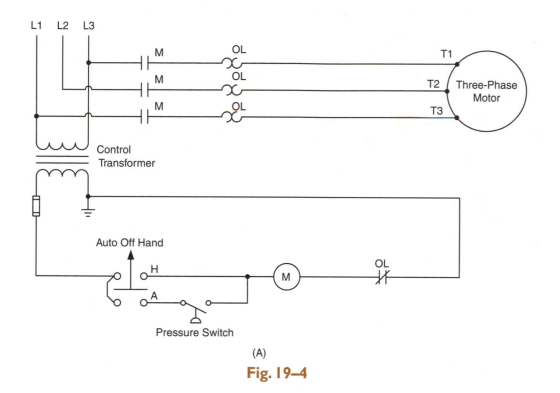

(A)

Fig. 19–4

UNIT 20

Multiple Push-Button Stations

Objectives

After studying this unit, the student should be able to

- Read and interpret diagrams using multiple push-button stations.
- Draw diagrams using multiple push-button stations.
- Connect multiple push-button stations.

The conventional three-wire, push-button control circuit may be extended by the addition of one or more push-button control stations. The motor may be started or stopped from a number of separate stations by connecting all start buttons in parallel and all stop buttons in series. The operation of each station is the same as that of the single push-button control in the basic three-wire circuit covered in Unit 18. Note in Figure 20–1 that pressing any stop button de-energizes the coil. Pressing any start button energizes the coil.

When a motor must be started and stopped from more than one location, any number of start and stop buttons may be connected. Another possible arrangement is to use only one start-stop station and several stop buttons at different locations to serve as emergency stops.

Multiple push-button stations are used to control conveyor motors on large shipping and receiving freight docks.

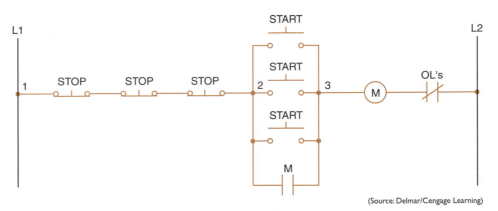

(Source: Delmar/Cengage Learning)

Fig. 20–1 Three-wire control using a momentary contact, multiple push-button station.

Study Questions

1. Using the schematic diagram in Figure 20–2, determine the number of control wires needed to control the three-phase motor. Use the terminal and wire identification shown in the schematic as well as proper power circuit identification.

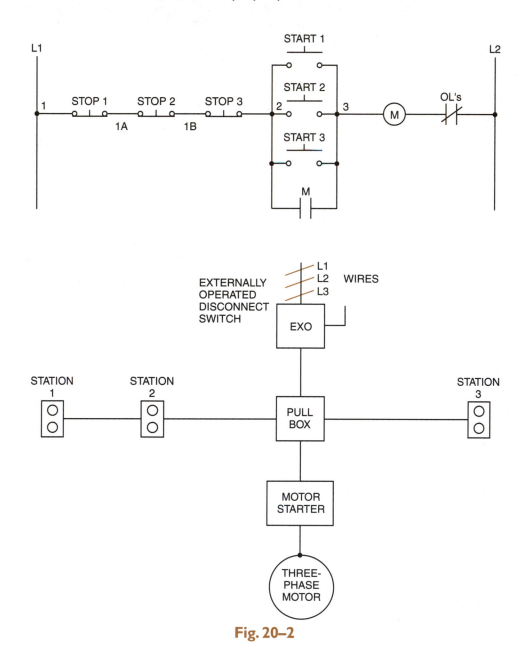

Fig. 20–2

Interlocking Methods
for Reversing Control

Objectives

After studying this unit, the student should be able to

- Explain the purpose of the various interlocking methods.
- Read and interpret wiring and line diagrams of reversing controls.
- Read and interpret wiring and line diagrams of interlocking controls.
- Wire and troubleshoot reversing and interlocking controls.

The direction of rotation of three-phase motors can be reversed by interchanging any two-motor leads to the line. If magnetic control devices are to be used, then reversing starters accomplish the reversal of the motor direction, Figure 21–1. Reversing starters wired to NEMA standards interchange lines L1 and L3, Figure 21–2. To do this, two contactors for the starter assembly are required—one for the forward direction and one for the reverse direction, Figure 21–3. A technique called *interlocking* is used to prevent the contactors from being energized simultaneously or closing together and causing a short circuit. There are three basic methods of interlocking.

MECHANICAL INTERLOCK

A mechanical interlocking device is assembled at the factory between the forward and reverse contactors. This interlock locks out one contactor at the beginning of the stroke of either contactor to prevent short circuits and burnouts.

The mechanical interlock between the contactors is represented in the elementary diagram of Figure 21–4 by the broken line between the coils. The broken line indicates the coils F and R cannot close contacts simultaneously because of the mechanical interlocking action of the device.

When the forward contactor coil (F) is energized and closed through the forward push button, the mechanical interlock prevents the accidental closing of coil R. Starter F is blocked by coil R in the same manner. The first coil to close moves a lever to a position that prevents the other coil from closing its contacts when it is energized. If an oversight allows the second coil to remain energized without closing its contacts, the excess current in the coil due to the lack of the proper inductive reactance will damage the coil.

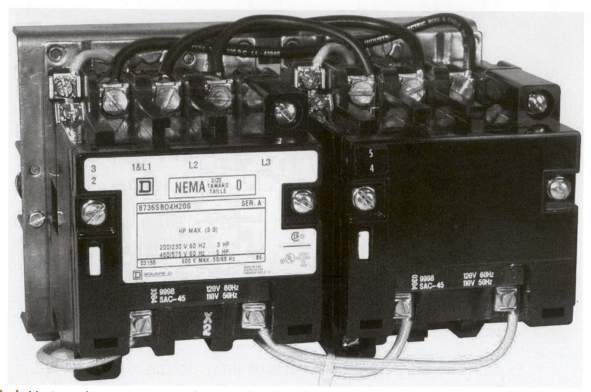

Fig. 21–1 Horizontal reversing starter shown without overload relay. *(Courtesy of Schneider Electric)*

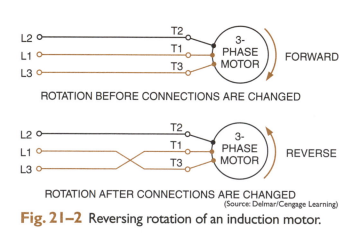

ROTATION BEFORE CONNECTIONS ARE CHANGED

ROTATION AFTER CONNECTIONS ARE CHANGED
(Source: Delmar/Cengage Learning)

Fig. 21–2 Reversing rotation of an induction motor.

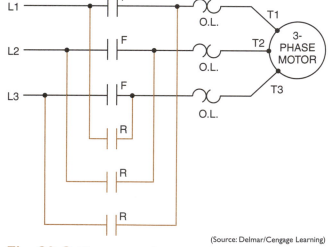

(Source: Delmar/Cengage Learning)

Fig. 21–3 Elementary diagram of a reversing starter power circuit.

Note in the elementary diagram of Figure 21–4 that the stop button must be pushed before the motor can be reversed.

Reversing starters are available in horizontal and vertical construction. A vertical starter is shown in Figure 21–5.

A mechanical interlock is installed on the majority of reversing starters in addition to the use of one or both of the following electrical methods: push-button interlock and auxiliary contact interlock.

PUSH-BUTTON INTERLOCK

Push-button interlocking is an electrical method of preventing both starter coils from being energized simultaneously.

When the forward button in Figure 21–6 is pressed, coil F is energized and the normally open (NO) contact F closes to hold in the forward contactor. Because the normally closed (NC) contacts are used in the forward and reverse push-button units, there is no need to press the

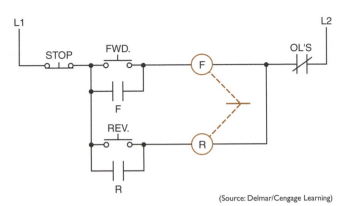

(Source: Delmar/Cengage Learning)

Fig. 21–4 Mechanical interlock between the coils prevents the starter from closing all contacts simultaneously. Only one contactor can close at a time.

Fig. 21–5 Vertical reversing motor starter. (*Courtesy Schneider Electric*)

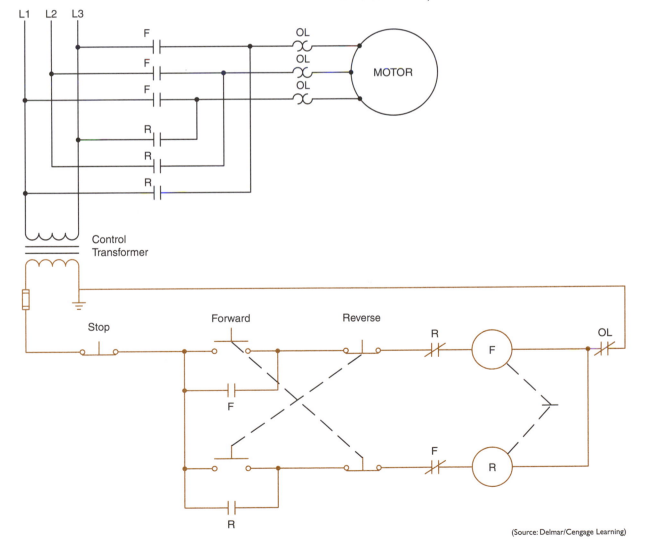

(Source: Delmar/Cengage Learning)

Fig. 21–6 Elementary diagram of the reversing starter shown in Figure 21–1. The mechanical, push-button, and auxiliary contact interlock are indicated.

stop button before changing the direction of rotation. If the reverse button is pressed while the motor is running in the forward direction, the forward control circuit is de-energized and the reverse contactor is energized and held closed.

Repeated reversals of the direction of motor rotation are not recommended. Such reversals may cause the overload relays and starting fuses to overheat; this disconnects the motor from the circuit. The driven machine may also be damaged. It may be necessary to wait until the motor has coasted to a standstill.

NEMA specifications call for a starter to be derated. That is, the next size larger starter must be selected when it is to be used for "plugging" to stop, or "reversing" at a rate of more than five times per minute.

Reversing starters consisting of mechanical and electrical interlocked devices are preferred for maximum safety.

AUXILIARY CONTACT INTERLOCK

Another method of electrical interlock consists of normally closed auxiliary contacts on the forward and reverse contactors of a reversing starter, Figure 21–6.

When the motor is running forward, a normally closed contact (F) on the forward contactor opens and prevents the reverse contactor from being energized by mistake and by closing. The same operation occurs if the motor is running in reverse. Electrical interlocks are usually mounted on the side of a motor starter. These are shown in Figure 21–7.

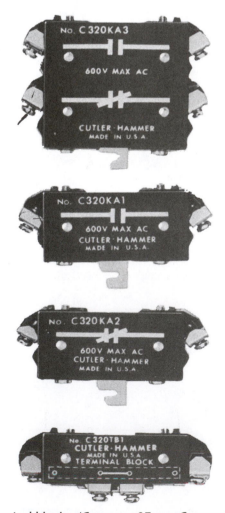

Fig. 21–7 Electrical interlocks and terminal blocks. *(Courtesy of Eaton Corporation)*

The term *interlocking* is also used generally when referring to motor controllers and control stations that are interconnected to provide control of production operations.

To reverse the direction of rotation of single-phase motors, *either* the starting *or* running winding motor leads are interchanged, but not both. A schematic diagram showing the connection for reversing the direction of rotation for a single-phase motor is shown in Figure 21–8. A wiring diagram of this connection is shown in Figure 21–9.

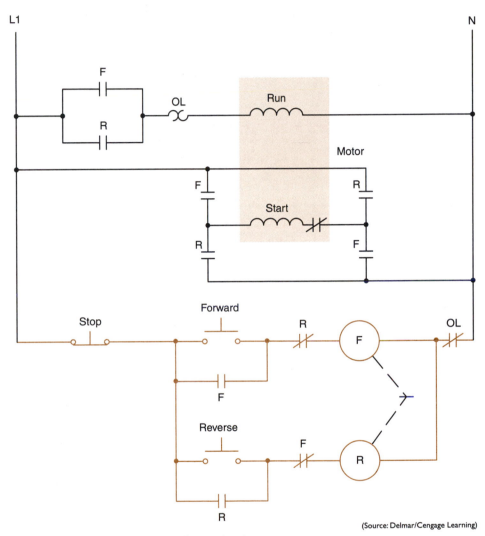

(Source: Delmar/Cengage Learning)

Fig. 21–8 Reversing the direction of rotation of a single-phase motor.

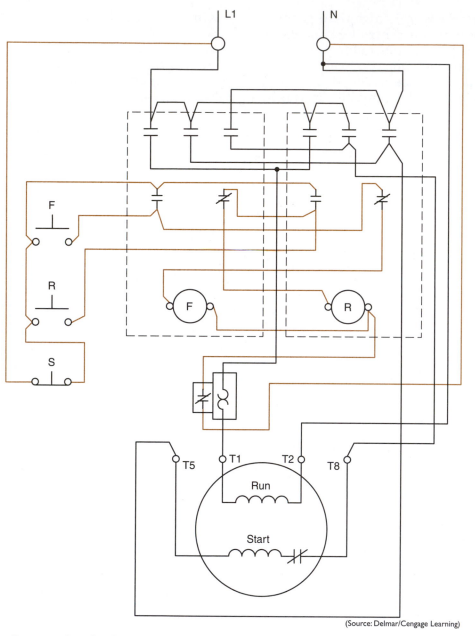

(Source: Delmar/Cengage Learning)

Fig. 21–9 Wiring diagram of single-phase motor reversing control.

Study Questions

1. How is a change in the direction of rotation of a three-phase motor accomplished?

2. What is the purpose of interlocking?

3. If a reversing control circuit contains push-button interlocking, what would happen if both the forward and reverse buttons were pushed at the same time?

4. How is auxiliary contact interlocking achieved on a reversing starter?

5. After the forward coil has been energized, is the auxiliary forward interlocking contact open or closed?

6. If a mechanical interlock is the only means of interlocking used, what procedure must be followed to reverse the direction of the motor? Assume the motor to be running forward.

7. If pilot lights are to indicate which coil is energized, where should they be connected? Additional auxiliary contacts are not to be used.

8. What is the sequence of the operations if the limit switches, shown in Figure 21–6, are used. (Limit switches were covered in Unit 13.)

9. What will happen in Figure 21–6 if limit switches are installed and the jumpers from terminals 6 and 7 to the coils are not removed?

10. In place of the push buttons in Figure 21–4, draw a heavy-duty selector switch for forward-reverse-stop control. Show the target table for this selector switch.

11. From the elementary drawing in Figure 21–10, determine the number and terminal identification of the wiring in each conduit in the conduit layout. Indicate your solutions in the same manner as the example given below the disconnect switch.

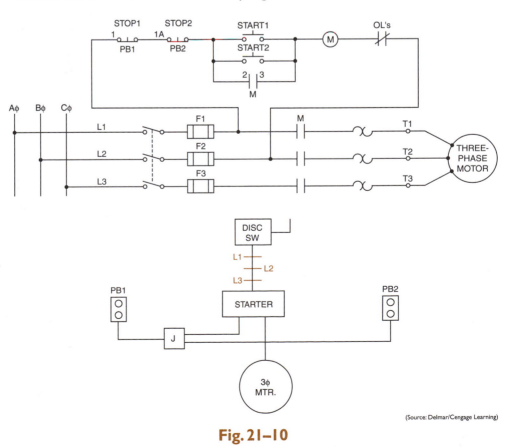

(Source: Delmar/Cengage Learning)

Fig. 21–10

12. Convert the *control circuit only,* Figure 21–11, from the wiring diagram to an elementary diagram. Include the limit switches (RLS, FLS) as operating in the control circuit.

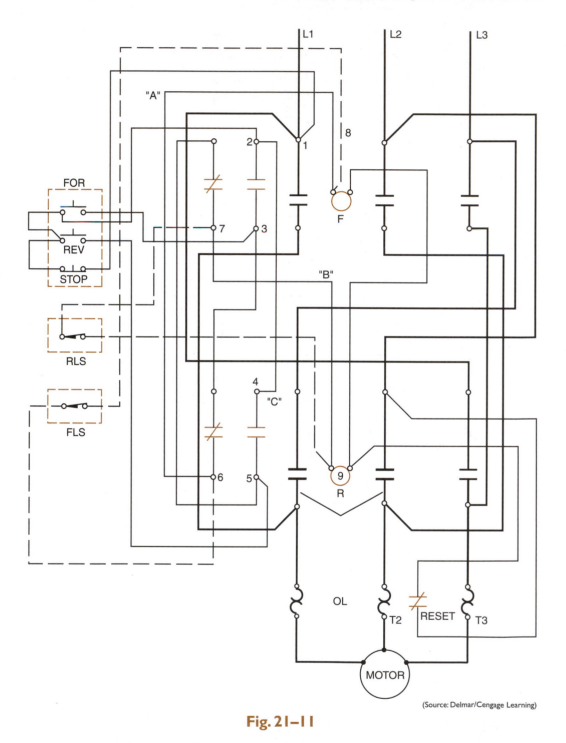

Fig. 21–11

(Source: Delmar/Cengage Learning)

13. Draw a standard duty selector switch to select either forward or reverse. Include a stop and run push button. Include proper interlocks.

UNIT 22

Sequence Control

Objectives

After studying this unit, the student should be able to

- Describe the purpose of starting motors in sequence.
- Read and interpret sequence control diagrams.
- Make the proper connections to operate motors in sequence.
- Troubleshoot sequence motor control circuits.
- Describe solid-state logic elements.
- Describe solid-state logic concept and control function.

Sequence control is the method by which starters are connected so that one cannot be started until the other is energized. This type of control is required whenever the auxiliary equipment associated with a machine, such as high-pressure lubricating and hydraulic pumps, must be operating before the machine itself can be operated safely. Another application of sequence control is in main or subassembly line conveyors.

The proper push-button station connections for sequence control are shown in Figure 22–1. Note that the control circuit of the second starter is wired through the maintaining contacts of the first starter. As a result, the second starter is prevented from starting until after the first starter is energized. If standard starters are used, the connection wire (X) must be removed from one of the starters.

If sequence control is to be provided for a series of motors, the control circuits of the additional

starters can be connected in the manner shown in Figure 22–2. That is, M3 will be connected to M2 in the same step arrangement by which M2 is connected to M1.

The stop button or an overload on any motor will stop all motors with this method.

AUTOMATIC SEQUENCE CONTROL

A series of motors can be started automatically with only one start-stop control station as shown in Figure 22–2 and Figure 22–3. When the lube oil pump, (M1) in Figure 22–3, is started by pressing the start button, the pressure must be built up enough to close the pressure switch before the main drive motor (M2) will start. The pressure switch also energizes a timing relay (TR). After a preset time delay, the contact (TR) will close and energize the feed motor starter coil (M3).

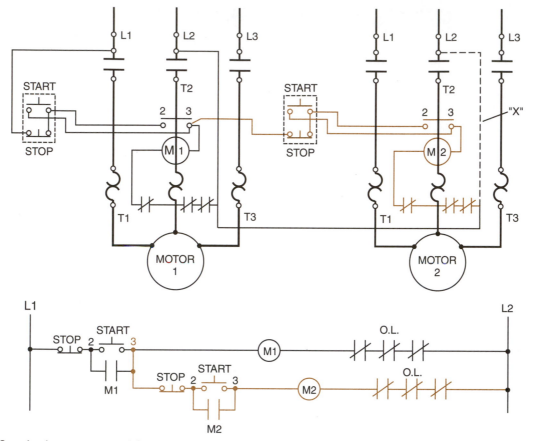

Fig. 22–1 Standard starters wired for sequence control. *(Courtesy of Rockwell Automation)*

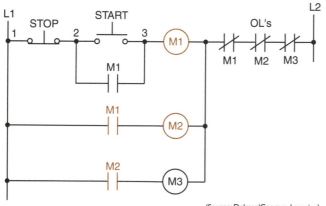

(Source: Delmar/Cengage Learning)

Fig. 22–2 Auxiliary contacts (or interlocks) used for automatic sequence control. Contact (M1) energizes coil (M2); contact (M2) energizes coil (M3).

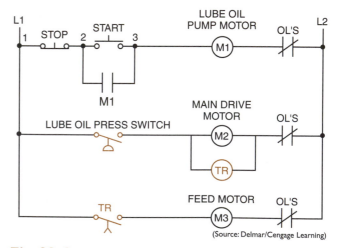

(Source: Delmar/Cengage Learning)

Fig. 22–3 Pilot devices used in an automatic sequence control scheme.

If the main drive motor (M2) becomes overloaded, the starter and timing relay (TR) will open. As a result, the feed motor circuit (M3) will be de-energized due to the opening of the contact (TR). If the lube oil pump motor (M1) becomes overloaded, all of the motors will stop. Practically any desired overload control arrangement is possible. Figure 22–4 reviews some common overload control arrangements for operating different machinery control schemes.

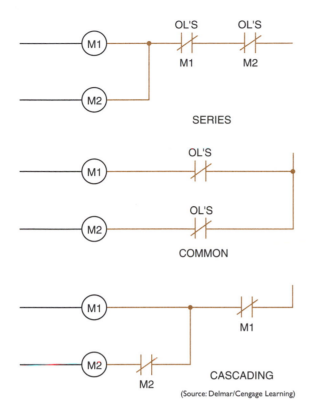

Fig. 22–4 Common overload connections.

Different Voltages

Figure 22–5 illustrates sequencing control connections using different operating voltages.

LOGIC CONCEPT AND CONTROL FUNCTION

The functions of electrical control systems and their associated devices can be grouped into three functions: *input, logic, output.* Table 22–1 shows the function, devices, and their use.

The *input,* or information, section of control consists of limit switches, pressure switches, push buttons, temperature sensors, and other control devices. These accessories receive and transmit information to the *logic,* or decision, section of the control, such as a relay or transistor. When information has been received, it must be assimilated and correlated in order to arrive at a decision to act or do something, the *output.*

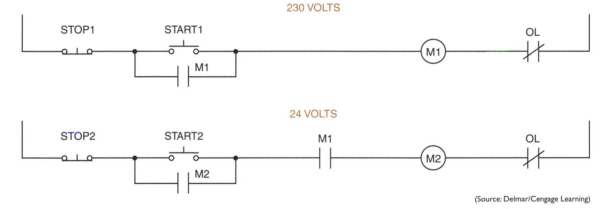

Fig. 22–5 Different voltage sequencing.

Table 22–1 Functions and Their Associated Devices

FUNCTION	DEVICE	USE
Input (Information)	Push buttons, limit switches, pressure switches, proximity switches, photocells, etc.	Sensing or information source
Logic (Decision)	Relays, electron tubes, magnetic amplifiers, hydraulic and pneumatic valves, transistors, etc.	Acts on information from input devices; makes decisions and provides desired output signals
Output (Action)	Contactors, starters, solenoids, power magnetic amplifiers, electronic and solid state contactors	Amplifies the logic output signals to the desired power level

Once a decision has been reached, the control must take action or have an output. Typical actions include energizing a solenoid, contactor or motor starter, an indicating light, or other electrical apparatus where something takes place.

Solid-state control performs the same decision-making function in a solid-state system that relays perform in a conventional magnetic system. Relays convert a single input from a push button to a coil into multiple outputs by means of a movable armature closing and opening contacts. Solid-state control logic elements convert a single input, or combination of inputs, into a single output by controlling a transistor.

The development of solid-state control has introduced new concepts and terminology to the electrical industry. This is true even though solid-state control functionally duplicates conventional magnetic control devices. It is in the area of performing the logic function that solid-state control is applied.

SOLID-STATE LOGIC ELEMENTS

Motor control circuits sometimes involve a special type of electronic logic called "gate logic." Gate logic generally involves the use of integrated circuits. There are five basic types of gates: AND, OR, INVERTER, NOR, and NAND. Gate logic differs from relay logic in several ways. Logic gates are solid-state devices, and as such they operate much faster, are not subject to vibration, and have no moving components. Another difference between gate logic and relay logic is that relays are single input (the coil), multi-output (the contacts) devices. Logic gates are multi-input, single output devices; they contain two or more inputs but only one output. There are also different sets of symbols used to depict logic gates. One set is the USASI standard, generally referred to as computer logic, and the other set is the NEMA standard.

The AND Gate

The AND gate requires all of its inputs to be in a high state or turned on (voltage applied to the input) before it will develop a high output. The USASI and NEMA symbols for a two-input AND gate are shown in Figure 22–6. The inputs are labeled A and B and the output is labeled Y. The AND gate equivalent logic is two normally open switches connected in series. In order to provide an output to the lamp, both switches must be changed from open to closed.

The OR Gate

The OR gate will have a high output if any or all of its inputs are high. The OR gate equivalent logic would be normally open switches connected in parallel. If either or both of the switches should change from open to closed, the lamp would turn on. The USASI and NEMA symbols for an OR gate are shown in Figure 22–7.

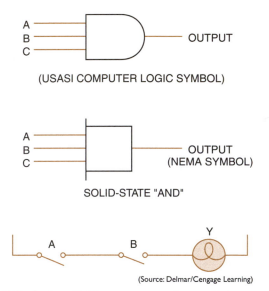

(USASI COMPUTER LOGIC SYMBOL)

(NEMA SYMBOL)

SOLID-STATE "AND"

(Source: Delmar/Cengage Learning)

Fig. 22–6 An AND gate must have all inputs high before it can have an output.

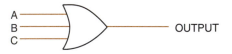

USASI (COMPUTOR LOGIC) **OR** GATE SYMBOL

NEMA LOGIC **OR** GATE SYMBOL

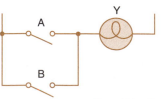

(Source: Delmar/Cengage Learning)

Fig. 22–7 An OR gate will have a high output if any or all of its inputs are high.

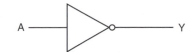

USASI (COMPUTER LOGIC) SYMBOL FOR AN INVERTER

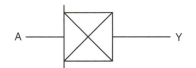

NEMA LOGIC SYMBOL FOR AN INVERTER

(Source: Delmar/Cengage Learning)

Fig. 22–8 The output of an INVERTER will always be the opposite of the input.

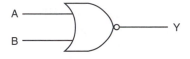

USASI (COMPUTER LOGIC) **NOR** GATE SYMBOL

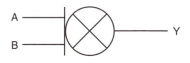

NEMA LOGIC **NOR** GATE SYMBOL

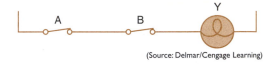

(Source: Delmar/Cengage Learning)

Fig. 22–9 A NOR gate will have a low input if any input is high.

The INVERTER

The INVERTER is a single-input, single output device. The output will always be inverted or opposite the input. If the input is high, the output will be low; if the input is low, the output will be high. In the USASI standard symbols, a circle is used to represent an inverted condition. The NEMA standard symbols use an "X" to represent an inverted input or output. The logic of an INVERTER is a normally closed contact connected in series with the output. As long as the contact is de-energized (not opened) the output will be high. When the input is energized (contact changed) the output will be low. The USASI and NEMA symbols for an INVERTER are shown in Figure 22–8.

The NOR Gate

A NOR gate is actually an abbreviation for "NOT OR GATE." The NOR gate is an OR gate with an inverted output. The logic of an OR gate is any high input will produce a high output. Conversely, the logic of the NOR gate is that any high input will produce a low output. An equivalent switch circuit for the NOR gate would be normally closed switches connected in series. If any switch should be energized or changed, there will be a low output. The USASI and NEMA logic symbols for a NOR gate are shown in Figure 22–9.

The NAND Gate

NAND is an abbreviation for "NOT AND." The NAND gate is the same as an AND gate with an INVERTER connected to the output.

The logic of an AND gate is that if any input is low, the output also will be low. The logic of a NAND gate is that if any input is low, the output will be high. The equivalent switch circuit for a NAND gate would be normally closed switches connected in parallel. The USASI and NEMA symbols for a NAND gate are shown in Figure 22–10.

The EXCLUSIVE OR Gate

Another type of gate that is derived from an OR gate is called the EXCLUSIVE OR gate. The

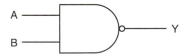

USASI (COMPUTER LOGIC) SYMBOL FOR A **NAND** GATE

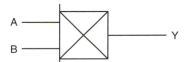

NEMA LOGIC SYMBOL FOR A **NAND** GATE

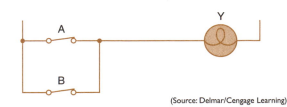

(Source: Delmar/Cengage Learning)

Fig. 22–10 Any low input will produce a high output for a NAND gate.

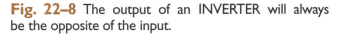

logic of a two-input EXCLUSIVE OR gate is that if either but not both of its inputs is high it will have a high output. If both inputs are low, the output will be low. If both inputs are high, the output will also be low. The USASI symbol and equivalent switch circuit for an EXCLUSIVE OR gate are shown in Figure 22–11.

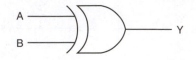

USASI SYMBOL FOR AN EXCLUSIVE **OR** GATE

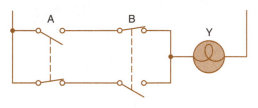

EXCLUSIVE **OR** GATE
(Source: Delmar/Cengage Learning)

Fig. 22–11 Exclusive OR gate.

Study Questions

1. Describe what is meant by sequence control.

2. Referring to the diagram in Figure 22–2, explain what will happen if the motor that is operated by coil "M" becomes overheated.

3. In Figure 22–2, what will happen if there is an overload on motor 2?

4. What is the sequence of operation in Figure 22–2?

5. Redraw Figure 22–3 so that if the feed motor (M3) is tripped out because of overload, it will also stop the main drive motor (M2).

6. Draw a symbol for a solid-state logic element "AND."

Jogging (Inching) Control Circuits

Objectives

After studying this unit, the student should be able to

- Define the process of jogging control.

- State the purpose of jogging controllers.

- Describe the operation of a jogging control circuit using a control relay.

- Describe the operation of a jogging control using a control relay on a reversing starter.

- Describe the operation of a jogging control using a selector switch.

- Connect jogging controllers and circuits.

- Recommend solutions for troubleshooting jogging controllers.

Jogging, or *inching,* is defined by NEMA as "the quickly repeated closure of a circuit to start a motor from rest for the purpose of accomplishing small movements of the driven machine." The term *jogging* is often used when referring to across-the-line starters; the term *inching* can be used to refer to reduced voltage starters. Generally, the terms are used interchangeably because they both prevent a holding circuit.

JOGGING CONTROL CIRCUITS

The control circuits covered in this unit are representative of the various methods that are used to obtain jogging.

Figure 23–1 is a line diagram of a very simple jogging control circuit. The stop button is held open mechanically, Figure 23–2. With the stop button held open, maintaining contact M cannot hold the coil energized after the start button is

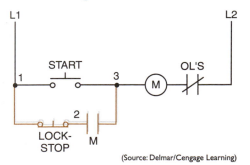

(Source: Delmar/Cengage Learning)

Fig. 23–1 Lock-stop push button in a jogging circuit.

Fig. 23–2 Oil-tight push-button operator with extra long button to accept padlock attachment (on right) to provide lockout on stop feature. *(Courtesy of Eaton Corporation)*

closed. The disadvantage of a circuit connected in this manner is the loss of the lock-stop safety feature. This circuit can be mistaken for a conventional three-wire control circuit, locked off for safety reasons, such as to keep a circuit or machine from being energized. A padlock should be installed for safety purpose.

If the lock-stop push button is used for jogging, it should be clearly marked for this purpose.

Figure 23–3 illustrates other simple schemes for jogging circuits. The normally closed push-button contacts on the jog button in Figure 23–3(B) are connected in series with the holding circuit contact on the magnetic starter. When the jog button is pressed, the normally open contacts energize the starter magnet. At the same time, the normally closed contacts disconnect the holding circuit. When the button is released, therefore, the starter immediately opens to disconnect the motor from the line. The action is similar in Figure 23–3(A). *A jogging*

attachment can be used to prevent the reclosing of the normally closed contacts of the jog button. This device ensures that the starter holding circuit is not reestablished if the jog button is released too rapidly. Jogging can be repeated by reclosing the jog button; it can be continued until the jogging attachment is removed.

CAUTION: If the circuits shown in Figure 23–3 are used without the jogging attachment mentioned, they are hazardous. A control station using such a circuit, less a jogging attachment, can maintain the circuit when the operator's finger is quickly removed from the button. This could injure production workers, equipment, and machinery. This circuit should not be used by responsible people committed to safety in the electrical industry.

JOGGING USING A CONTROL RELAY

When a jogging circuit is used, the starter can be energized only as long as the jog button is depressed. This means the machine operator has instantaneous control of the motor drive.

The addition of a control relay to a jogging circuit provides even greater control of the motor drive. A control relay jogging circuit is shown in Figure 23–4. When the start button is pressed, the control relay is energized and a holding circuit is formed for the control relay and the starter magnet. The motor will now run. The jog button is connected to form a circuit to the starter magnet. This circuit is independent of the control relay. As a result, the jog button can be pressed to obtain the jogging or inching action.

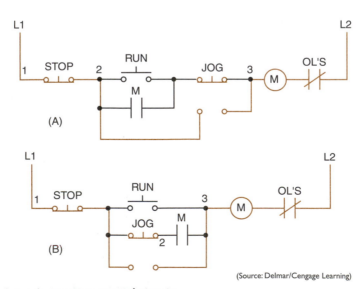

(Source: Delmar/Cengage Learning)

Fig. 23–3 Line diagrams of simple jogging control circuits.

Other typical jogging circuits using control relays are shown in Figure 23–5. In Figure 23–5(A), pressing the start button energizes the control relay. In turn, the relay energizes the starter coil. The normally open starter interlock and relay contact then form a holding circuit around the start button. When the jog button is pressed, the starter coil is energized independently of the relay and a holding circuit does not form. As a result, a jogging action can be obtained.

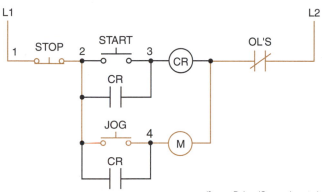

(Source: Delmar/Cengage Learning)

Fig. 23–4 Jogging is achieved with added use of control relay.

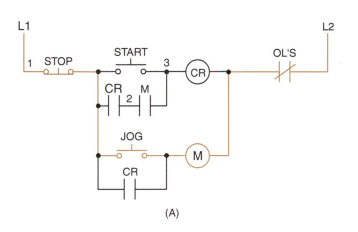

(A)

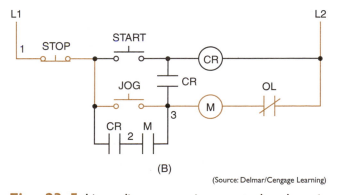

(B)

(Source: Delmar/Cengage Learning)

Fig. 23–5 Line diagrams using control relays in typical installations.

Jogging with a Control Relay on a Reversing Starter

The control circuit shown in Figure 23–6 permits the motor to be jogged in either the forward or the reverse direction while the motor is at standstill or is rotating in either direction. Pressing either the start-forward button or the start-reverse button causes the corresponding starter coil to be energized. The coil then closes the circuit to the control relay, which picks up and completes the holding circuit around the start button. While the relay is energized, either the forward or the reverse starter will also remain energized. If either jog button is pressed, the relay is de-energized and the closed starter is released. Continued pressing of either jog button results in a jogging action in the desired direction.

Jogging with a Selector Switch

The use of a selector switch in the control circuit to obtain jogging requires a three-element control station with start and stop controls and a selector switch. A standard-duty, two-position selector switch is shown connected in the circuit in Figure 23–7. The starter maintaining circuit is disconnected when the selector switch is placed in the jog position. The motor is then

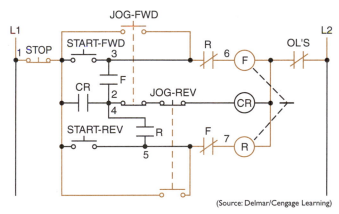

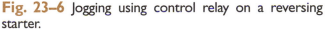

(Source: Delmar/Cengage Learning)

Fig. 23–6 Jogging using control relay on a reversing starter.

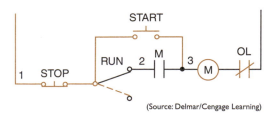

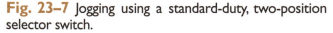

(Source: Delmar/Cengage Learning)

Fig. 23–7 Jogging using a standard-duty, two-position selector switch.

inched with the start button. Figure 23–8 is the same circuit as that shown in Figure 23–7 with the substitution of a heavy-duty, two-position selector switch.

The use of a selector push button to obtain jogging is shown in Figure 23–9. In the jog position, the holding circuit is broken, and jogging is accomplished by depressing the push button.

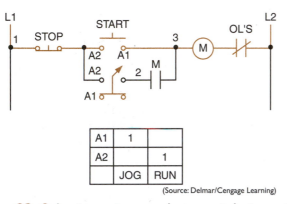

A1		
A2		1
	JOG	RUN

(Source: Delmar/Cengage Learning)

Fig. 23–8 Jogging using a selector-switch jog with start button.

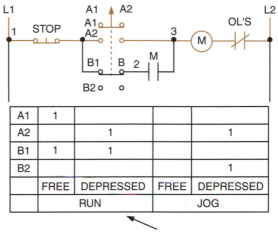

	FREE	DEPRESSED	FREE	DEPRESSED
A1	1			
A2		1		1
B1	1	1		
B2				1
	RUN		JOG	

(Source: Delmar/Cengage Learning)

Fig. 23–9 Jogging using selector push button.

Study Questions

1. What is the safety feature of a lock-stop push button?

2. In Part (A) of Figure 23–3, what will happen if both the run and jog push buttons are closed?

3. What will happen if the start and jog push buttons of the circuit shown in Figure 23–4 are pushed at the same time?

4. In Figure 23–6, what will happen if both jog push buttons are pushed momentarily?

5. Draw an elementary control diagram of a reversing starter. Use a standard-duty selector switch with forward, reverse, and stop push buttons with three methods of interlocking.

UNIT 24

Time-Delay, Low-Voltage Release Relay

Objectives

After studying this unit, the student should be able to

- Describe the purpose of a time-delay, low-voltage release relay.

- Describe the construction and operation of a time- delay, low-voltage release relay.

- Make proper connections to insert this type of relay in a circuit that contains a motor starter and a push-button station.

- Read and interpret control diagrams of low-voltage release relay.

There is a possibility of injury to an operator or damage to machinery when the power suddenly resumes after a prolonged voltage failure. Therefore, it is desirable to provide some means of preventing motors from restarting in such cases. One method of preventing motor restart is known as low-voltage protection or three-wire control. However, three-wire control is not practical because momentary line voltage fluctuations occur in many localities, or the voltage actually fails for only a few seconds. Remember that with three-wire control, each time a voltage failure occurs the motors must be restarted manually. This is a time-consuming procedure and, in some instances, can cause damage to material in process or production.

One solution to this problem is to use a time-delay, low-voltage release device, Figure 24–1. When this device is used with a magnetic starter and a momentary contact push-button station, the motor is automatically reconnected to the

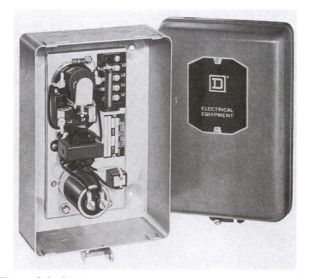

Fig. 24–1 Time-delay, low-voltage release relay. (Courtesy of Schneider Electric)

power lines after a voltage failure of short duration. If the voltage failure exceeds the time-delay setting of the low-voltage release device, or if the stop button is pressed, the motor will not restart automatically; it must be restarted by pressing the start button. As a result, the use of a time-delay, low-voltage release device provides the safety of three-wire control. In addition, it eliminates the inconvenience of a loss of production time where momentary voltage failures are common.

CONSTRUCTION AND OPERATION

The time-delay, low-voltage release device consists of a single-pole, normally open control relay, an electrolytic capacitor, a rectifier, two resistors, and a control transformer, Figure 24–2. Resistor R1 is connected in parallel with the coil of the control relay CR1 to provide a time delay of approximately 2 seconds. Removal of this resistor provides a time delay of approximately 4 seconds.

When used with a magnetic starter and a momentary contact, start-stop push-button station, the coil of the low-voltage release relay is connected to the secondary of the transformer through the rectifier and a resistor. Current flows through the relay coil at all times. The value of resistor R2, however, is such that the current flowing through the coil is too low to allow relay pickup.

When the start button is pressed, resistor R2 is cut out, full voltage is applied to the relay coil, and the relay is energized. (When the start button is released, the resistor is again in series with the relay coil. The relay, however, is still energized because less current is required to hold the relay once it is picked up.) A relay contact 1CR is wired in series with the operating coil 1M of the magnetic starter. This contact closes and allows the operation of the magnetic starter.

If the line voltage is reduced or fails completely, the electrolytic capacitor C1, which is charged through the rectifier, discharges through the control relay coil to keep the coil

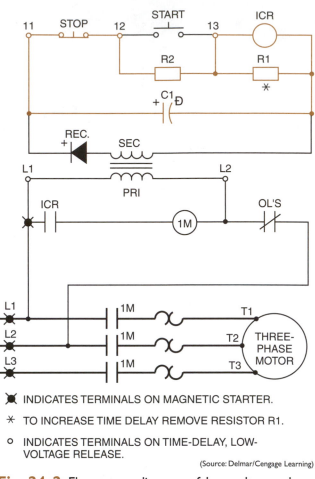

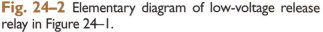

✖ INDICATES TERMINALS ON MAGNETIC STARTER.

✳ TO INCREASE TIME DELAY REMOVE RESISTOR R1.

○ INDICATES TERMINALS ON TIME-DELAY, LOW-VOLTAGE RELEASE.

(Source: Delmar/Cengage Learning)

Fig. 24–2 Elementary diagram of low-voltage release relay in Figure 24–1.

energized. The time required for the capacitor to discharge is a function of the resistance in the circuit and the capacitance of the device.

If the line voltage returns to approximately 85 percent of its normal value before the capacitor is discharged, the magnetic starter automatically recloses and restarts the motor. If the voltage does not return to normal before the capacitor is discharged, the control relay opens. As a result, to restart the motor the push button must be pressed when power is returned.

At any time, pushing the stop button causes the control relay to be de-energized. The starter circuit then opens immediately.

Study Questions

1. In the event of line voltage fluctuations or failure, how is the starter maintained in a closed position?

2. How is this principle applied to a DC starter?

3. Under what condition will the motor restart automatically after a voltage failure?

4. Explain in detail the purpose of a time-delay, low-voltage release relay assembly installed in a motor control circuit.

AC Reduced Voltage Starters

The Motor and Starting Methods

Objectives

After studying this unit, the student should be able to

- Describe the most important factors to consider when selecting motor starting equipment.
- State why reduced current starting is important.
- Describe typical starting methods.
- Identify squirrel cage induction motors.
- Describe how a squirrel cage motor functions.

There are two reasons for the use of reduced voltage starting:

1. To reduce the high starting current drawn by the motor
2. To reduce the starting torque provided by the motor

The simplicity, ruggedness, and reliability of squirrel cage induction motors have made them the standard choice for AC, all-purpose, constant-speed motor applications. Several types of motors are available; therefore, various kinds of starting methods and control equipment are also obtainable.

THE MOTOR

The Revolving Field

The squirrel cage motor consists of a frame, a stator, and a rotor. The *stator*, or stationary portion, carries the stator windings, Figure 25–1 (center). The *rotor* is a rotating member,

Figure 25–1 (bottom) that is constructed of steel laminations mounted rigidly on the motor shaft. The rotor winding consists of many copper, or aluminum, bars fitted into slots in the rotor. The bars are connected at each end by a closed continuous ring. The assembly of the rotor bars and end rings resembles a squirrel cage. This similarity gives the motor its name, *squirrel cage motor*.

For a three-phase motor, the stator frame has three windings. The stator for a squirrel cage motor never has fewer than two poles. The stator windings are connected to the power source. When a 60-hertz current flows in the stator winding, a magnetic field is produced. Because of the three-phase alternating current and the displacement of each phase winding, this field circles the rotor at a speed equal to the number of revolutions per minute (r/min or rpm) divided by the number of pairs of stator poles. Therefore, on 60 hertz, a motor having one pair of two poles will run at 3600 r/min; a four-pole motor (two pairs of two poles each) will run at 1800 r/min

(3600/2). The revolving stator magnetic field induces current in the short-circuited rotor bars. The induced current in the squirrel cage then has a magnetic field of its own. The two fields interact, with the rotor field following the stator rotating field, thereby establishing a torque on the motor shaft. The current induced has its largest value when the rotor is at a standstill. The current then decreases as the motor comes up to speed. In designing motors for specific applications, changing the resistance and reactance of the rotor will alter the characteristics of the motor. For any one rotor design, however, the characteristics are fixed. There are no external connections to the rotor. A cutaway view of an assembled squirrel cage motor is shown in Figure 25–2.

Locked Rotor Currents

The locked rotor current and the resulting torque are factors that determine if the motor can be connected across the line or if the current must be reduced to obtain the required performance. Locked rotor currents for different motor types vary from 2½ to 10 times the full-load current of the motor. Some motors, however, have even higher inrush currents.

The Induction Motor at Start

Figure 25–3 illustrates the behavior of the current taken by an induction motor at various speeds. First, note that the starting current is high compared with the running current. In addition, the starting current remains fairly constant at this high value as the motor speed increases. The current then drops sharply as the

Fig. 25–1 Squirrel cage induction motor frame, stator, and rotor. Note the cooling blades on the rotor. *(Courtesy of Emerson Motors)*

Fig. 25–2 Cutaway view of a squirrel cage motor. *(Courtesy of Emerson Motors)*

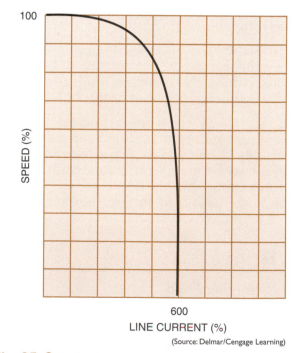

Fig. 25–3 Induction motor current at various speeds.

(Source: Delmar/Cengage Learning)

motor approaches its full rated speed. Because the motor heating rate is a function of I² (copper loss), this rate is high during acceleration. For most of the acceleration period, the motor can be considered to be in the locked condition.

No-Load Rotor Speed

The induced current in the rotor gives rise to magnetic forces, which cause the rotor to turn in the direction of rotation of the stator field. The motor accelerates until the necessary speed is reached to overcome windage and friction losses. This speed is referred to as the *no-load speed*. The motor never reaches synchronous speed because a current will not be induced in the rotor under these conditions and thus the motor will not produce a torque. *Torque* refers to the twisting or turning of the motor shaft. (The rotor bars of the squirrel cage must be cut by the rotating magnetic AC field to produce a torque.)

Speed Under Load

As the rotor slows down under load, the speed adjusts itself to the point where the forces exerted by the magnetic field on the rotor are sufficient to overcome the torque required by the load. *Slip* is the term given to the difference between the speed of the magnetic field and the speed of the rotor.

The slip necessary to carry the full load depends on the characteristics of the motor. In general, the following situations are true:

1. The higher the inrush current, the lower the slip at which the motor can carry full load, and the higher the efficiency.
2. The lower the value of inrush current, the higher the slip at which the motor can carry full load, and the lower the efficiency.

An increase in line voltage causes a decrease in the slip, and a decrease in line voltage causes an increase in the slip. In either case, sufficient current is induced in the rotor to carry the load. A decrease in the line voltage causes an increase in the heating of the motor. An increase in the line voltage decreases the heating. In other words, the motor can carry a larger load. The slip at rated load may vary from 3 percent to 20 percent for different types of motors.

Variation of Torque Requirements

Different loads have different torque requirements. This must be kept in mind when considering the starting torque required and the rate

of acceleration most desirable for the load. In general, a number of motors will satisfy the load requirements of an installation under normal running conditions. However, it may be more difficult to use a motor that will perform satisfactorily during both starting and running. Often, it is necessary to decide which is the more important factor to consider for a particular application. For example, a motor may be selected to give the best starting performance, but there may be a sacrifice in the running efficiency. In another case, to obtain a high running efficiency, it may be necessary to use a motor with a high-current inrush. For these and other examples, the selection of the proper starter to overcome the objectionable features is an important consideration in view of the high cost of energy today.

The machine to which the motor is connected may be started at no load, normal load, or overload conditions. Many industrial applications require that the machine be started when it is not loaded. Thus, the only torque required is that necessary to overcome the inertia of the machine. Other applications may require that the motor be started while the machine it is driving is subjected to the same load it handles during normal running. In this instance, the starting requirements include the ability to overcome both the normal load and the starting inertia.

Important factors in providing the proper starting equipment include using a starter that satisfies the horsepower rating of the motor, and connecting the motor directly across the line. Another factor is that the motor itself must meet the torque requirements of the industrial application. Actually, the starting equipment may be selected to provide adequate control of the torque after the motor is selected.

Controlling Torque

The most common method of starting a polyphase, squirrel cage induction motor is to connect the motor directly to the plant distribution system at full voltage. In this case a manual or a magnetic starter is used. From the standpoint of the motor itself, this is a perfectly acceptable practice. As a matter of fact, it is probably the most desirable method of starting this type of motor.

Overload protective devices have reached such a degree of reliability that the motor is given every opportunity to make a safe start. The application of a reduced voltage to a motor in an attempt to prevent overheating during acceleration is generally wasted effort. The accelerating time will increase and correctly sized overload elements may still trip.

Reduced voltage starting minimizes the shock on the driven machine by reducing the starting torque of the motor. A high torque, applied suddenly, with full-voltage starting, may cause belts to slip and wear or may damage gears, chains, or couplings. The material being processed, or conveyed, may be damaged by the suddenly applied jerk of high torque. By reducing the starting voltage, or current, at the motor terminals, the starting torque is decreased.

Reduced Voltage, Reduced Current, Reduced Torque

The category of reduced voltage methods generally includes all starting methods that deviate from standard, line voltage starting. Not all of these starting schemes reduce the voltage at the motor terminals. Even reduced voltage starters reduce the voltage only to achieve either the reduction of line current or the reduction of starting torque. The reduction of line current is the most commonly desired result.

You should note one important point: *When the voltage is reduced to start a motor, the current is also reduced, and so is the torque that the machine can deliver.* Regardless of the desired result (either reduced current or reduced torque), remember that the other will always follow.

If this fact is kept in mind, it is apparent that a motor that will not start a load on full voltage cannot start that same load at reduced voltage or reduced current conditions. Any attempt to use a reduced voltage or current scheme will not be successful in accelerating troublesome loads. The very process of reducing the voltage and current will further reduce the available starting torque.

Need for Reduced Current Starting

The most common function of reduced voltage starting devices is to reduce, or in some way modify, the starting current of an induction motor. In other words, the rate of change of the starting current is confined to prescribed limits, or there is a predetermined current-time picture that the motor presents to the supply wiring network.

A current time picture for an entire area is maintained and regulated by the public power utility serving the area. The power company attempts to maintain a reasonably constant voltage at the points of supply so that lamp flicker will not be noticeable. The success of the power company in this attempt depends on the generating capacity to the area; transformer and line-loading conditions and adequacies; and the automatic voltage regulating equipment in use. Voltage regulation also depends on the sudden demands imposed on the supply facilities by residential, commercial, and industrial customers. Transient overloading of the power supply may be caused by (1) sudden high surges of reactive current from large motors on starting, (2) pulsations in current taken by electrical machinery driving reciprocating compressors and similar machinery, (3) the impulse demands of industrial x-ray equipment, and (4) the variable power factor of electric furnaces. All of these demands are capable of producing voltage fluctuations.

Therefore, each of these particularly difficult loads is regulated in some way by the power utility. The utility requires the use of some form of reduced voltage, reduced current method and helps its customers determine the best method.

Power company rules and regulations vary between individual companies and the areas served. The following list gives some commonly applied regulations. (All of the possible restrictions on energy usage are not given.) An installation may be governed by just one of these restrictions, or two or more rules may be combined. Regulations include

1. A maximum number of starting amperes, either per horsepower or per motor.
2. A maximum horsepower for line starting. A limit in percent of full-load current is set for anything above this value.
3. A maximum current in amperes for particular feeder size. It is up to the user to determine if the motor will conform to the power company requirements.
4. A maximum rate of change of line current taken by the motor; for example, 200 amperes per half-second.

It should be apparent that it is very important for the electrician to understand the behavior of an induction motor during the start-up and acceleration periods. Such an understanding enables you to select the proper starting method to conform to local power company regulations. Even though several starting methods may appear to be appropriate, a careful examination of the specific application will usually indicate the one best method for motor starting.

TYPICAL STARTING METHODS

The most common methods of starting polyphase squirrel cage motors include:

■ *Full-voltage starting:* A hand-operated or automatic starting switch throws the motor directly across the line.

■ *Primary-resistance starting:* A resistance unit connected in series with the stator reduces the starting current.

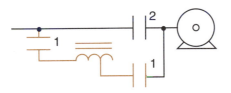

■ *Autotransformer or compensator starting:* Manual or automatic switching between the taps of the autotransformer gives reduced voltage starting.

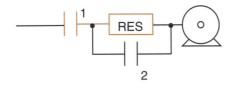

■ *Impedance starting:* Reactors are used in series with the motor.

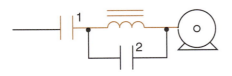

■ *Star-delta starting:* The stator of the motor is star connected for starting and delta connected for running.

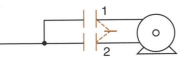

■ *Part winding starting:* The stator windings of the motor are made up of two or more circuits; the individual circuits are connected to the line in series for starting and in parallel for normal operation.

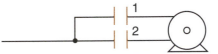

Some solid-state motor starters include many desirable features such as a power saver. This reduces motor voltage under light or no-load conditions to keep motors cooler and save energy. Starting currents are adjustable and contain adjustable current sensors for overload protection. They provide for reduced voltage and gradual starting of motors.

■ *Solid-state electronic control:* Control of current or acceleration time is achieved by gating silicon controlled rectifiers with the AC half cycle.

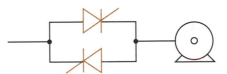

Of these methods, the two most fundamental ways of starting squirrel cage motors are full-voltage starting and reduced voltage starting. Once again, full-voltage starting can be used where the driven load can stand the shock of starting and objectionable line disturbances are not created. Reduced voltage starting may be required if the starting torque must be applied gradually or if the starting current produces objectionable line disturbances.

Study Questions

1. List five commonly used starting methods.

2. When can full-voltage starting be used?

3. Name the simplest, most rugged, and reliable AC motor. Describe how it operates.

4. Describe the term *slip*.

5. When the voltage is reduced to start a motor, what happens to the
   a. torque?
   b. current?

6. Why is it more advisable to start some machines at reduced torque?

Primary Resistor-Type Starters

Objectives

After studying this unit, the student should be able to

- State why reduced current starting is important.

- Describe the construction and operation of primary resistor starters.

- Interpret and draw diagrams for primary resistor starters.

- Connect squirrel cage motors to primary resistor starters.

- Troubleshoot electrical problems on primary resistor starters.

PRIMARY RESISTOR-TYPE STARTERS

A simple and common method of starting a motor at reduced voltage is used in primary resistor-type starters. In this method, a resistor is connected in series in the lines to the motor, Figure 26–1. Thus, there is a voltage drop across the resistors and the voltage is reduced at the motor terminals. Reduced motor starting speed and current are the result. As the motor accelerates, the current through the resistor decreases, reducing the voltage drop $(E = IR)$ and increasing the voltage across the motor terminals. A smooth acceleration is obtained with gradually increasing torque and voltage.

The resistance is disconnected when the motor reaches a certain speed. The motor is then connected to run on full-line voltage, Figure 26–2. The introduction and removal of resistance in the motor starting circuit may be accomplished manually or automatically.

Primary resistor starters are used to start squirrel cage motors in situations where limited torque is required to prevent damage to driven machinery. These starters are also used with limited current inrush to prevent excessive power line disturbances.

It is desirable to limit the starting current in the following cases:

1. When the power system does not have the capacity for full-voltage starting

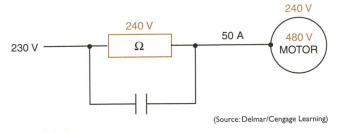

(Source: Delmar/Cengage Learning)

Fig. 26–1 Reduced voltage starting.

2. When full-voltage starting may cause serious line disturbances, such as in lighting circuits, electronic circuits, the simultaneous starting of many motors, or when the motor is distant from the incoming power supply

In these situations, reduced voltage starters may be recommended for motors with ratings as small as 5 horsepower.

Reduced voltage starting must be used for driving machinery, which must not be subjected to a sudden high starting torque and the shock of sudden acceleration. Among typical applications are those where belt drives may slip or where large gears, fan blades, or couplings may be damaged by sudden starts.

Automatic primary resistor starters may use one or more than one step of acceleration, depending on the size of the motor being controlled. These starters provide smooth acceleration with out the line current surges normally experienced when switching autotransformer types of reduced voltage starters.

Primary resistor starters provide closed transition starting. This means the motor is never disconnected from the line from the moment it is first connected until the motor is operating at full line voltage. This feature may be important in wiring systems sensitive to voltage changes. Primary resistor starters do consume energy, with the energy being dissipated as heat. However, the motor starts at a much higher power factor than with other starting methods.

Special starters are required for very high inertia loads with long acceleration periods or where power companies require that current surges be limited to specific increments at stated intervals.

Primary Resistor-Type Reduced Voltage Starter

Figure 26–3 and Figure 26–4 illustrate an automatic, primary-resistor-type reduced voltage starter with two steps of acceleration. The circuit consists of resistors connected in series with a three-phase squirrel cage induction motor.

A control transformer is used to reduce the line voltage, supplying the motor to 120 volts to operate the control circuit. When the start button is pressed, a circuit is completed to the coil of M starter, causing all M contacts to close. The three M load contacts connect the motor to the line through the series resistors. The three resistors limit the amount of current supplied to the motor. Current flowing through the resistors produces a voltage drop across them. The resistor voltage drop causes a voltage drop across the motor terminals also. The motor starts with reduced voltage applied to it. As the motor accelerates, the current flow to the motor decreases, causing less voltage drop across the resistors and more voltage to be applied to the motor.

When M coil is energized, the two M auxiliary contacts also close. One is used as a holding

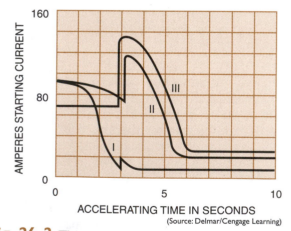

Fig. 26–2 The curves illustrate how a primary resistance starter reduces the starting current of a 10-hp, 230-volt motor under three different load conditions. Curve I is at a light load and Curve II is at a heavy load. In either case, the running switch closes after 3 seconds. With a heavy load, an increase in the starting time reduces the second current inrush; the motor will reach a higher speed before it is connected to full line voltage. In Curve III the motor cannot start on the current allowed through the resistance; it comes up to speed only after connection to full line voltage.

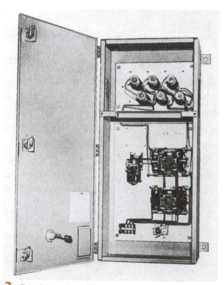

Fig. 26–3 Reduced voltage, primary resistance-starter. Resistors are in the top ventilated system. (Courtesy of Eaton Corporation)

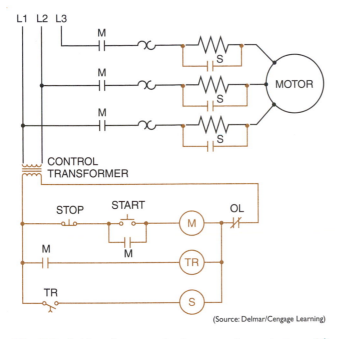

Fig. 26–4 Line diagram of primary resistor starter with two-step acceleration.

connected in parallel with the resistors close and short circuit the resistors. The motor is now connected directly to the power line.

When the stop button is pressed, M starter coil de-energizes, causing all M contacts to return to their open position. The three M load contacts open and disconnect the motor from the power line. The M auxiliary contact connected in series with TR coil opens and causes on-delay timer TR to de-energize. This causes TR contact to open and de-energize S coil. When the three S load contacts reopen, the circuit is back in its original de-energized state.

For maximum operating efficiency, push buttons and other pilot devices are usually located on the driven machine or in easy reach of the operator. The starter is located near the motor to keep the power circuit wiring as short as possible. Only three small connecting wires are necessary between the starter and the pilot devices. A motor can be operated from any of several remote locations if a number of push buttons or pilot switches are used with one magnetic starter, such as on a conveyor system.

AC primary resistor starters are available for use on single-phase and three-phase reversing operations. They are also available with multipoints of acceleration.

contact, and the other energizes the coil of on-delay timer TR. At the end of the timing period, TR contacts close and energize the coil of S contactor. When S coil energizes, the S load contacts

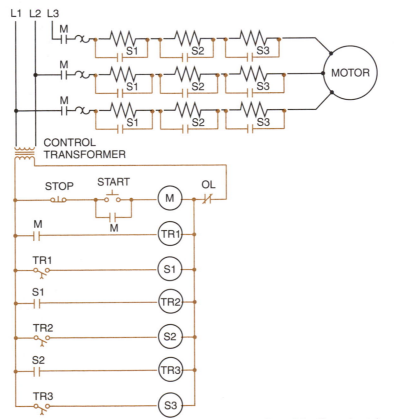

(Source: Delmar/Cengage Learning)

Fig. 26–5 Schematic diagram of a four-point primary resistor starter.

Four-Point Resistor Starter

Figure 26–5 illustrates a resistor starter with four points of acceleration. When the start button is pressed, M coil energizes and causes all M contacts to close. The three M load contacts connect the motor to the power line through the resistors connected in series with the motor. At this point, the resistors provide maximum voltage drop and current limit to the motor. The motor is now in its first or lowest speed. When M auxiliary contact connected in series with TR1 on-delay timer coil closed, timer TR1 began its time sequence. At the end of its time sequence, TR1 contact closes and energizes the coil of S1 contactor coil. This causes the three S1 load contacts to close and shunt out the first set of resistors in the motor circuit. Motor current is now limited by two series resistors instead of three. More current is supplied to the motor, causing it to accelerate. When S1 coil was energized, S1 auxiliary contact closed and energized the coil of on-delay timer TR2. After the preset time for TR2 expires, TR2 contact will close and energize the coil of S2 contactor. S2 load contacts close and shunt out the second set of resistors. The motor now has only one set of resistors connected in series with it. The motor advances to third speed. The S2 auxiliary contact connected in series with the coil of TR3 on-delay time closes and energizes TR3 coil. At the end of its time sequence, contact TR3 closes and energizes the coil of S3 contactor. All S3 load contacts close and shunt out the last of the resistors connected in series with the motor. The motor is now connected directly to the power line and operates in its fourth or highest speed.

Figure 26–6 shows two types of resistor banks used to start small motors. The resistors, shown in Figure 26–6(A), consist of resistance wire wound around porcelain bases and imbedded in refractory cement. The other resistor, Figure 26–6(B), has a ribbon-type construction and is used on larger motor starters. This type of resistor consists of an alloy ribbon formed in a zig-zag shape. The formed ribbon is supported between porcelain blocks, which have recesses for each bend of the ribbon. The ribbons are not stressed and do not have a reduced cross section, because they are bent on the flat and not on the edge. Any number of units can be combined vertically or horizontally in series or in parallel.

Some starters have graphite compression disk resistors and are used for starting polyphase squirrel cage motors, Figure 26–7.

A primary resistance-type starter has the following features:

- simple construction
- low initial cost
- low maintenance
- smooth acceleration in operation
- continuous connection of the motor to the line during the starting period
- a high power factor

These starters should *not* be used for starting *very heavy* loads because of their low starting torque. These starters are said to have a low starting economy because the starting resistors dissipate electrical energy.

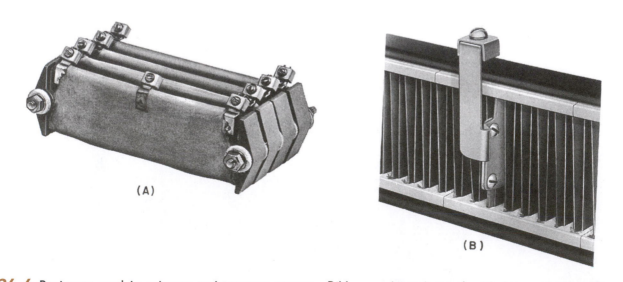

(A)   (B)

Fig. 26–6 Resistors used in primary resistor-type starters. Ribbon resistor is used with larger starter. (*Courtesy of Schneider Electric*)

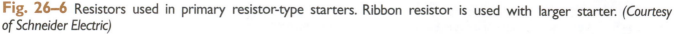

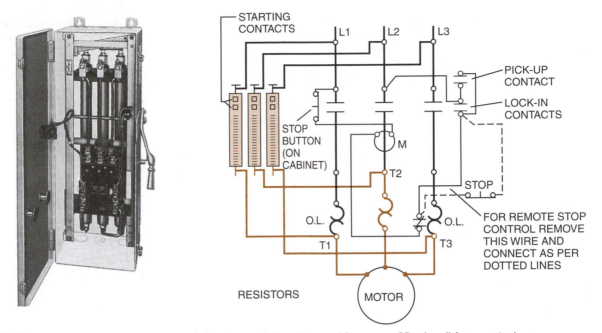

Fig. 26–7 Manual resistance starter with "carbon pile" resistors. *(Courtesy of Rockwell Automation)*

Study Questions

1. What is the purpose of inserting resistance in the stator circuit during starting?

2. Why is the power not interrupted when the motor makes the transition from start to run?

3. If the starter is to function properly and a timing relay is used, where must the coil of the relay be connected?

4. How may additional steps of acceleration be added?

5. What is meant by the low or poor starting economy of a primary resistor starter?

6. How does this starter provide smooth acceleration and a gradual increase of torque?

UNIT 27

Autotransformer Starters

Objectives

After studying this unit, the student should be able to

- Describe the construction and operation of autotransformer starters.

- Draw and interpret diagrams for autotransformer starters.

- Connect squirrel cage motors to autotransformer starters.

- Define what is meant by open transition and closed transition starting.

- Troubleshoot electrical problems on autotransformer starters.

Autotransformer reduced voltage starters are similar to primary resistor starters in that they are used primarily with AC squirrel cage motors to limit the inrush current or to lessen the starting strain on driven machinery, Figure 27–1. This type of starter uses autotransformers between the motor and the supply lines to reduce the motor starting voltage. Taps are provided on the autotransformer to permit the user to start the motor at approximately 50 percent, 65 percent, or 80 percent of line voltage.

Most motors are successfully started at 65 percent of line voltage. In situations where this value of voltage does not provide sufficient starting torque, the 80 percent tap is available. If the 50 percent starting voltage creates excessive line drop to the motor, the 65 percent taps are available. This way of changing the starting

voltage is not usually available with other types of starters. The starting transformers are inductive loads; therefore, they momentarily affect the power factor. They are suitable for long starting periods, however.

Autotransformer Starters

To reduce the voltage across the motor terminals during the accelerating period, an autotransformer-type starter generally has two autotransformers connected in open delta. During the reduced voltage starting period, the motor is connected to the taps on the auto transformer. With the lower starting voltage, the motor draws less current and develops less torque than if it were connected to the line voltage, Figure 27–2.

Fig. 27–1 Reduced voltage autotransformer starter, size 3. *(Courtesy of Eaton Corporation)*

An adjustable time-delay relay controls the transfer from the reduced voltage condition to full voltage. A current-sensitive relay may be used to control the transfer to obtain current-limiting acceleration.

Figure 27–2(A) shows the power circuit for starting the motor with two autotransformers. Figure 27–2(B) shows the circuit for starting a motor with three autotransformers.

To understand the operation of the auto-transformer starter more clearly, refer to the schematic diagram shown in Figure 27–3. When the start-button is pressed a circuit is completed to the coil of control relay CR, causing all CR contacts to close. One contact is employed to hold CR coil in the circuit when the start-button is released. Another completes a circuit to the coil of TR timer, which permits the timing sequence to begin. The CR contact connected in series with the normally closed TR contact supplies power to the coil of S (start) contactor. The fourth CR contact permits power to be connect to R (run) contactor when the normally open timed TR contact closes.

When the coil of S contactor energizes, All S contacts change position. The normally closed S contact connected in series with R coil opens to prevent both S and R contactors from ever being energized at the same time. This is the same interlocking used with reversing starters. When the S load contacts close the motor is connected to the power line through the auto-transformers. The autotransformers supply 65% of the line voltage to the motor. This reduced voltage produces less in-rush current during starting and also reduces the starting torque of the motor.

When the time sequence for TR timer is completed, both TR contacts change position. The normally closed TR contact opens and disconnects contactor S from the line causing all S contacts to return to their normal position. The normally open TR contact closes and supplies power through the now closed S contact to coil R. When contactor R energizes, all R contacts change position. The normally closed

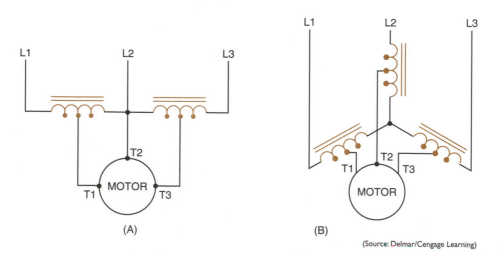

(A)

(B)

(Source: Delmar/Cengage Learning)

Fig. 27–2 Power circuit connections showing two and three autotransformers used for reduced voltage starting.

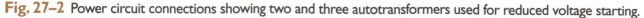

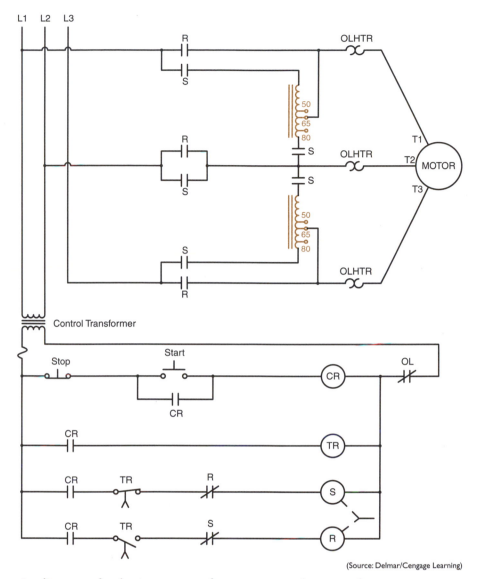

Fig. 27–3 Schematic diagram of a basic autotransformer starter. Autotransformer starters provide the greatest amount of starting torque per ampere of starting current than any other type of reduced voltage starter.

R contact connected in series with S coil opens to provide interlocking for the circuit. The R load contacts closed and connect the motor to full voltage.

When the stop-button is pressed, control relay CR de-energizes and opens all CR contacts. This disconnects all other control components from the power line and the circuit returns to its normal position. A wiring diagram for this circuit is shown in Figure 27–4.

Starting compensators (autotransformer starters) using a five-pole starting contactor are classified as open transition starters. The motor is disconnected momentarily from the line during the transfer from the start to the run conditions.

Closed transition connections are usually found on standard size 6 and larger starters. For the closed transition starter, Figure 27–5, the starting contactors consist of a three-pole (S2) and a two-pole (S1) contactor operating independently of each other. During the transfer from start to run, the two-pole contactor is open and the three-pole contactor remains closed. The motor continues to accelerate with the autotransformer serving as a reactor. With this type of starter, the motor is not disconnected from the line during the transfer period. Thus, there is less line disturbance and a smoother acceleration.

The transformers are de-energized while in the running position. This is done to conserve electrical energy and to extend their life.

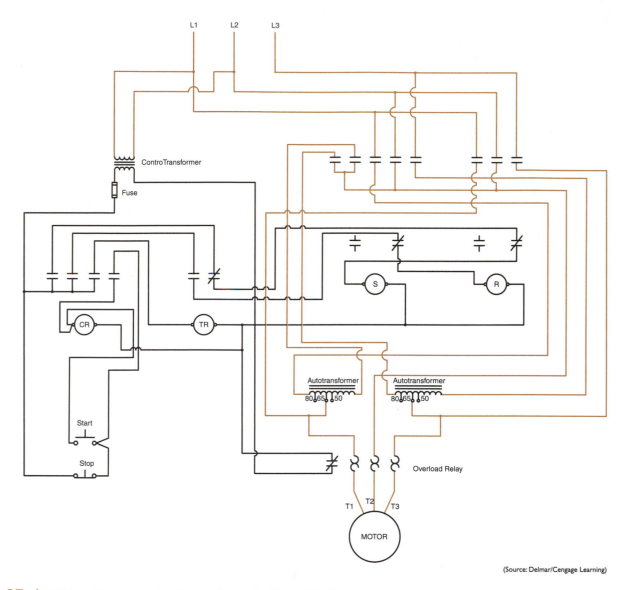

(Source: Delmar/Cengage Learning)

Fig. 27–4 Wiring diagram of the circuit shown in Figure 27–3.

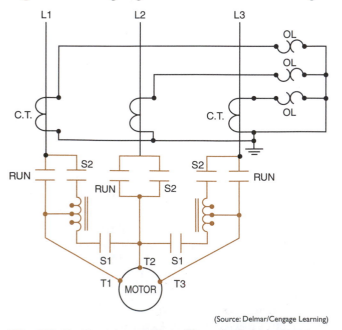

The (CT) designations in Figure 27–5 indicate current transformers. These transformers are used on large motor starters to step down the current so that a conventional overload relay and heater size may be used. Magnetic overload relays are used on large reduced voltage starters also.

(Source: Delmar/Cengage Learning)

Fig. 27–5 Closed transition (Korndorfer) connection.

Study Questions

1. Why is it desirable to remove the autotransformers from the line when the motor reaches its rated speed?

2. What is meant by an "open transition" from start to run? Why is this condition objectionable at times when used with large horsepower motors?

3. Which of the following applies to an autotransformer starter with a five-pole starting contactor: open transition or closed transition? Locate in Figure 27–3.

4. How are reduced voltages obtained from autotransformer starters?

5. Assume the motor is running. What happens when the stop button is pressed and then the start button is pressed immediately?

6. What is a disadvantage of starting with autotransformer coils rather than with resistors?

7. What is one advantage of using an autotransformer starter?

UNIT 28

Part Winding Motor Starters

Objectives

After studying this unit, the student should be able to

- Describe the construction and operation of a part winding motor starter.

- Draw and interpret diagrams for part winding motor starters.

- List the advantages and disadvantages of two-step part winding starters.

- Connect motors to part winding starters.

- Describe the difference between part winding motors and dual-voltage motors.

- Troubleshoot part winding motor starters.

Part winding motors are similar in construction to standard squirrel cage motors. However, part winding motors have two identical windings that may be connected to the power supply in sequence to produce reduced starting current and reduced starting torque. These motor windings are intended to operate in parallel. Because only half of the windings are connected to the supply lines at startup, the method is described as *part winding*. (Many, but not all, dual-voltage, 230/460-volt motors are suitable for part winding starting at 230 volts.) By bringing out leads from each winding, the motor manufacturer enables the windings to be connected in parallel, external to the motor, with the starter (refer to Figure 5–13).

Part winding starters, Figure 28–1, are designed to be used with squirrel cage motors having two separate and parallel stator windings. In the part winding motor, these windings may be Y connected or delta connected, depending on the motor design. Part winding starters are not suitable for use with delta-wound, dual-voltage motors.

Part winding motors are used to drive centrifugal loads such as fans, blowers, or centrifugal pumps. They are also used for other loads where a reduced starting torque is necessary. This type of motor is also used where the full-voltage starting current will produce objectionable voltage drops in the distribution feeders or where power company restrictions require a reduced starting current. Using a part winding starter to start a motor does not necessarily reduce the maximum starting current. Instead, incremental starting is obtained. Neither one of the two windings has the thermal capacity to operate

Fig. 28–1 Part-winding-type, reduced current starter. *(Courtesy of Eaton Corporation)*

alone for more than a few seconds. Therefore, unless the motor accelerates to practically full speed on the one winding, the TOTAL current drawn may approach that of line voltage starting.

This type of starting has many applications in air-conditioning systems. This is due to the increased capacity built into these systems and the necessity of limiting both the current and torque on starting.

TWO-STEP STARTING

A part winding starter, Figure 28–2, has two main contactors. Each contactor controls one winding of the part winding motor. By energizing the contactors in sequence, first one motor

winding is energized, and a short time later the second winding is connected in parallel with the first. As stated previously, the motor is started with one winding energized, giving rise to the name of the starting method part winding starting.

In a two-step part winding starter, Figure 28–3, pressing the start button energizes the start contactor (S). As a result, half of the motor windings are connected to the line and a timing relay (TR) is energized. After a time delay of approximately 5 seconds, the timer contacts (TR) close to energize the "run" contactor. This contactor connects the second half of the motor windings to the line in parallel with the first half of the windings. Note that the control circuit is maintained with an instantaneous normally open contact (TR) operated by the timing relay (TR) and the starting contact (S).

Pressing the stop button or tripping any overload relay disconnects both windings from the line. When starting on one winding, the motor draws approximately two-thirds of the normal locked rotor current and develops approximately one-half of the normal locked rotor torque.

A two-step part winding starter has certain obvious advantages:

1. It is less expensive. Most other starting methods require additional voltage reducing elements such as transformers, resistors, or reactors.
2. It uses only half-size contactors.
3. It provides closed transition starting.

The two-step part winding starter also has disadvantages:

1. The fixed starting torque is poor.
2. The starter is almost always an incremental start device. It is unsuitable for long-starting, high-inertia, loads.

Fig. 28–2 Part winding starter, two step. *(Courtesy of Siemens Energy & Automation)*

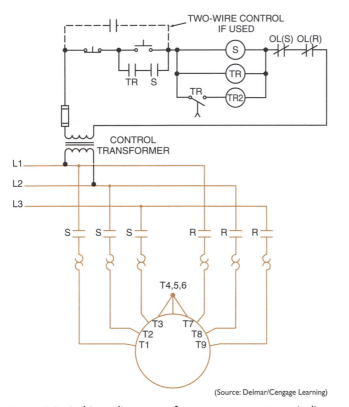

Fig. 28–3 Line diagram of a two-step, part winding starter. When dual-voltage, Y-connected motors are used on the lower voltage only, terminals T4, T5, and T6 are connected to form the center of the second Y.

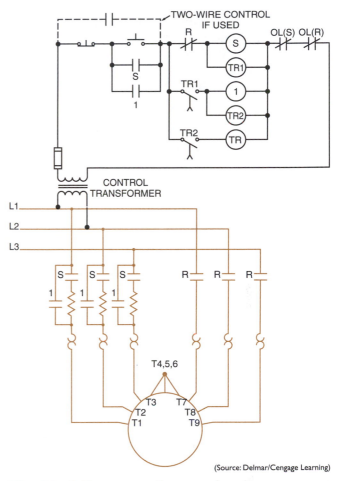

Fig. 28–4 Elementary diagram of a three-step, part winding motor starter.

Part winding starters cannot be used in as many applications as other, more expensive types of reduced voltage starters.

THREE-STEP STARTING

The thermal capacity of the motor limits the length of acceleration on the first winding to approximately 5 seconds, depending on the connected load. In many instances, the motor will not begin to accelerate until the second winding is connected.

Three-step starting is similar to two-step starting, except that when the first contactor closes, the first winding is connected to the line through a resistor in each phase, Figure 28–4. After a time delay of approximately 2 seconds, this resistor is shunted out and the first winding is connected to the full voltage. After another time delay of about 2 seconds, the run contactor closes to connect both windings to the line voltage, Figure 28–5. The resistors are designed to provide approximately 50 percent of the line voltage to the starting winding. Thus,

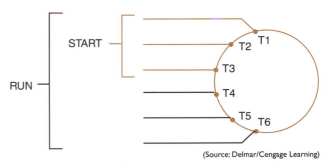

Fig. 28–5 Part winding, single-voltage star or delta motor terminal markings.

the motor starts with three approximately equal increments of starting current.

OVERLOAD PROTECTION

Motor overload protection is provided by six overload relays, three in each motor winding. Because the windings are identical, each winding should have the same full-load current. In addition, each phase of the start winding will carry a current identical to that of the corresponding phase in the run winding. Thus, if three overload relays are connected in three phases in the run winding and three other relays are connected in three different phases in the start winding, then the effect of full three-phase protection is obtained.

When selecting heater elements for the overload relays, the electrician should remember that each of the relays carries only half of the motor current. The elements, therefore, must be selected on the basis of one-half of the full-load current of the motor, unless otherwise specified.

The current requirements in part winding starters permit the use of smaller contactors with *smaller overload elements*. As a result, because the starting fuses must be proportionally smaller to protect these elements, dual-element fuses are required. When in doubt regarding installation and protection problems, local electrical codes and ordinances should always be consulted.

Study Questions

1. How can a 230/460-volt motor be used as a part winding motor?

2. Which voltage may be applied to a 230/460-volt motor if used with a part winding starter? Why?

3. If the nameplate of a dual-voltage motor connected to a part winding starter reads 15/30 amperes, which amperage value is used as the basis for the selection of the overload heating elements?

4. If a timing relay fails to cut in the run starter, what percentage of the full-load current can the motor carry safely?

5. When starting on one winding, a motor draws approximately what percentage of the normal locked rotor current?

6. How is a third step gained in part winding starting?

7. Who determines the compliance of an installation to applicable electrical codes with regard to adequate overload or starting and short-circuit protection?

UNIT 29

Automatic Starters for Star-Delta Motors

Objectives

After studying this unit, the student should be able to

- Identify terminal markings for a star-delta motor and motor starter.
- Describe the purpose and function of star-delta starting.
- Troubleshoot star-delta motor starters.
- Connect star-delta motors and starters.

A commonly used means of reducing inrush currents without the need of external devices is star-delta motor starting (sometimes called wye-delta starting). Figure 29–1 shows a typical star-delta starter.

Star-delta motors are similar in construction to standard squirrel cage motors. However, in star-delta motors, both ends of each of the three windings are brought out to the terminals. If the starter used has the required number of properly wired contacts, the motor can be started in star and run in delta.

The motor must be wound in such a manner that it will run with its stator windings connected in delta. The leads of all of the windings must be brought out to the motor terminals for their proper connection in the field.

APPLICATIONS

The primary applications of star-delta motors are for driving centrifugal chillers of large, central air-conditioning units for loads such as

Fig. 29–1 Star-delta starter.

(Source: Delmar/Cengage Learning)

fans, blowers, pumps, or centrifuges, and for situations where a reduced starting torque is necessary. Star-delta motors also may be used where a reduced starting current is required. Because all of the stator winding is used and there are no limiting devices such as resistors or autotransformers, star-delta motors are widely used on loads having high inertia and a long acceleration period.

The speed of a star-delta, squirrel cage induction motor depends on the number of poles of the motor and the supply line frequency (hertz). Because both of these values are constant, the motor will run at approximately the same speed for either the star or delta connection. The inrush and line current are less when the motor is connected in star than when it is connected in delta. The winding current is less than the line current when the motor is connected in delta. That is, the inrush and line current in the star connection are one-third the values of these quantities in the delta connection. Starting on the higher current delta connection would bump the voltage and create line disturbances.

OVERLOAD PROTECTION SELECTION

Three overload relays are furnished on star-delta starters. These relays are wired so that they carry the motor winding current, Figure 29–2. This means that the relay units must be selected on the basis of the winding current, not the delta-connected full-load current. If the motor nameplate indicates only the delta-connected full-load current, divide this value by 1.73 to obtain the winding current, which is used as the basis for selecting the motor winding protection. A schematic showing the entire connection of the motor stator winding for a star-delta starter is shown in Figure 29–3.

OPERATION

Open transition starting for a star-delta starter is shown in Figure 29–4. As indicated in the line diagram on the right, the automatic transfer from star to delta is accomplished by a pneumatic timer. The timer is operated by the movement of the armature of one of the contactors. Operating the push-button station start button energizes the contactor S. The main contacts of contactor S connect three of the motor leads together (T4, T5, and T6) to form a star. At about the same time, the normally open, control

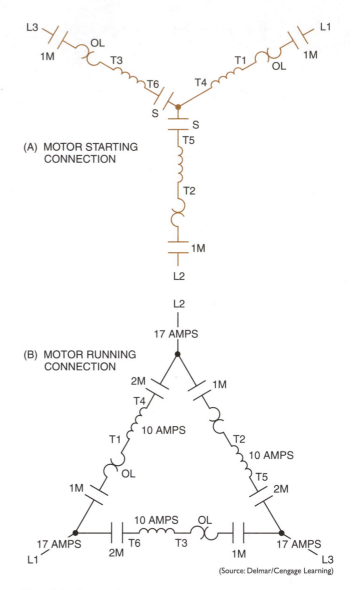

(A) MOTOR STARTING CONNECTION

(B) MOTOR RUNNING CONNECTION

(Source: Delmar/Cengage Learning)

Fig. 29–2 Elementary diagrams of motor power circuits of Figure 29–3. Controller connects motor in wye on start and in delta for run. Note that the overload relays are connected in the motor winding circuit, not in the line. Note also that the line current is higher than the phase winding current in the diagram for the delta connection (B). Winding current is the same as the line in diagram (A).

contact (S) of the same contactor energizes another contactor (1M) and maintains itself. Because the pneumatic timer is attached to contactor (1M), the motor is connected to the line in star and the timing period is started. When the timing period is complete, the first contactor (S) is de-energized. As a result, normally closed interlock S is closed, and contactor 2M is energized to connect the motor in delta. The motor then runs in the delta-connected configuration. This start-run scheme is called *open transition* because there is a moment in which the motor circuit is open between the opening of the power contacts (S) and the closing of the contacts (2M).

One advantage of this starting method is that it does not require accessory voltage reducing equipment. A star-delta starter has the disadvantage of open circuit transition, but it does give a larger starting torque per ampere of line current than a part winding starter.

CLOSED TRANSITION STARTING

Figure 29–5 shows a modification of Figure 29–4. In Figure 29–5, resistors maintain continuity to the motor to avoid the difficulties associated with the open circuit form of transition between start and run.

With closed transition starting, the transfer from the star to delta connections is made without disconnecting the motor from the line. When the transfer from star to delta is made in open transition starting, the starter momentarily disconnects the motor and then reconnects it in delta. Although an open transition is satisfactory in many cases, some installations may require closed transition starting to prevent power line disturbances. A size 6 closed transition starter is shown in Figure 29–6. Closed transition starting is achieved by adding a three-pole contactor and three resistors

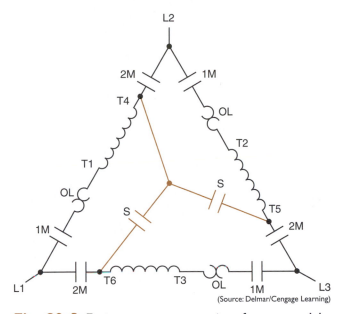

(Source: Delmar/Cengage Learning)

Fig. 29–3 Entire stator connection for a star-delta starter.

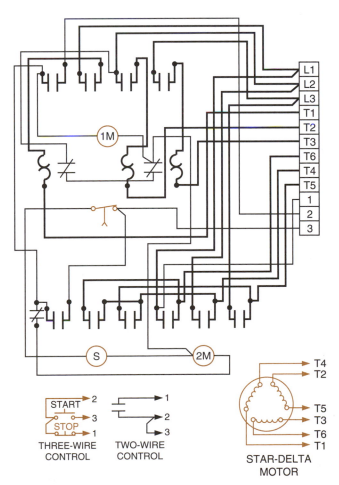

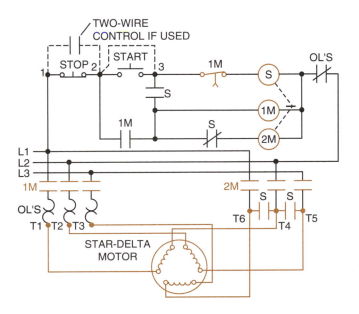

(Source: Delmar/Cengage Learning)

Fig. 29–4 Wiring diagram (left) and line diagram of star-delta starter (right).

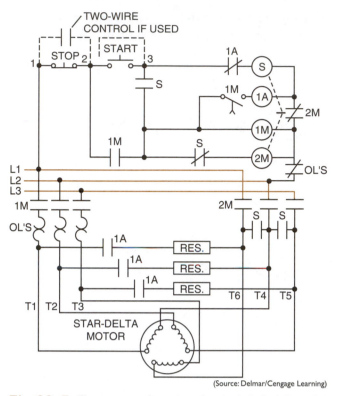

(Source: Delmar/Cengage Learning)

Fig. 29–5 Elementary diagram of sizes 1, 2, 3, 4, 5, and 6 star-delta starters with transition starting.

Fig. 29–6 Size 6 combination wye-delta (closed transition) motor starter. *(Courtesy of General Electric Co.)*

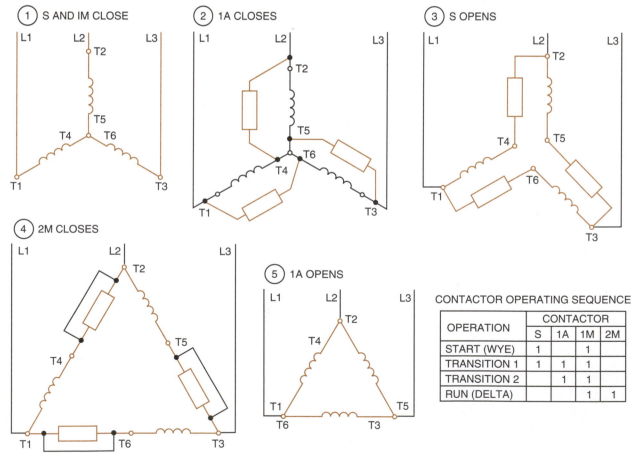

OPERATION	CONTACTOR			
	S	1A	1M	2M
START (WYE)	1		1	
TRANSITION 1	1	1	1	
TRANSITION 2		1	1	
RUN (DELTA)			1	1

CONTACTOR OPERATING SEQUENCE

(Source: Delmar/Cengage Learning)

Fig. 29–7 Motor connection and operating sequence for the star-delta closed transition starting.

to the starter circuit. The connections are made as shown in the closed transition schematic diagram, Figure 29–5. The contactor is energized only during the transition from star to delta. It keeps the motor connected to the power source through the resistors during the transition period, Figure 29–7. There is a reduction in the incremental current surge, which results from the transition. The balance of the operating sequence of the closed transition starter is similar to that of the open transition star-delta motor starter.

A single method of reduced current starting may not achieve the desired results because the motor starting requirements are so involved, the restrictions so stringent, and the needs are so conflicting. It may be necessary to use a combination of starting methods before satisfactory performance is realized. For special installations, it may be necessary to design a starting system to fit the particular conditions.

Study Questions

1. Indicate the correct terminal markings for a star-delta motor on Figure 29–8.

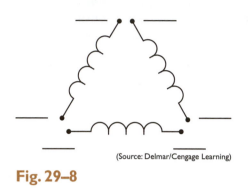

(Source: Delmar/Cengage Learning)

Fig. 29–8

2. What is the principal reason for using star-delta motors?

3. In Figure 29–5, which contactor closes the transition?

4. The closed transition contactor is energized only on transfer from star to delta. How is this accomplished?

5. If a delta-connected, six-lead motor nameplate reads "Full Load Current 170 Amperes," on what current rating should the overload relay setting or the selection of the heater elements be based?

UNIT 30

AC Solid-State Reduced Voltage Controller

Objectives

After studying this unit, the student should be able to

- Describe and draw how a solid-state controller is connected to an existing electromagnetic starter.

- Describe additional features of a solid-state controller.

- Describe electrical noise and how to help prevent it.

- Discuss the function of SCRs in reduced voltage motor starters.

- Install a solid-state controller.

THE NEED FOR SOLID-STATE CONTROL

Electromechanical devices such as magnetic relays, stepping switches, and timers, have been used for many years to provide decision or logic circuits for the electrical industry. For example, these devices are used to provide interlocking methods and sequencing control. Their continued wide usage can be expected because they are proven devices, backed by years of engineering experience. Their advantages include simple construction, flexibility in use, numerous contact combinations, low cost per point switched, easy availability, and ability to carry and break relatively large currents.

Solid-state, or static, control has no moving parts (except accessories). It provides a method for electrically controlling machines and different processes with logic elements and accessories. By means of semiconductor switching devices, such as transistors, they can manipulate electrical signals without any mechanical movement. Logic elements perform the same function in solid-state control systems that relays do in conventional electromagnetic control systems.

The development of solid-state control was *not* originally intended to replace conventional relays or to render them obsolete. Rather, solid-state switching devices were developed to help meet the more critical demands for reliability, life, and speed for controllers used in automated industries. In automated process lines, not only is it necessary to control individual machines, but also to coordinate material transfer between the machines and to perform many interlocking functions. In such complex circuitry, the reliability of each component becomes increasingly important.

Increased industrial production requires higher duty cycles and creates the need for a device whose life is independent of the number of operations. In addition, many industries (such as chemical, petroleum, and food) impose difficult environmental conditions on electrical controls that must be overcome. Solid-state control is often the best method to achieve these objectives because these modules are sealed units.

Solid-state starters are an effective choice in reduced voltage applications. Aside from not having moving power circuit parts, their great advantage is tailoring the output control closer to the needs of the load, because many elements are field adjustable. These features include precise starting and stopping time, precise overcurrent protection, power factor control, single phasing protection, current limiting, transient protection, and more. A particularly valuable option is a power factor corrector that permits significant energy savings when motors are operating at very low loadings. Solid-state starters are available in ranges up to several hundred horsepower.

SOLID-STATE REDUCED VOLTAGE MOTOR STARTING

Although the details of how a solid-state reduced voltage starter works are complex, the basic principles are easy to grasp. Reduced voltage starting is accomplished by reducing the applied voltage to the motor, which allows the motor to start at a reduced current. The motor then accelerates to full speed at a lower torque level. This technique reduces the high current inrush and the mechanical shock to driven equipment produced by across-the-line motor starting. A solid-state, reduced voltage motor starter uses silicon controlled rectifiers (SCRs) for power control. SCRs are indicated by an arrowhead pointing in the direction of conventional current flow, Figures 30–1(A) and (B). SCRs block reverse current flow, like diodes. They also block forward current flow until they receive a turn-on signal, or gate pulse. Once an SCR is turned on (gated), it does not stop forward current flow until the current drops below a certain level. Full-wave control employs two SCRs in each phase by connecting them in inverse parallel. This connection is shown in Figure 30–2. In this connection scheme, SCR 1 controls the voltage when it is positive in the sine wave. SCR 2 controls the voltage when it is negative. For three-phase motors, the configuration shown in Figure 30–3 is used.

Control of current or acceleration time is achieved by gating the SCR at different times within the half-cycle. If the gate pulse is applied early in the half-cycle, the output is high. If the gate pulse is applied late in the half-cycle, only a small part of the waveform is passed through and the output is low. Thus, by controlling the SCR's output voltage, the motor acceleration characteristics can be controlled. These effective

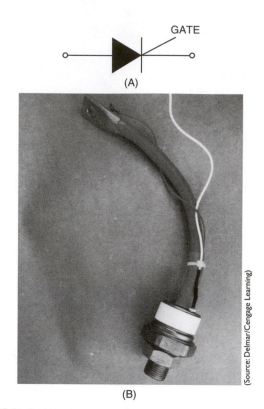

Fig. 30–1 (A) Symbol for silicon-controlled rectifier (SCR). (B) Typical power SCR.

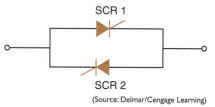

(Source: Delmar/Cengage Learning)

Fig. 30–2 Inverse parallel SCRs per one phase.

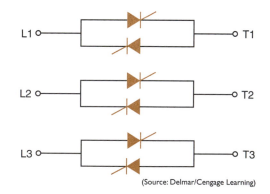

(Source: Delmar/Cengage Learning)

Fig. 30–3 A three-phase SCR connection.

waveforms are shown in Figure 30–4. Feedback to the controller is supplied by current transformers in each phase on the line side of the starter, Figure 30–5. Should there be any abnormal changes in the line current through these current transformers, it is detected in the secondary circuit and fed back to a control circuit that adjusts for smooth and timed acceleration. Timed acceleration is also available from a tachometer generator that is attached to the

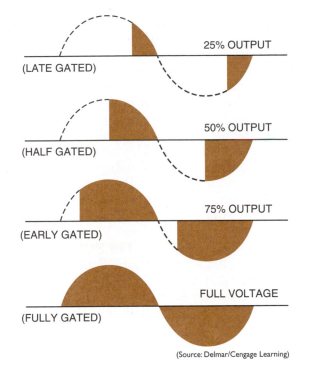

25% OUTPUT

(LATE GATED)

50% OUTPUT

(HALF GATED)

75% OUTPUT

(EARLY GATED)

FULL VOLTAGE

(FULLY GATED)

(Source: Delmar/Cengage Learning)

Fig. 30–4 Outputs from differently gated SCRs.

(Source: Delmar/Cengage Learning)

Fig. 30–5 Current transformers used for feedback control and metering.

motor shaft. This provides additional feedback to the controller, Figure 30–6.

In addition, solid-state controllers contain solid-state overload relays that monitor both the controller and the motor for overcurrent

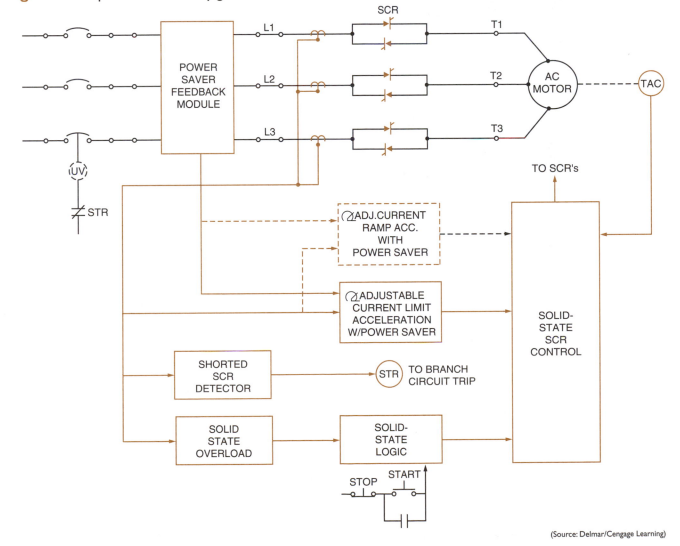

(Source: Delmar/Cengage Learning)

Fig. 30–6 Block diagram of a reduced voltage starter.

conditions. Power-saving modules are available that detect idling or low-power periods of operation. The motor terminal voltage is then reduced to minimize electrical losses during these operating conditions. A solid-state, reduced voltage motor starter is shown in Figure 30–7. Motor and remote control terminal connections are made at a marked terminal block.

Fig. 30–7 Solid-state, reduced voltage, three-phase starter. *(Courtesy of Eaton Corporation)*

Solid-state starters are available in many different types, incorporating different features. As indicated in the elementary diagram in Figure 30–8, the starter includes both start and run contactors. The start contacts are in series with the SCRs. The run contacts are in parallel with the series combination of SCRs and the start contacts. When the starter is energized, the start contacts close. The motor acceleration is now controlled by phasing-on the SCRs. When the motor reaches full speed, the run contacts close, connecting the motor directly across the line. At this point, the SCRs are turned off and the start contacts open. Thus, under full-speed running conditions, the SCRs are out of the circuit. This eliminates SCR power dissipation during the run cycle.

With the starter in the de-energized state, the start contacts are open. This condition isolates the motor from the lines and the SCR circuits. This open circuit condition protects against SCR misfiring or damage from overvoltage transients, or any electrical noise.

SOLID-STATE CONTROL WITH EXISTING MAGNETIC DEVICES

In addition to the greater noise immunity provided by integrated electronic circuits, the logic functions (such as the relays, contactors, and starters previously described) include additional filtering and buffering to limit the effect of transient, electrical "spikes," which may exceed the safe voltage noise margin. Refer to Figure 30–9.

The integration and use of solid-state relays for industrial control poses several problems. When they are used to operate electromagnetic equipment, replacing relays and contactors,

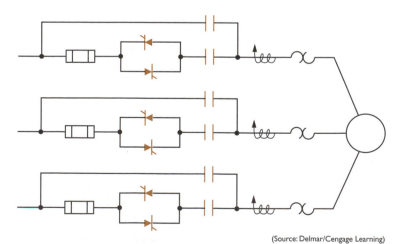

(Source: Delmar/Cengage Learning)

Fig. 30–8 Power circuit of SCRs with contactors.

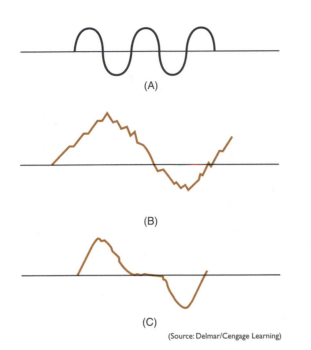

(A)

(B)

(C)

(Source: Delmar/Cengage Learning)

Fig. 30–9 (A) Good computer-grade AC power. (B) Objectionable line noise. (C) Bad temporary blackouts.

Fig. 30–10 Voltage surge suppressor. *(Courtesy of Siemens Energy & Automation)*

they are susceptible to damage by these transient voltages. They cannot be used simply to replace a relay that fails because of contact arcing. Although the relay contacts may have reduced life, the solid-state relay will have no life unless adequately protected.

Surge suppressors are installed on magnetic device coils, such as relays, contactors, and motor starters. A voltage surge suppressor is shown in Figure 30–10. The leads of the suppressor are connected to the coil terminals. This further limits voltage noise and overvoltage spikes produced by the starter coil when the coil circuit is opened.

Reduced voltage solid-state starting units can be connected into existing magnetic starter installations for standard three-phase, AC induction motors as shown in Figures 30–11(A) and (B). This series incorporates an automatic energy controller with a reduced voltage starter to maximize the efficiency of the motor duty cycle. It can be used on any motor load or electrical system that benefits from reduced voltage soft start or where power can be saved. Energy savings are accomplished by sensing the load and by reducing the voltage applied to the motor when it is running at less than full-load conditions. The result is lower motor amperage and voltage (resulting in true kilowatt savings).

In addition to being a reduced voltage starter and energy saver, such a unit is also a motor monitor. It features adjustable current limit, phase loss detection, adjustable start, undervoltage protection, and no-voltage phase reversal. It also has a three-phase sensing overload current trip that acts on an inverse time function similar to a thermal overload relay. The unit also features SCR overtemperature protection. Figure 30–12 shows a graph of the adjustable linear timed acceleration of a solid-state starter. This option starts the motor with a linear speed versus time acceleration. The starting is adjustable, thereby altering the current ramp. Some units may be adjusted for across-the-line starting as well. The current is still monitored so it does not exceed a preset level while starting. A locked rotor or overload would activate this preset trip level. A motor-mounted tachometer output is used for feedback. Sometimes current transformers are used.

Nearly all of these functions are illuminated with electronic LEDs. LEDs are similar to pilot lights on push-button control stations. They are color-coded and serve as fault indicators and advise of condition status. The color coding of LEDs typically is as follows:

GREEN	Control voltage present; or running	
AMBER	Starting; or energy saver	
RED	Fault, including:	
	Shorted SCR; phase loss	GREEN
	SCR overtemperature	AMBER
	Stalled motor	AMBER

This may be seen in Figure 30–13.

ENERGY SAVING

When a motor is lightly loaded or idling, it requires less magnetic flux. An energy-saving option will reduce the magnetic field strength

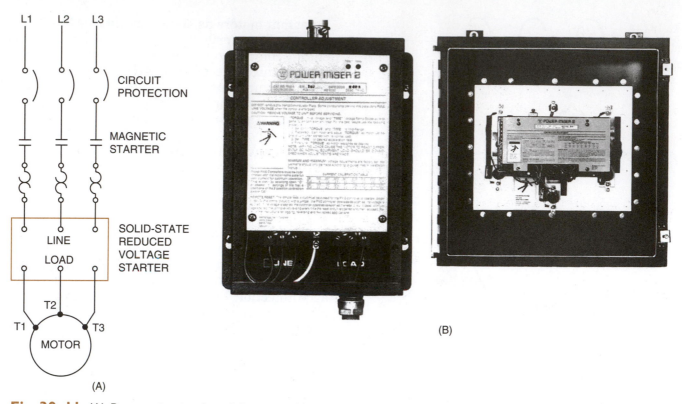

Fig. 30–11 (A) Power circuit of a solid-state reduced voltage starter connected into an existing motor installation. (B) This controller senses the motor load and feeds back the information, which is compared to a reference. As a result, the voltage can be continuously changed to reduce the current and voltage under light-load conditions. This energy saving feature results in lower power consumption and increased line power factor. *(Courtesy of Rockwell Automation)*

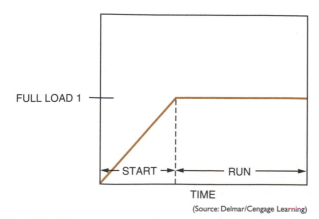

(Source: Delmar/Cengage Learning)

Fig. 30–12 Smooth, stepless acceleration of an induction motor.

Fig. 30–13 LEDs show the condition status of the motor starter. *(Courtesy of Allen Bradley)*

by reducing the voltage to the motor during these periods. When the field is reduced, the motor will draw less current. The energy-saving option, including a microcomputer, is activated after the motor is up to speed. It automatically applies more voltage when the load increases.

APPLICATION

The solid-state reduced voltage starter is suited for starting most industrial loads, including fans, compressors, pumps, machine tools, and conveyors. This starter should provide complete control of the starting current and motor torque.

There are some general precautions, which should be considered in any application. Because of their greater complexity, failures will be more difficult to find. A more qualified maintenance person or an electronic technician may be required to service the equipment. Installation should be no more difficult than electromagnetic equipment, as long as the installer follows the wiring diagram and the terminal markings.

Study Questions

1. List some advantages of solid-state control.

2. What element is the heart of the solid-state starter?

3. How is smooth acceleration achieved with a reduced voltage starter?

4. How are solid-state controllers connected into existing electromagnetic starters?

5. What are some additional features of a solid-state controller not commonly found on an electromagnetic starter?

6

Three-Phase, Multispeed Controllers

Controllers for Two-Speed, Two-Winding (Separate Winding) Motors

Objectives

After studying this unit, the student should be able to

- Identify terminal markings for two-speed, two-winding motors and starters.

- Describe the purpose and function of two-speed, two-winding motor starters.

- Describe how to obtain different speeds for AC squirrel cage motors.

- Connect two-speed, two-winding controllers and motors.

- Recommend solutions for troubleshooting these motors and controllers.

THE MOTOR

The squirrel cage induction motor is the most widely used industrial motor. It is simple and less expensive than most other motors. Essentially, it is a constant-speed motor. Some models, however, are made to operate at several fixed speeds. To do this, the motor manufacturer changes the number of poles for which the stator is wound. The operating principle on which the squirrel cage motor is based calls for a rotor revolving within a stator; this is due to the action of a rotating magnetic field. The speed of the revolving field is determined by the frequency of the alternating current (expressed in hertz) supplied by the alternator and by the number of poles on the stator. The formula defining speed in rpm is

$$\text{Speed in rpm} = \frac{60 \times \text{Hertz}}{\text{Pairs of Poles}}$$

The metric abbreviation r/min (revolutions per minute) refers to the synchronous speed of the motor. At the synchronous speed, the rotor speed equals the rotating magnetic field speed. Practically, however, this synchronous speed is not achieved due to slip, or the lag, of the rotor behind the rotating stator field. The phrase *pairs of poles* is used to indicate that motor poles are always in pairs; a motor never has an odd number of poles.

Using the formula for speed, the synchronous speed of a two-pole motor supplied by 60-hertz electrical energy is determined as follows:

$$\frac{60 \times 60}{1} = 3600 \text{ rpm}$$

The motor will run just below the synchronous speed due to slip. The design of the motor will determine the percentage of slip. This value is not the same for all motors.

CONTROLLERS

Line-voltage, ac multispeed starters, Figure 31–1, are designed to provide control for squirrel cage motors operating on two, three, or four different constant speeds, depending on their construction. The use of an automatic starter and a control station gives greater operating efficiency. In addition, both the motor and the machines are protected against improper sequencing or rapid speed change. Protection against motor overload at each speed is necessary.

Motors with separate windings have a different winding for each speed required. Although this type of motor is slightly more complicated in construction, and thus more expensive, the controller is relatively simple. Assuming the

Fig. 31–1 Magnetic controller for two-speed, two-winding (separate winding) motor. *(Courtesy of Schneider Electric)*

frequency is constant, two-winding motors will operate on each winding at the speed for which they are wound.

In the control arrangement in Figure 31–2, note that the motor can be started in either speed. However, the stop button must be pressed to transfer from high speed to the low speed. The starter shown in the figure is provided with a mechanical interlock between the L and H starters. Auxiliary contact interlocks are also provided.

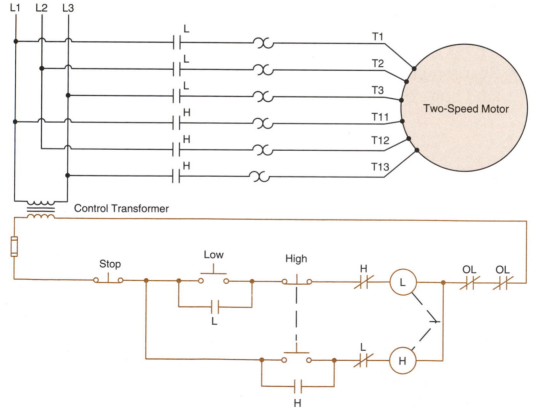

(Source: Delmar/Cengage Learning)

Fig. 31–2 Schematic diagram of two-speed, two-winding three-phase motor.

Figure 31–3 illustrates push-button interlocking through the use of combination push buttons. The figure also demonstrates that a transfer can be made to either speed without touching the stop button. However, rapid and continued speed transfers may activate the overload relays. To prevent this, the control scheme can be equipped with time-delay relays that will provide a time lag between the speed changes.

When the motor is operating, the inactive winding must be open circuited. This is done to prevent circulating current due to transformer action between the idle winding and the energized winding.

Figure 31–4 illustrates an automatic speed control for a two-speed, two-winding three-phase motor. The circuit is so designed that a certain amount of time must pass before the motor can

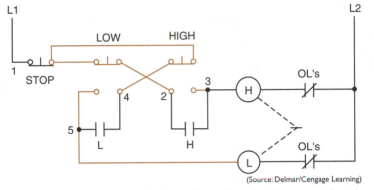

(Source: Delmar/Cengage Learning)

Fig. 31–3 Elementary diagram of push-button interlock and transfer to either speed without stopping.

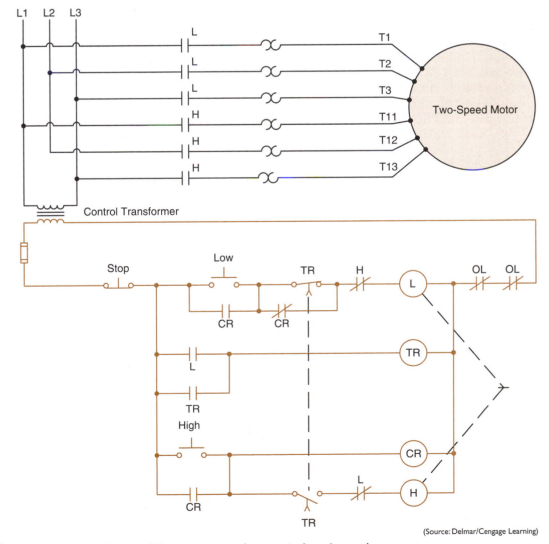

(Source: Delmar/Cengage Learning)

Fig. 31–4 Automatic speed control for a two-speed, two-winding three-phase motor.

be changed from low to high speed. Assume that timer TR is set for a delay of 10 seconds. If the low-speed push button is pressed, a circuit is completed through the normally closed CR contact and the normally closed H contact to motor starter L. When the L starter energizes, the motor starts in low speed and the normally open L auxiliary contact closes to energize the coil of timer TR. At the same time, the normally closed L auxiliary contact connected in series with H starter coil opens. After a delay of 10 seconds, the two TR timed contacts change position. The normally closed TR contact connected in series with L starter coil opens, but the circuit is maintained through the normally closed CR contact. The normally open TR contact connected in series with H starter coil closes, but a circuit is not completed to the coil because of the normally open CR contact and the normally closed L auxiliary contact that is now open.

If high speed is desired, the high push button can be pressed, energizing control relay CR and causing all CR contacts to change position. The normally closed CR contact opens and interrupts the power to L starter coil. This permits the normally closed L auxiliary contact connected in series with H coil to reclose. The normally open CR contact connected in parallel with the high push button closes and provides a current path to both the CR coil the H starter coil through the now closed TR contact and

closed L contact. The stop button must be pressed before the motor can be operated in low speed again.

If the motor is not running and the high push button is pressed, CR control relay energizes, causing all CR contacts to change position. The normally open CR contact connected in parallel with the low-speed push button closes and provides a current path through the closed TR contact and closed H contact to starter coil L. The motor starts in low speed and the normally open L auxiliary contact connected in series with TR timer coil closes. When TR time energizes, the normally open TR instantaneous contact closes to maintain the circuit when the L contact reopens. After a 10-second time delay, both TR timed contacts change position. The normally closed TR contact connected in series with L starter coil opens and disconnects the starter from the power line. Note that because the normally closed CR contact in now open, a current path does not exist around the TR contact. When L starter de-energizes, the normally closed L auxiliary contact connected in series with H coil recloses, completing a current path through the closed CR contact, and closed TR timed contact. The motor will continue to operate in high speed until the stop button is pressed or an overload occurs. Connections for two-speed, separate winding, three-phase motors are shown in Figure 31–5.

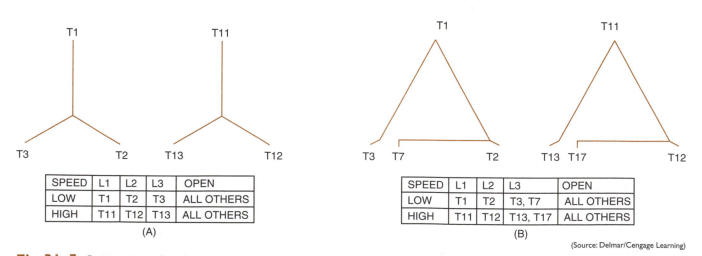

SPEED	L1	L2	L3	OPEN
LOW	T1	T2	T3	ALL OTHERS
HIGH	T11	T12	T13	ALL OTHERS

(A)

SPEED	L1	L2	L3	OPEN
LOW	T1	T2	T3, T7	ALL OTHERS
HIGH	T11	T12	T13, T17	ALL OTHERS

(B)

(Source: Delmar/Cengage Learning)

Fig. 31–5 Connections for three-phase, two-speed, two-winding (separate winding) motors (A) star and (B) delta.

Study Questions

1. Excluding slip, at what speed (r/min) will an eight-pole, 60-hertz, 240-volt motor run?

2. What is meant by the description "separate winding, two-speed motor"?

3. What will happen if an operator makes too many rapid transitions from low to high speed and back on a two-speed motor?

4. Why are there four motor contacts for high and low speeds in Figure 31–4?

5. In Figure 31–4, what does the broken line between both time-delay contacts (TR) mean?

6. When the high button is depressed in Figure 31–4, what prevents the high-speed starter from closing immediately?

7. Which coils are energized immediately when the high push button is closed in Figure 31–4?

8. At what speed can a four-pole, 240-volt, 5-hp, 50-hertz motor operate?

9. What is meant by the "compelling action" produced by an accelerating relay?

10. Is it necessary to press the stop button when changing from a high speed to a low speed in Figure 31–4? Why or why not?

Two-Speed, One-Winding (Consequent Pole) Motor Controller

Objectives

After studying this unit, the student should be able to

- Identify terminal markings for two-speed, one-winding motors and controllers.

- Describe the purpose and function of two-speed, one-winding motor starters and motors.

- Connect two-speed, one-winding controllers and motors.

- Connect a two-speed starter with reversing controls.

- Recommend solutions for troubleshooting these motors and controllers.

Certain applications require the use of a squirrel cage motor having a winding arranged so that the number of poles can be changed by reversing some of the currents. If the number of poles is doubled, the speed of the motor is cut approximately in half.

The number of poles can be cut in half by changing the polarity of alternate pairs of poles, Figure 32–1. The polarity of half the poles can be changed by reversing the current in half the coils, Figure 32–2.

If a stator field is laid flat (as in Figure 32–1) the established stator field must move the rotor twice as far in Part (B) as in Part (A) and in the same amount of time. As a result, the rotor must travel faster. The fewer the number of poles established in the stator, the greater is the speed in r/min of the rotor.

A three-phase squirrel cage motor can be wound so that six leads are brought out, Figure 32–3. By making suitable connections with these leads, the windings can be connected in series delta or parallel wye, Figure 32–4. If the winding is such that the series delta connection gives the high speed, then the horsepower rating is the same at both speeds. If the winding is such that the series delta connection gives the low speed and the parallel wye connection gives the high speed, then the torque rating is the same at both speeds.

Consequent pole motors have a single winding for two speeds. Extra taps can be brought

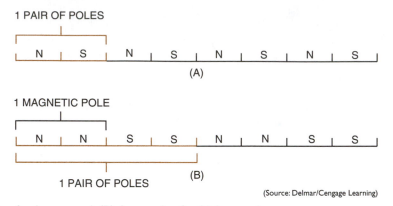

1 PAIR OF POLES

| N | S | N | S | N | S | N | S |

(A)

1 MAGNETIC POLE

| N | N | S | S | N | N | S | S |

1 PAIR OF POLES

(B)

(Source: Delmar/Cengage Learning)

Fig. 32–1 (A) Eight poles for low speed; (B) four poles for high speed.

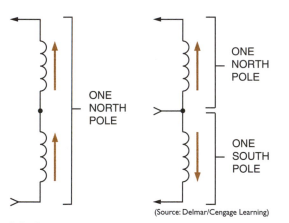

ONE NORTH POLE

ONE NORTH POLE

ONE SOUTH POLE

(Source: Delmar/Cengage Learning)

Fig. 32–2 The number of poles is doubled by reversing current through half a phase. Two speeds are obtained by producing twice as many consequent poles for low-speed operation as for high speed.

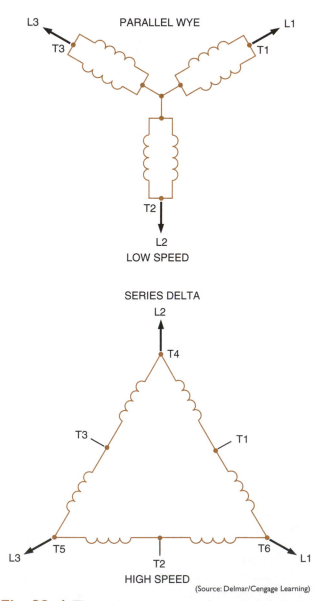

PARALLEL WYE

L3   T3   T1   L1

T2

L2

LOW SPEED

SERIES DELTA

L2

T4

T3   T1

L3   T5   T2   T6   L1

HIGH SPEED

(Source: Delmar/Cengage Learning)

Fig. 32–4 Three-phase, two-speed, one-winding constant horsepower motor connections made by motor controller.

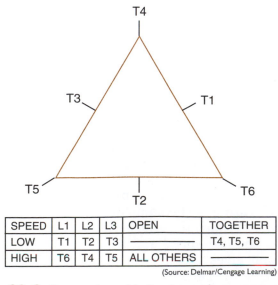

T4

T3   T1

T5   T2   T6

SPEED	L1	L2	L3	OPEN	TOGETHER
LOW	T1	T2	T3	—	T4, T5, T6
HIGH	T6	T4	T5	ALL OTHERS	—

(Source: Delmar/Cengage Learning)

Fig. 32–3 Connection table for three-phase, two-speed, one-winding constant horsepower motor.

from the winding to permit reconnection for a different number of stator poles. The speed range is limited to a 1:2 ratio, such as 600–1200 rpm or 900–1800 rpm.

Two-speed, consequent pole motors have one reconnectable winding. However, three-speed, consequent pole motors have two windings, one of which is reconnectable. Four-speed consequent pole motors have two reconnectable windings.

Referring back to the motor connection table in Figure 32–3, note that for low-speed operation, T1 is connected to L1; T2 to L2; T3 to L3; and T4, T5, and T6 are connected together. For high-speed operation, T6 is connected to L1; T4 to L2; T5 to L3; and all other motor leads are open.

Figures 32–5(A) and (B) illustrates the circuit for a size 1, selective multispeed starter connected for operation with a reconnectable,

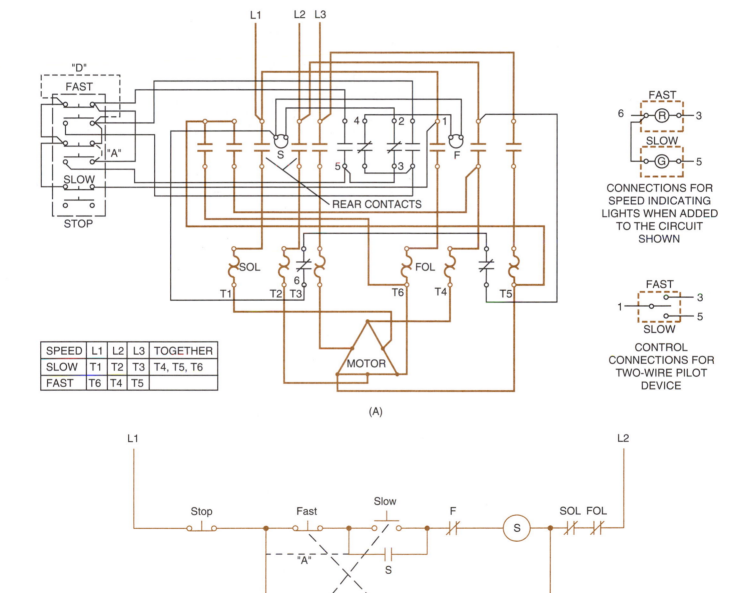

SPEED	L1	L2	L3	TOGETHER
SLOW	T1	T2	T3	T4, T5, T6
FAST	T6	T4	T5	

(A)

(B)

(Source: Delmar/Cengage Learning)

Fig. 32–5 Wiring diagram (A) and line diagram (B) of an AC, full-voltage, two-speed magnetic starter for single-winding (reconnectable pole) motors.

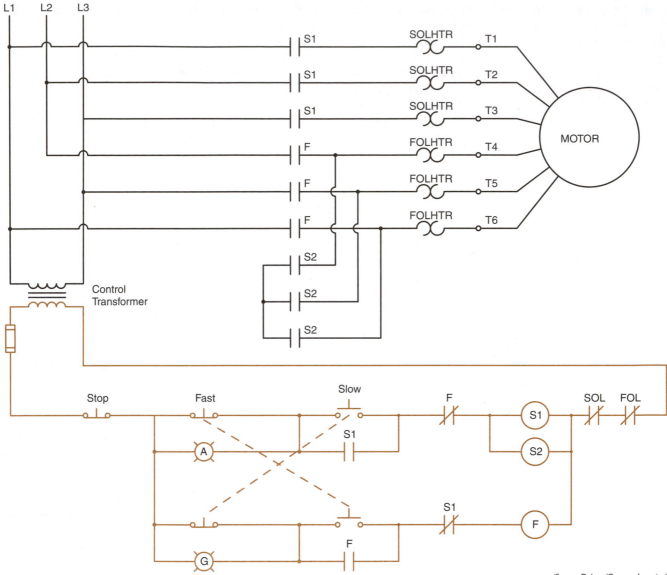

Fig. 32–6 A circuit for controlling a two speed consequent pole motor using contactors with three load contacts.

(Source: Delmar/Cengage Learning)

constant horsepower motor. The control station is a three-element, fast-slow-stop station connected for starting at either the fast or slow speed. The speed can be changed from fast to slow or slow to fast without pressing the stop button between changes. If equipment considerations make it desirable to stop the motor before changing speeds, this feature can be added to the control circuit by making connections "D" and "A," shown in Figure 32–5 by the dashed lines. Adding these jumper wires eliminates push-button interlocking in favor of stopping the motor between speed changes. This feature may be desirable for some applications.

Connections for the addition of indicating lights or a two-wire pilot device instead of the control shown are also given.

The circuit illustrated in Figure 32–5(A) requires the use of a contactor with five load contacts. Circuits of this type are often constructed using contactors that contain three load contacts because they are more readily available. A schematic diagram of a circuit of this type is shown in Figure 32–6. The circuit permits the motor speed to be changed without pressing the stop button. According to the chart in Figure 32–5(A), terminals T4, T5, and T6 must be connected together for low speed operation. Two contactors will be employed when low speed is selected. The contactors will be identified as S1 and S2. The fast speed contactor will be identified as F. A separate overload relay for each speed is required because the full load current for each speed will be different. Pilot lamps

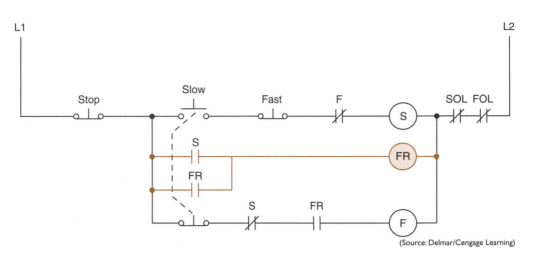

(Source: Delmar/Cengage Learning)

Fig. 32–7 Two-speed control circuit using a compelling relay.

have been added to the circuit to indicate motor speed. An amber colored lamp will indicate slow speed and a green lamp will indicate fast speed.

A compelling-type control scheme is shown in Figure 32–7. These connections mean that the operator must start the motor at the slow speed. This controller cannot be switched to the fast speed until after the motor is running.

When the slow button is pressed, the slow-speed starter S and the control relay FR are energized. Once the motor is running, pressing the fast button causes the slow-speed starter to drop out. The high-speed starter is picked up through the normally closed interlock contacts S of the slow-speed starter and through the normally open contacts of control relay FR. The normally open contacts of control relay FR will now close.

If the fast button is pressed in an attempt to start the motor, nothing will happen because the normally open contacts of control relay FR will prevent the high-speed starter from energizing. When the fast button is pressed, it breaks a circuit but does not make a circuit. This scheme is another form of sequence starting.

MISTAKEN REVERSAL CAUTION

When multispeed controllers are installed, the electrician should check carefully to ensure that the phases between the high- and low-speed windings are not accidentally reversed. Such a phase reverse will also reverse the direction of motor rotation. The driven machine should remain disconnected from the motor until an operational inspection is complete. The upper oil inspection plugs (pressure plugs) in large gear reduction boxes should be removed. Failure to remove these plugs may create broken casings if the motor is reversed accidentally.

Machines can be damaged if the direction of rotation is changed from that for which they are designed. In general, the correct rotational direction is indicated by arrows on the driven machine.

TWO-SPEED STARTER WITH REVERSING CONTROLS

Figure 32–8 is a diagram of a two-speed, reversing controller. The desired speed, high or low, is determined by the setting of a two position switch labeled Low and High. The load circuit consists of forward and reverse contactors as well as a contactor that contains three load contacts for high speed and a contactor that contains five load contacts for low speed. The circuit provides automatic acceleration to high speed in either the forward or reverse direction. The circuit operates as follows:

- The direction of rotation is determined by pressing the forward or reverse push button. The forward-reverse circuit is interlocked electrically and mechanically in the same manner as a standard forward-reverse control as discussed in Unit 21.
- If the selector switch is set in the high position, the motor will start in low speed in the direction selected because a current path exists through the normally closed TR contact, and normally closed H contact to L contactor coil. The L and H contactors are electrically and mechanically interlocked in the same manner as the forward-reverse contactors.
- Since the selector switch is set in the high position, a current path also exists to the coil of timer TR.
- After some period of time, the normally closed TR contact opens and de-energizes

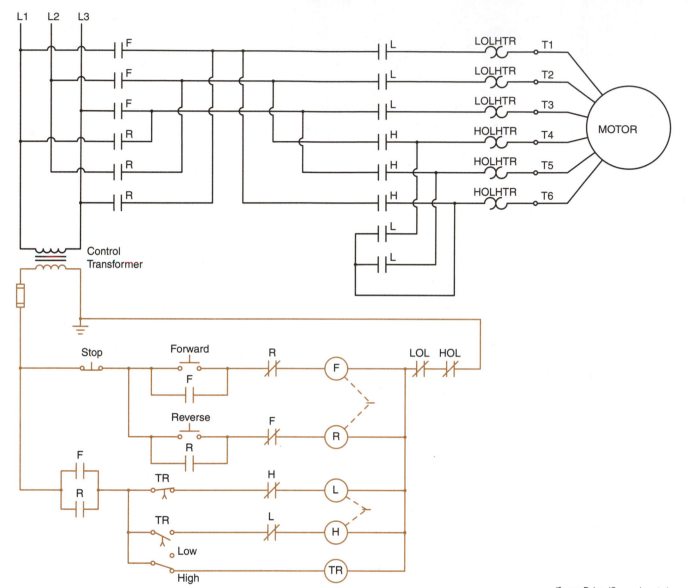

(Source: Delmar/Cengage Learning)

Fig. 32–8 A forward-reverse two-speed control circuit.

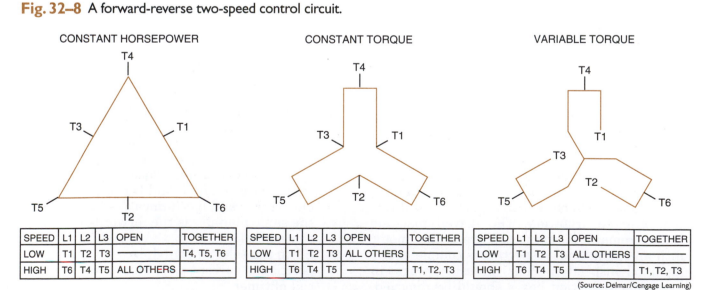

SPEED	L1	L2	L3	OPEN	TOGETHER
LOW	T1	T2	T3	———	T4, T5, T6
HIGH	T6	T4	T5	ALL OTHERS	———

SPEED	L1	L2	L3	OPEN	TOGETHER
LOW	T1	T2	T3	ALL OTHERS	———
HIGH	T6	T4	T5	———	T1, T2, T3

SPEED	L1	L2	L3	OPEN	TOGETHER
LOW	T1	T2	T3	ALL OTHERS	———
HIGH	T6	T4	T5	———	T1, T2, T3

(Source: Delmar/Cengage Learning)

Fig. 32–9 Typical motor connection arrangement for three-phase, two-speed, one-winding motors. These connections conform to NEMA and ASA standards. (All possible arrangements are not shown.)

contactor coil L. The normally open TR contact closes and provides a current path through the now closed L contact to contactor coil H causing the motor to accelerate to high speed.

■ The motor will remain in the selected speed until the stop push button is pressed.

■ If the selector switch is set in the low position, a current path is never established to TR timer and the motor will continue to operate in low speed until the stop push button is pressed.

■ An overload in either the low or high speed setting will de-energize the control circuit.

Study Questions

1. The rotating magnetic field of a two-pole motor travels 360 electrical degrees. How many mechanical degrees does the rotor travel?

2. How can two speeds be obtained from a one-winding motor?

3. What will happen if terminals T5 and T6 in Figure 32–3 are interchanged on high speed?

4. Is the motor operating at high speed or low speed when it is connected in series delta? In parallel wye?

5. Describe the operation of the line diagram in Figure 32–5(B) by adding jumper "D" only.

6. Describe the operation of Figure 32–5(B) by adding jumper wire "A" to the original circuit.

7. In Figure 32–7, the motor cannot be started in high speed. Why?

8. Draw a schematic diagram of a two-speed control circuit using a standard-duty selector switch for the high and low speeds. Add a red pilot light for fast speed and a green pilot light for low speed. Omit "A" and "D" jumper wires.

Four-Speed, Two-Winding (Consequent Pole) Motor Controller

Objectives

After studying this unit, the student should be able to

- Identify terminal markings for four-speed, two-winding (consequent pole) motors and controllers.

- Describe the purpose and function of four-speed, two-winding motor starters and motors.

- Describe the purpose and function of a compelling relay, an accelerating relay, and a decelerating relay.

- Connect and troubleshoot four-speed, two-winding controllers and motors.

Four-speed, consequent pole, squirrel cage motors have two reconnectable windings and two speeds for each winding.

OPERATIONAL SEQUENCE

Standard starters generally are connected so that the operator can start the motor from rest at any speed, or change from a lower speed to a higher speed. Before change is made from a higher to a lower speed, the stop button must be pressed. This arrangement protects the motor and the driven machinery from the excessive line current and shock that results when a motor running at a high speed is reconnected to run at a lower speed.

Multispeed starters are provided with mechanical and electrical interlocking to avoid the possibility of short-circuiting the line or connecting more than one speed winding at the same time.

COMPELLING RELAYS

Standard starters are not equipped with compelling relays. Such relays ensure that the motor starts at the lowest speed or that acceleration is accomplished in a series of steps. (These devices are described in previous units covering multispeed control.) However, when the type of motor used or the characteristics of the load involved make it necessary to use a particular starting sequence, there are three types of relays available to accomplish the function required: compelling relays, accelerating relays, and decelerating relays.

Form 1 Compelling Relay

When this type of relay is used, the motor must be started at low speed before a higher speed can be selected. The motor can be started only by pressing the low-speed push button. If any other push button is pressed, nothing will happen. This arrangement ensures that the motor first moves the load only at low speed. The stop button must be pressed before a change can be made from a higher speed to a lower speed.

Form 2 Accelerating Relays

When the starter is equipped with a Form 2 accelerating relay, the final speed is determined by the button pressed. The motor is started at low speed and is then accelerated automatically through successive speed steps until the selected speed is reached. Definite time intervals must elapse between each speed change. Individual timing relays are provided for each interval. Each relay is adjustable. The stop button must be pressed before a change can be made from a higher speed to a lower speed.

Decelerating Relays

This type of relay is similar in action to a Form 2 accelerating relay. The difference is that it controls a stepped *deceleration* from a high speed to a lower speed. Definite time intervals must elapse between each speed change.

Tremendous strains are placed on both the motor and the driven machinery when a change from a higher speed to a lower speed is made and the motor is not allowed to slow down to the desired speed. Such strain is avoided if a definite, adjustable time interval is provided between speeds when the motor is decelerating.

Using three-wire, push-button control and decelerating relays, it is possible to select any speed less than that at which the motor is running by pressing the proper push button. Once this push button is pressed, it de-energizes the contactor driving the motor at the higher speed and energizes the timing relays. After a preset time period, the timing relays energize the contactor to drive the motor at the lower speed. As succeedingly lower speeds are selected, there is an increase in the elapsed time between disconnecting the higher speed winding and connecting the lower speed winding.

When the two-wire control devices, such as pressure or float switches, are used to control motors at various speeds, decelerating relays should always be used. The only exception to this statement is in the event that both the motor manufacturer and the machine manufacturer approve the intended application and agree that decelerating relays are not necessary.

Figure 33–1 is an elementary diagram of a four-speed controller that can be used on a two-winding motor. This circuit is for a standard starter arrangement that is not equipped with compelling relays. (Remember that such relays permit starting only at the lowest speed.) This controller has electrical interlocks that prevent an operator from changing to different speeds without pressing the stop button. A definite time interval must elapse between each speed change on deceleration. The motor should be allowed to slow to the speed desired before a transfer to a lower speed is made.

In the motor connection table in Figure 33–1, note that the windings are connected alternately during acceleration. If the first winding is 6 and 12 poles and the second winding is 4 and 8 poles, then the successive pole connections for acceleration from low speed to high speed are 12, 8, 6, and 4. Figure 33–2 shows common motor connections for four-speed, three-phase, two-winding motors; all have different speeds. Note that the idle winding remains open circuited on given speeds. In this way an unnecessary electrical load is not imposed on the energized winding. This load would result from transformer action that creates circulating currents in the secondary, idle winding.

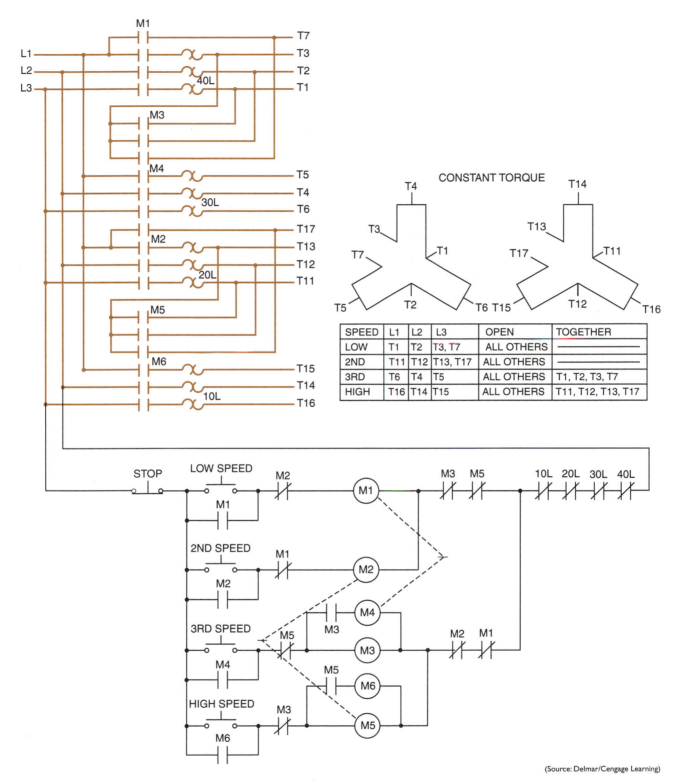

SPEED	L1	L2	L3	OPEN	TOGETHER
LOW	T1	T2	T3, T7	ALL OTHERS	————
2ND	T11	T12	T13, T17	ALL OTHERS	————
3RD	T6	T4	T5	ALL OTHERS	T1, T2, T3, T7
HIGH	T16	T14	T15	ALL OTHERS	T11, T12, T13, T17

(Source: Delmar/Cengage Learning)

Fig. 33–1 Elementary diagram of a four-speed, two-winding controller.

CONSTANT HORSEPOWER

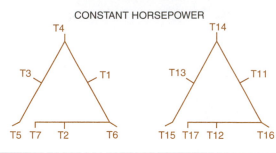

SPEED	L1	L2	L3	OPEN	TOGETHER
LOW	T1	T2	T3	ALL OTHERS	T4, T5, T6, T7
2ND	T6	T4	T5, T7	ALL OTHERS	————
3RD	T11	T12	T13	ALL OTHERS	T14, T15, T16, T17
HIGH	T16	T14	T15, T17	ALL OTHERS	————

(A)

CONSTANT HORSEPOWER

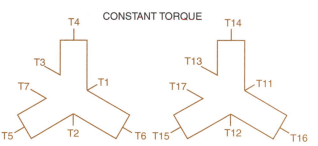

SPEED	L1	L2	L3	OPEN	TOGETHER
LOW	T1	T2	T3	ALL OTHERS	T4, T5, T6, T7
2ND	T11	T12	T13	ALL OTHERS	T14, T15, T16, T17
3RD	T6	T4	T5, T7	ALL OTHERS	————
HIGH	T16	T14	T15, T17	ALL OTHERS	————

(B)

CONSTANT TORQUE

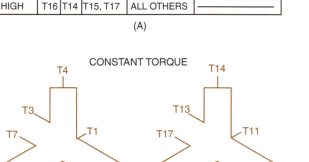

SPEED	L1	L2	L3	OPEN	TOGETHER
LOW	T1	T2	T3, T17	ALL OTHERS	————
2ND	T6	T4	T5	ALL OTHERS	T1, T2, T3, T7
3RD	T11	T12	T13, T17	ALL OTHERS	————
HIGH	T16	T14	T15	ALL OTHERS	T11, T12, T13, T17

(C)

CONSTANT TORQUE

SPEED	L1	L2	L3	OPEN	TOGETHER
LOW	T1	T2	T3, T17	ALL OTHERS	————
2ND	T11	T12	T13, T17	ALL OTHERS	————
3RD	T6	T4	T5	ALL OTHERS	T1, T2, T3, T7
HIGH	T16	T14	T15	ALL OTHERS	T11, T12, T13, T17

(D)

VARIABLE TORQUE

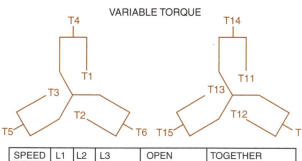

SPEED	L1	L2	L3	OPEN	TOGETHER
LOW	T1	T2	T3	ALL OTHERS	————
2ND	T6	T4	T5	ALL OTHERS	T1, T2, T3
3RD	T11	T12	T13	ALL OTHERS	————
HIGH	T16	T14	T15	ALL OTHERS	T11, T12, T13

(E)

VARIABLE TORQUE

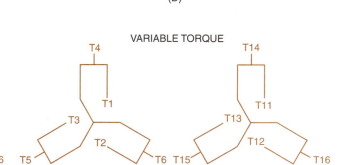

SPEED	L1	L2	L3	OPEN	TOGETHER
LOW	T1	T2	T3	ALL OTHERS	————
2ND	T11	T12	T13	ALL OTHERS	————
3RD	T6	T4	T5	ALL OTHERS	T1, T2, T3
HIGH	T16	T14	T15	ALL OTHERS	T11, T12, T13

(F)

(Source: Delmar/Cengage Learning)

Fig. 33–2 Typical arrangements for four-speed, three-phase, two-winding motors.

Study Questions

1. When a four-speed, two-winding motor is running, why is it desirable to allow the idle winding to remain open circuited?

2. In Figure 33–1, what is the function of this particular arrangement of electrical interlocks?

3. Why are definite time intervals required between each speed change on deceleration?

4. What is the primary reason for using a compelling relay?

5. What is the purpose of an accelerating relay?

6. When is it most important to use decelerating relays for multispeed installations?

7. Why are different motor connections shown for what appears to be the same motor in the various diagrams in Figure 33–2?

8. How many windings are required for three-speed motors?

Wound Rotor (Slip Ring) Motor Controllers

Wound Rotor Motors and Manual Speed Control

Objectives

After studying this unit, the student should be able to

- Identify terminal markings for wound rotor (slip ring) motors and controllers.

- Describe the purpose and function of manual speed control and wound rotor motor applications.

- Explain the difference between two-wire and three-wire control for wound rotor motors.

- Connect wound rotor motors with manual speed controllers.

- Recommend solutions to troubleshoot problems with these motors.

The AC three-phase wound rotor, or slip ring, induction motor was the first AC motor that successfully provided speed control characteristics. This type of motor was an important factor in successfully adapting alternating current for industrial power applications. Because of their flexibility in specialized applications, wound rotor motors and controls are widely used throughout industry to drive conveyors for moving materials, hoists, grinders, mixers, pumps, variable speed fans, saws, and crushers. Advantages of this type of motor include maximum utilization of driven equipment, better coordination with the overall power system, and reduced wear on mechanical equipment. The wound rotor motor has the added features of high starting torque and low starting current. These features give the motor better operating characteristics for applications requiring a large motor or where the motor must start under load. This motor is especially desirable where its size is large with respect to the capacity of the transformers or power lines.

The phrase "wound rotor" actually describes the construction of the rotor. In other words, it is wound with wire. When the rotor is installed in a motor, three leads are brought out from the rotor winding to solid conducting slip rings, Figure 34–1. Carbon brushes ride on these rings and carry the rotor winding circuit out of the motor to a speed controller. Unlike the squirrel cage motor, the induced current can be varied in the wound rotor motor. As a result, the motor speed can also be varied, Figure 34–2. (Wound rotor motors have stator windings identical to those used in squirrel cage motors.) The controller

Fig. 34–1 Rotor of an 800-hp wound rotor induction motor. *(Courtesy of Electric Machinery Company, Inc.)*

varies the resistance (and thus the current) in the rotor circuit to control the acceleration and speed of the rotor once it is operating.

Resistance is introduced into the rotor circuit when the motor is started or when it is operating at slow speed. As the external resistance is eliminated by the controller, the motor accelerates.

Generally, wound rotor motors under load are not suitably started with the rings shunted. If such a motor is started in this manner, the rotor resistance is so low that starting currents are too high to be acceptable. In addition, the starting torque is less than if suitable resistors are inserted in the slip ring circuit. By inserting resistance in the ring circuit, starting currents are decreased and starting torque is increased.

Motors for school and classroom use are provided with one stator and several interchangeable rotors. These rotors can be used with the *same stator* and bearing end brackets. The selection of available rotors includes the following types: *squirrel cage, wound,* and *synchronous.*

A control for a wound rotor motor consists of two separate elements: (1) a means of connecting the primary or stator winding to the power lines and (2) a mechanism for controlling the current in the secondary or rotor circuit. For this reason, wound rotor controllers are sometimes called *secondary resistor starters.*

Secondary resistor starters and *speed regulators* have a sliding contact (faceplate) design. This regulator has stationary contacts connected into the resistor circuit. Movable contacts slide over the stationary contacts (from left to right) and cut out of the rotor circuit successive steps of resistance. As a result, there is an increase in the speed of the motor. The entire resistance is removed from the rotor circuit when the extreme right position of the stationary contacts is reached. At this point, the motor is operating at normal speed.

Basically, there are two types of manual controllers, *starters* and *regulators.* The resistors used in starters are designed for starting duty only. This means that the operating lever must be moved to the full *ON* position. The lever must *not* be left in any intermediate position. The resistors used in *regulators* are designed for continuous duty. As a result, the operating lever in regulators can be left in any speed position.

When wound rotor motors are used as adjustable speed drives, they are operated on a continuous basis with resistance in the rotor circuit. In this case, the speed regulation of the motor is changed, and the motor operates at less than the full-load speed.

Sliding contact wound rotor starters and regulators do not have a magnetic primary contactor. They control only the secondary (rotor) circuit of the motor. A separate magnetic contactor, automatic starter, or circuit breaker is required for the primary circuit. If a primary magnetic contactor is used, electrical interlocks on the movable contact arm will control the operation of the contactor.

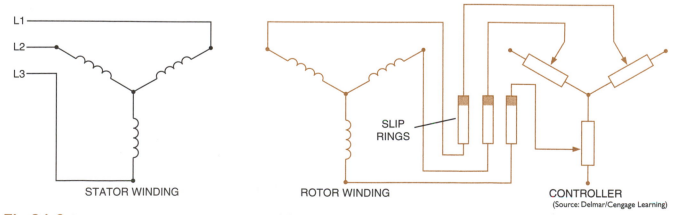

Fig. 34–2 Power circuit: stator, rotor, and controller.

STATOR WINDING

SLIP RINGS

ROTOR WINDING

CONTROLLER
(Source: Delmar/Cengage Learning)

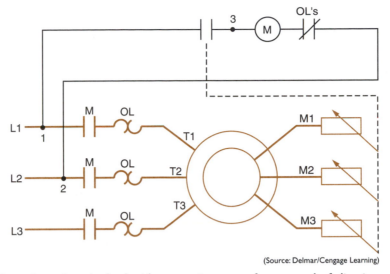

Fig. 34–3 Manual speed regulator interlocked with magnetic starter for control of slip ring motor (two-wire control).

When starting wound rotor motors, the operating lever is moved to the first running position to insert the full value of resistance in the secondary circuit, Figure 34–3. This action also operates the electrical normally open contact and completes the circuit to the magnetic primary contactor. This contactor connects the primary circuit, motor stator, to the line by closing the normally open contact (two-wire control). As the operating lever moves to the right, more and more resistance is cut out of the circuit. At the extreme right-hand position of the lever, the motor is running at full speed. Normally open or normally closed electrical interlocks control the primary magnetic switch and ensure that sufficient resistance is used in the rotor circuit for starting.

The use of two-wire control (normally open contact) for starting provides low-voltage release. This means that power to the motor can be interrupted if the line voltage drops to a low value or fails completely. When the voltage returns to its normal value, the motor is restarted automatically. For some applications, this feature is not always desirable.

The use of three-wire control for starting (normally closed control contact), Figure 34–4

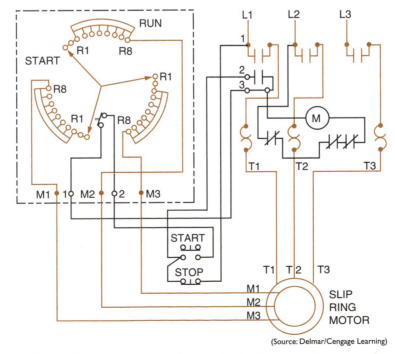

Fig. 34–4 Wiring diagram of manual speed regulator interlocked with magnetic starter for control of slip ring motor (three-wire control).

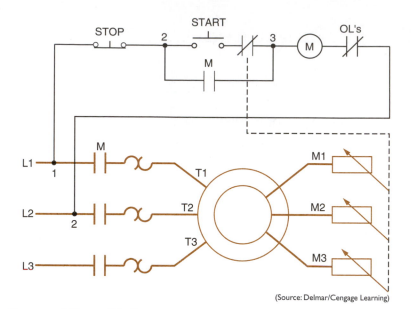

(Source: Delmar/Cengage Learning)

Fig. 34–5 Elementary diagram of Figure 34–4.

and Figure 34–5, means that low-voltage protection is provided. The motor is disconnected from the line in the event of voltage failure. To restart the motor when the voltage returns to its normal value, the normal starting procedure must be followed. To reverse the motor, any two of the three-phase motor leads may be interchanged.

Drum controllers, Figure 34–6, may be used in place of faceplate starters. The drum controller, however, is an independent component; it is separate from the resistors and is used for switching or controlling external resistors. As in the faceplate starter, a starting contact controls a line voltage, across-the-line starter. Generally, the electrician must make the connections on the job site.

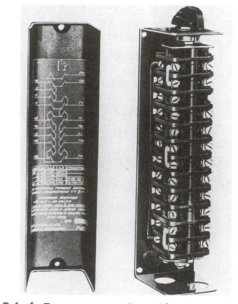

Fig. 34–6 Drum controller. *(Courtesy of Rockwell Automation)*

Study Questions

1. What characteristic of wound rotor motors led to their wide use for industrial applications?

2. By what means is the rotor coupled to the stator?

3. What other name is given to the wound rotor motor?

4. Does increased resistance in the rotor circuit produce low or high speed?

5. What two separate elements are used to control a wound rotor motor?

6. What is the difference between a manual faceplate starter and a manual faceplate regulator?

UNIT 35

Push-Button Speed Selection

Objectives

After studying this unit, the student should be able to

- Describe the purpose of push-button speed selection of wound rotor motors.

- Describe what happens during acceleration and deceleration of a motor used with a push-button speed selector control.

- Connect wound rotor motors and push-button speed selector controls.

- Recommend troubleshooting solutions for wound rotor problems.

Magnetic controllers for wound rotor motors consist of (1) a magnetic starter that connects the primary or stator winding to the power line and (2) one or more accelerating contactors that cut out steps of resistance in the secondary or rotor circuit. The number of accelerating contactors depends on the desired number of speeds for the motor. Enough steps of speed are used to ensure a smooth acceleration and to keep the inrush or starting current within practical limits. The circuit shown in Figure 35–1 permits three steps of speed. The motor must be started in low speed and progress to second speed and then to high speed. If the motor is operating in high speed, a slower speed can be selected by pushing the appropriate speed push button.

The motor is started by pressing the low-speed push button and energizing coil M, causing power to be supplied to the motor's stator winding through load contacts M. The motor starts slowly with full resistance in the secondary circuit. When the second, or medium-speed, push-button is pressed, contactor S energizes and closes the two S load contacts. This shorts the first bank of resistors out of the circuit and the motor accelerates to a higher speed. When the high-speed push button is pressed, contactor H energizes and shorts all the resistance out of the rotor circuit and the motor accelerates to its highest speed.

If it is desirable to go from high to medium speed, pressing the medium-speed push button causes the normally closed medium-speed push button to open and de-energize contactor coil H, causing the H load contacts to open and one bank of resistors to be connected back in the rotor circuit. This additional resistance in

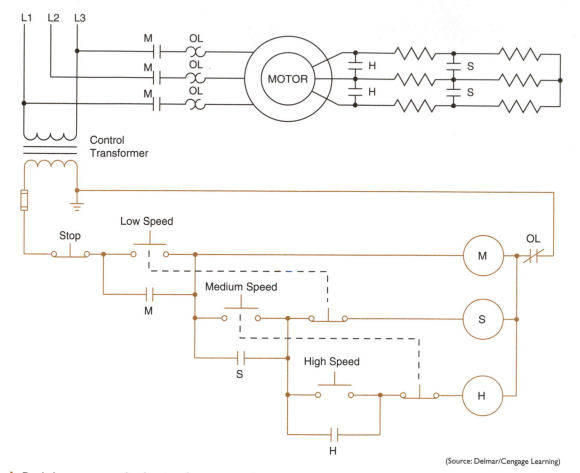

Fig. 35–1 Push-button speed selection for a wound rotor motor.

(Source: Delmar/Cengage Learning)

the rotor circuit will cause the motor to slow down. A similar change can be made from medium to low by pressing the low-speed push button. The speed can also be changed from medium to low or high to low by pressing the low-speed push button.

This method of push-button speed selection is achieved by a relatively simple, inexpensive control installation. The resistors should have

a high enough power rating to operate at any speed.

A disadvantage of this method is that the motor and the driven machine can be accelerated without allowing sufficient time between steps for the rotor to gain its maximum speed for each step of acceleration. The desired time delay can be achieved by adding compelling relays to the circuit to prevent too rapid an acceleration.

Study Questions

1. Can the motor in Figure 35–1 be started in medium or high speed? Why?

2. Must the stop button be pressed between speeds on acceleration or deceleration? Why?

3. When the motor is being accelerated, must each speed push button be closed in succession? Why?

4. Is it possible to transfer from high speed to low speed without pressing the medium push button? Why?

5. Do all contactors remain energized when the motor is in high speed?

6. Acceleration resistors cannot be used for continuous duty. Why is this true and what may happen if an attempt is made to use them for continuous duty?

7. Draw a line diagram of a three-speed acceleration control that uses compelling timing relays to prevent too rapid an acceleration.

Automatic Acceleration for Wound Rotor Motors

Objectives

After studying this unit, the student should be able to

- State the advantages of using controllers to provide automatic acceleration of wound rotor motors.

- Identify terminal markings for automatic acceleration controllers used with wound rotor motors.

- Describe the process of automatic acceleration using reversing control.

- Describe the process of automatic acceleration using frequency relays.

- Connect wound rotor motor and automatic acceleration and reversal controllers using push buttons and limit switches.

- Recommend troubleshooting solutions for problems with using wound rotor motors for acceleration.

Secondary resistor starters used for the automatic acceleration of wound rotor motors consist of (1) an across-the-line starter for connecting the primary circuit to the line and (2) one or more accelerating contactors to shunt out resistance in the secondary circuit as the rotor speed increases. The secondary resistance consists of banks of three uniform wye sections. Each section is to be connected to the slip rings of the motor. The wiring of the accelerating starters and the design of the resistor sections are meant for starting duty only. This type of controller cannot be used for speed regulation. The current inrush on starters with two steps of acceleration is limited by the secondary resistors to a value of approximately 250 percent at the point of the initial acceleration. Resistors on starters with three or more steps of acceleration limit the current inrush to 150 percent at the point of initial acceleration. Resistors for acceleration are generally designed to withstand one 10-second accelerating period in each 80 seconds of elapsed time, for a duration of one hour without damage.

The operation of accelerating contactors is controlled by a timing device. This device provides timed acceleration in a manner similar to the operation of primary resistor starters. Normally, the timing of the steps of acceleration is controlled by adjustable accelerating relays. When these timing relays are properly adjusted, all starting periods are the same regardless of variations in the starting load. This automatic timing feature eliminates the danger

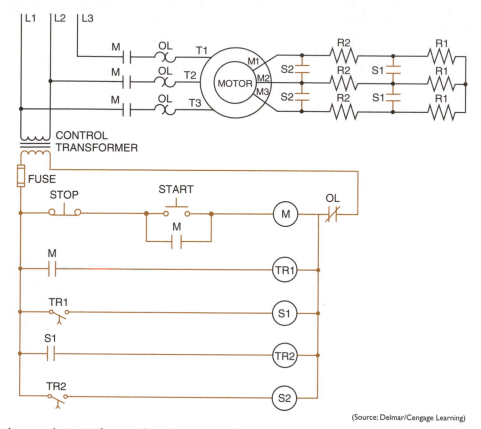

(Source: Delmar/Cengage Learning)

Fig. 36–1 Typical control circuit for accelerating a wound rotor motor with three steps of speed.

of an improper startup sequence by an inexperienced machine operator.

The circuit shown in Figure 36–1 is a typical control circuit used to accelerate a wound rotor motor with timing relays. When the start button is pressed, M starter is energized and all M contacts close. At this point, resistor banks R1 and R2 are connected in the rotor circuit. This provides the lowest speed for the motor. When M auxiliary contact connected in series with the coil of timer TR1 closes, the timer begins its time count. At the end of the set period, timed contacts TR1 close and energize the coil of S1 contactor. This causes the S1 contacts connected between resistor banks R1 and R2 to close, shorting out resistor bank R1. The motor now accelerates to the second speed.

The normally open S1 auxiliary contact connected in series with timer coil TR2 also closes, permitting the timer to start. At the end of the time period, timed contacts TR2 close and energize coil S2. This causes the S2 contacts connected between the motor and resistor bank R2 to close and short out all of the resistance in the rotor circuit. The motor now accelerates to the third and highest speed.

AUTOMATIC ACCELERATION WITH REVERSING CONTROL

Automatic acceleration can be obtained in either direction of rotation with the addition to the circuit of reversing contactors and push buttons. The wiring of these devices is shown in Figure 36–2.

As described for a squirrel cage motor, a wound rotor motor can be reversed by interchanging any two stator leads.

The motor may be started in either direction of rotation—at low speed with the full secondary resistance inserted in the circuit. For either direction of rotation, the timing relay TR is energized by the normally open auxiliary contacts (F or R). Coil TR activates the normally open, delay-in-closing contact TR. Coil S is energized when contact TR times out and closes and removes all of the resistors from the circuit to achieve maximum motor speed. The primary contactors are interlocked with the push buttons, normally closed interlocking contacts R and F, and the mechanical devices. The connections that occur if a limit switch is used are shown by the dashed lines

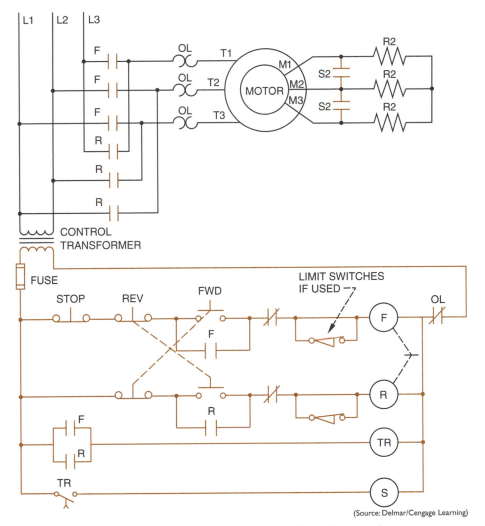

Fig. 36–2 Typical elementary diagram of a starter with two points of acceleration for a reversing wound rotor motor.

(Source: Delmar/Cengage Learning)

in Figure 36–2. The motor will stop when the limit switch is struck and opened. In this situation, it is necessary to restart the motor in the opposite direction with the push button. As a result, lines 1 and 3 on the primary side will be interchanged. Of course if limit switches are used, the jumper wire must be removed to avoid shunting out the operation of the limit switches.

AUTOMATIC ACCELERATION USING FREQUENCY RELAYS

Definite timers or compensated timers may be used to control the acceleration of wound rotor motors. Definite timers, which usually consist of pneumatic or static relays, are set for the highest load current and remain at the same setting regardless of the load. The operation of a compensated timer is based on the applied load. In other words, the motor will be allowed to accelerate faster for a light load

and slower for a heavy load. The frequency relay is one type of compensating timer. This relay uses the principle of electrical resonance in its operation.

When a 60-hertz AC, wound rotor motor is accelerated, the frequency induced in the secondary circuit decreases from 60 hertz at zero speed to 2 or 3 hertz at full speed, Figure 36–3. The voltage between the phases of the secondary circuit decreases in the same proportion from zero speed to full-speed operation. At zero speed, the voltage induced in the rotor is determined by the ratio of the stator and rotor turns. This action is similar to the operation of a transformer. The frequency, however, is the same as that of the line supply. As the rotor accelerates, the magnetic fields induced in it almost match the rotating magnetic field of the stator. As a result, the number of lines of force cut by the rotor is decreased, causing a decrease in the frequency and voltage of the rotor. The rotor never becomes fully synchronized with the rotating field. This is due to the slip

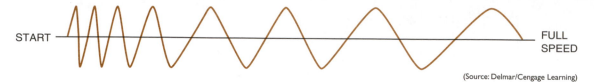

START   FULL
                                                                                                          SPEED

(Source: Delmar/Cengage Learning)

Fig. 36–3 Rotor frequency decreases as the motor approaches full speed.

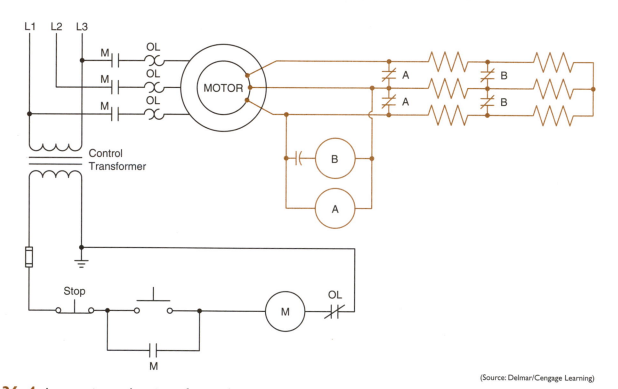

(Source: Delmar/Cengage Learning)

Fig. 36–4 Automatic acceleration of wound rotor motor using a simplified frequency relay system.

necessary to achieve the relative motion required for induction and the operation of the rotor. The percentage of slip determines the value of the secondary frequency and voltage. If the slip is 5 percent, then the secondary frequency and voltage are 5 percent of normal.

Figure 36–4 illustrates a simplified frequency relay system operated by push-button starting. This system has two contactor coils connected in parallel (A and B) and a capacitor connected in series with coil B. A three-step automatic acceleration results from this arrangement. When the motor starts, full voltage is produced across coils A and B, causing normally closed contacts A and B to open. The full resistance is connected across the secondary of the motor. As the motor accelerates, the secondary frequency decreases. This means that coil B drops out and contacts B close to decrease the resistance in the rotor circuit, resulting in continued acceleration of the motor. The capacitor depends on the frequency of an alternating current. As the motor continues to accelerate, coil A drops out and closes contacts A, accelerating the motor further. Because the

normally closed contacts are used, the secondary of the motor cannot be shunted out completely. If the secondary could be completely removed from the circuit, the electron flow would take the path of least resistance, resulting in no energy being delivered to coils A and B on starting.

The controllers for large crane hoists have a resistance, capacitance, and inductance control circuit network that is independent of the secondary rotor resistors.

Frequency relays have a number of advantages, including

1. Positive response.
2. Operating current drops sharply as the frequency drops below the point of resonance.
3. Accuracy is maintained because this type of relay operates in a resonant circuit.
4. Simple circuit.
5. Changes in temperature and variations in line voltage do not affect the relay.
6. An increase in motor load prolongs the starting time.

Study Questions

1. Are the secondary resistors connected in three uniform wye or delta sections?

2. Do secondary resistors on starters with three or more steps of acceleration have more or less current inrush than those with two steps of acceleration?

3. Does reversing the secondary rotor leads mean that the direction of rotation will reverse?

4. If one of the secondary resistor contacts (S2) fails in Figure 36–1, what will happen?

5. In Figure 36–2, how many different interlocking conditions exist? Name them.

6. Referring to Figure 36–4, why is it not possible to remove all of the resistance from the secondary circuit?

7. If frequency relays are used in starting, why is the starting cycle prolonged with an increase in motor load?

8. If there is a locked rotor in the secondary circuit, what will be the value of the frequency?

9. Why is it necessary to remove the jumpers in Figure 36–2 if limit switches are used?

10. Referring to Figure 36–2, why must the push buttons be used to restart if the limit switches are used?

UNIT 37

Automatic Speed Control for Wound Rotor Motors

Objectives

After studying this unit, the student should be able to

- Identify terminal markings for wound rotor motors used with automatic speed controllers.

- Describe the purpose and function of automatic speed controllers for wound rotor motors.

- Connect wound rotor motors and controllers for automatic speed control.

- Recommend solutions for troubleshooting problems with these motors.

Automatic speed control of a wound rotor motor can be obtained with the use of pilot devices. It is possible to control acceleration and deceleration and maintain selected speeds. The line diagram of a typical controller using pilot devices to provide automatic speed control is shown in Figure 37–1.

Assume that the wound rotor motor in Figure 37–1 is coupled to a fluid pump in a liquid-controlled system. The operation of such a system is described as follows: To maintain the liquid level automatically, the selector switch is placed in the automatic position. As the liquid rises, the master float switch (MFS) closes the circuit to the control switch. As the fluid continues to rise, float switch FS1 energizes control relay CR1. CR1 closes the main starter contacts M to start the motor in slow speed and energize

timing relay T1. If the motor speed is too slow to permit proper fluid delivery, the changing liquid level in the tank eventually closes the third float switch (FS2). CR2 is then energized through the now closed contacts of T1 to operate the first accelerating contactor (1A). In addition, the first bank of resistance is removed from the circuit, and the second time-delay relay (T2) is energized. This process continues until a motor speed is reached that maintains the liquid in the tank at a constant level. If the control selector switch is placed on manual, the float switches are bypassed and the motor must start with all of the resistance inserted in the secondary circuit. The motor must follow the preset timing sequence until all resistance is removed to obtain the maximum operation from the pump.

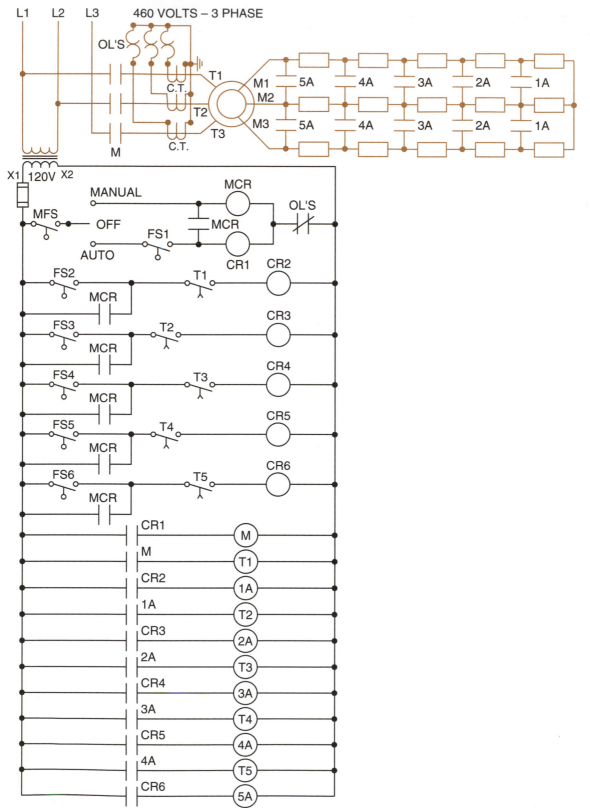

(Source: Delmar/Cengage Learning)

Fig. 37–1 Line diagram of wound rotor motor controller using pilot devices to provide automatic speed control. It includes control of acceleration and deceleration and maintenance of selected speeds.

Study Questions

1. Referring to Figure 37–1, can the motor be started in the manual position when there is no fluid in the tank? Why?

2. Assuming that this system is installed in an industrial mixing or processing tank and that the tank input decreases, what actions will be initiated by the control system?

3. How does the connection of the overload relay contacts cause all other control relays and contactors to drop out?

Solid-State Adjustable Speed Controller for AC Wound Rotor Motors

Objectives

After studying this unit, the student should be able to

- Describe slip power recovery and the method used to achieve it.
- Compare the solid-state method of control with resistor methods.
- Describe how speed control is accomplished electronically.
- Connect power factor correction capacitors into a motor circuit.

SOLID-STATE ADJUSTABLE SPEED CONTROLLER

An ideal solid-state adjustable speed controller is a modern, energy-saving method of controlling motor speed. It provides stepless, smooth adjustable speed control of AC wound rotor induction motors. It also recovers slip loss power when the motor runs at less than rated speed. This means that the controller will operate a wound rotor induction motor on less electrical power from the AC power line as compared with resistors, liquid rheostats or any reactor controls, and fluid or magnetic clutches.

The solid-state secondary speed controller is ideally suited for any application where a motor must run at less than its rated maximum speed, such as in water and waste water pumping and various fan applications. Speed control is accomplished by controlling motor rotor current, thereby controlling the motor torque.

Escalating energy costs have made efficiency an important consideration in determining which adjustable speed controller is best suited for a specific application.

SLIP POWER RECOVERY

A slip power recovery controller consists of a three-phase, full-wave diode rectifier, which converts the motor secondary voltage to a DC voltage, a filter reactor to smooth the rectification ripple, and a line commutated thyristor/inverter (DC to AC) synchronized to the power line to return motor slip loss power, Figure 38–1.

This adjustable controller permits stepless speed control of the motor. On reaching full speed, motor secondary shorting thyristors are automatically energized and the thyristor inverter is turned off. This eliminates reactive

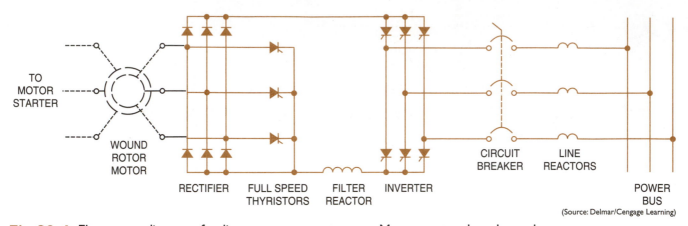

Fig. 38–1 Elementary diagram of a slip power recovery system. Motor starters have been shown.

current returned to the power system during full-speed operation.

Electronic circuitry is arranged on plug-in circuit cards, or modules, and are located on a plug-in rack. Electronic cards, or modules, are coated to protect against damage due to moisture and a corrosive environment. Card-edge connectors are generally gold coated to ensure reliable contact between the card and the rack terminals. Cards and modules are also keyed to prevent in-correct insertion. LEDs are provided to indicate the operating status to assist in troubleshooting.

An instantaneous overcurrent circuit detects a fault condition, activates circuitry electronically to clear the fault condition, and allows the controller to resume normal operation. All controller startup adjustments are located on a drive adjust card. Electronic circuits are located on the remaining cards. Ideally,

these are replaceable without readjustment of the controller.

POWER FACTOR CORRECTION CAPACITORS

Power factor correction is recommended for all slip power recovery systems. During the operation of a slip power recovery control system, motor secondary current is flowing back into the power line. As the motor speed increases, the current becomes more reactive and if not corrected would result in a very poor power factor. The correct method of connecting the power factor correction capacitors in the circuit is shown in Figure 38–2.

Figure 38–3 is a line diagram showing a total operating system. Starting resistors are required for reduced speed range systems only.

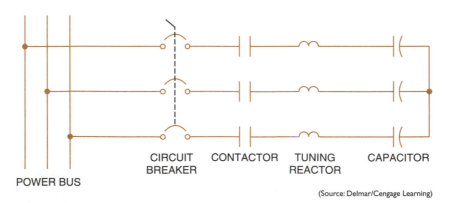

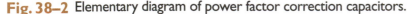

Fig. 38–2 Elementary diagram of power factor correction capacitors.

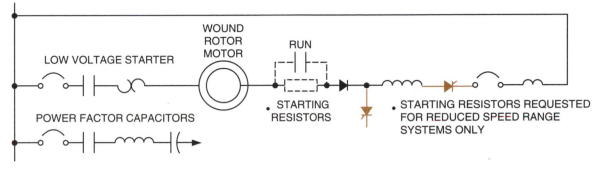

(Source: Delmar/Cengage Learning)

Fig. 38–3 Line diagram of solid-state control with power factor correction.

Study Questions

1. If someone told you that the solid-state power recovery controller is a perpetual motion machine, what would your reaction be?

2. How do solid-state adjustable speed controllers compare with secondary resistor control?

3. Basically, how is speed control accomplished?

4. Why are capacitors connected into the power circuit?

Synchronous Motor Controls

Synchronous Motor Operation

Objectives

After studying this unit, the student should be able to

- Describe the operation and applications of a synchronous motor.

- Describe lagging and leading power factor and the causes of each.

- Describe how the use of a synchronous motor improves the efficiency of an electrical system having a lagging power factor.

- Identify a brushless synchronous motor.

One of the distinguishing features of the synchronous motor is that it runs without slip at a speed determined by the frequency of the connected power source and the number of poles it contains. This type of motor sets up a rotating field through stator coils energized by alternating current. (This action is similar to the principle of an induction motor.) An independent field is established by a rotor energized by direct current through slip rings mounted on the shaft. The rotor has the same number of coils as the stator. At running speed, these fields (north and south) lock into one another magnetically so that the speed of the rotor is in step with the rotating magnetic field of the stator. In other words, the rotor turns at the synchronous speed. Variations in the connected load do not cause a corresponding change in speed, as they would with the induction motor.

The rotor, Figure 39–1, is excited by a source of direct current so that it produces alternate north and south poles. These poles are then attracted by the rotating magnetic field in the stator. The rotor must have the same number of poles as the stator winding. Every rotor pole, north and south, has an alternate stator pole, south and north, with which it can synchronize.

Fig. 39–1 Rotor of 300-hp, 720-rpm synchronous motor. (*Courtesy of Electric Machinery Company, Inc.*)

The rotor has DC field windings to which direct current is supplied through collector rings (slip rings). The current is provided from either an external source or a small DC generator connected to the end of the rotor shaft.

The magnetic fields of the rotor poles are locked into step with—and pulled around by—the revolving field of the stator. Assuming that the rotor and stator have the same number of poles, the rotor moves at the stator frequency (in hertz) actually produced by the generator supplying the motor.

Synchronous motors are constructed almost exactly like alternators. They differ only in those features of design that may make the motor better adapted to its particular purpose.

A synchronous motor cannot start without help because the DC rotor poles at rest are alternately attracted and repelled by the revolving stator field. Therefore, an induction squirrel cage, or starting winding is embedded in the pole faces of the rotor. This is called an *armortisseur* (ah-more-ti-sir) *winding*.

This starting winding resembles a squirrel cage winding. The induction effect of the starting winding provides the starting, accelerating, and pull-up torques required. The winding is designed to be used only for starting and for damping oscillations during running. It cannot be used like the winding of the conventional squirrel cage motor. It has a relatively small cross-sectional area and will overheat if the motor is used as a squirrel cage induction motor.

The slip is equal to 100 percent at the moment of starting. Thus, when the AC rotating magnetic field of the stator cuts the rotor windings, which are stationary at startup, the induced voltages produced may be high enough to damage the insulation if precautions are not taken.

If the DC rotor field is either connected as a closed circuit or connected to a discharge resistor during the starting period, the resulting current produces a voltage drop that is opposed to the generated voltage. Thus, the induced voltage at the field terminals is reduced. The squirrel cage winding is used to *start* the synchronous motor in the same way it is used in the squirrel cage induction motor. When the rotor reaches the maximum speed to which it can be accelerated as a squirrel cage motor (about 95 percent or more of the synchronous speed), direct current is applied to the rotor field coils to establish north and south rotor poles. These poles then are attracted by the poles on the stator. The rotor then accelerates until it locks into synchronous motion with the stator field.

Synchronous motors are used for applications involving large, slow-speed machines with steady loads and constant speeds. Such applications include compressors, fans and pumps, many types of crushers and grinders, and pulp, paper, rubber, chemical, flour, and metal-rolling mills, Figure 39–2.

POWER FACTOR CORRECTION BY SYNCHRONOUS MOTOR

A synchronous motor converts AC electrical energy into mechanical power. In addition, it also provides power factor correction. It can operate at a leading power factor or at unity. In very rare occasions, it can operate at a lagging power factor.

The power factor is of great concern to industrial users of electricity with respect to energy conservation. Power factor is the ratio of the actual power, being used in a circuit, expressed in watts or kilowatts, to the power apparently being drawn from the line, expressed in volt-amperes or kilovolt-amperes (kVA). The kVA value is obtained by multiplying a voltmeter reading and an ammeter reading of the same

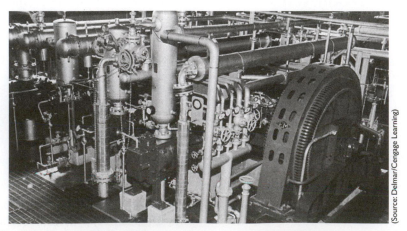

Fig. 39–2 A 2000-hp, 300-rpm synchronous motor driving compressor.

(Source: Delmar/Cengage Learning)

circuit or equipment. Inductance within the circuit will cause the current to lag the voltage.

When the values of the apparent and actual power are equal or in phase, the ratio of these values is 1:1. In other words, when the voltage and amperage are in phase, the ratio of these values is 1:1. This is the case of pure resistive loads. The unity power factor value is the highest power factor that can be obtained. The higher the power factor, the greater is the efficiency of the electrical equipment.

AC loads generally have a lagging power factor. As a result, these loads burden the power system with a large reactive load. Refer to the induction motor in Figure 39–3. A synchronous motor should be operated at unity, or at its greatest leading power factor for the most efficient operation. This helps supply reactive power where required in the electrical system. Synchronous motors are operated to offset the low-power factor of the other loads on the same electrical system. Therefore, the synchronous motor works to improve the power factor of the

electrical system, while turning large, slow-speed machines.

BRUSHLESS SYNCHRONOUS MOTORS

Another method for applying excitation current to the rotor is with a brushless exciter. Rotors of this type contain a separate three-phase alternator winding mounted on the rotor shaft, Figure 39–4. Electromagnets are mounted on each side of the three-phase winding. The external direct-current supply provides power for the electromagnets, Figure 39–5. The amount of induced voltage in the three-phase winding is controlled by the strength of the magnetic field of the electromagnets. The output of the three-phase alternator winding is also connected to a three-phase bridge rectifier mounted on the rotor shaft, Figure 39–6. The bridge rectifier converts the three-phase AC voltage into DC voltage. The DC voltage supplies the excitation current for the rotor. The amount of rotor excitation is still

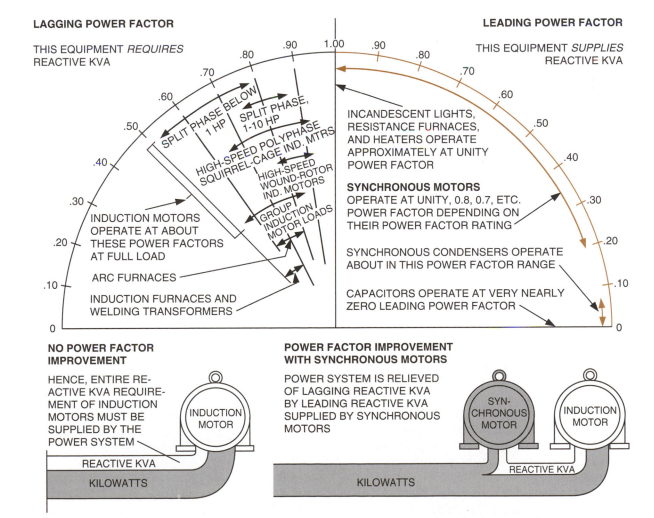

Fig. 39–3 Power factor operation of various devices may be improved through the use of synchronous motors. *(Courtesy of Electric Machinery Company, Inc.)*

Main Rotor Winding

Brushless Exciter Winding

(Source: Delmar/Cengage Learning)

Fig. 39–4 A brushless exciter contains a separate three-phase winding on the rotor shaft.

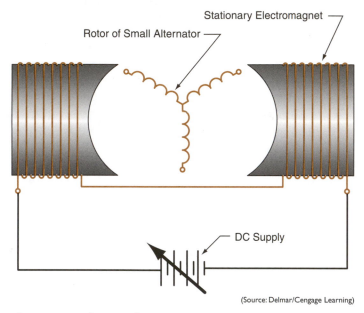

Stationary Electromagnet

Rotor of Small Alternator

DC Supply

(Source: Delmar/Cengage Learning)

Fig. 39–5 The brushless exciter uses stationary electromagnets.

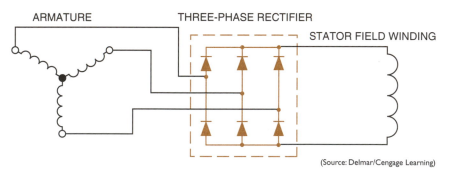

ARMATURE   THREE-PHASE RECTIFIER   STATOR FIELD WINDING

(Source: Delmar/Cengage Learning)

Fig. 39–6 Basic brushless exciter circuit.

controlled by the external DC source, but there are no brushes or slip rings that can cause mechanical problems. The elimination of slip rings and brushes gives the synchronous motor equipped with a brushless exciter several outstanding advantages:

- Brush sparking is eliminated, reducing safety hazards in some areas.
- Field control and excitation are provided by a static system, requiring much less maintenance.
- Field excitation is automatically removed whenever the motor is out of step. Automatic resynchronization can be achieved whenever it is practical.

The rotor of a large synchronous motor with a brushless exciter winding is shown in Figure 39–7. More is written on this subject in Unit 42.

Fig. 39–7 The rotor of a large synchronous motor with a brushless exciter. (*Courtesy of GE Industrial Systems, Fort Wayne, Indiana*)

Study Questions

1. Why is it necessary that the rotor and stator have an equal number of poles?

2. What is the effect of the starting winding of the synchronous motor on the running speed?

3. What are typical applications of synchronous motors?

4. Explain how the rotor magnetic field is established.

5. A loaded synchronous motor cannot operate continuously without DC excitation on the rotor. Why?

6. Why must a discharge resistor be connected in the field circuit for starting?

7. Depending on their power factor ratings, what is the range of the leading power factor at which synchronous motors operate?

8. At what power factor do incandescent lights operate?

9. At what power factor do high-speed, wound rotor motors operate?

Select the *best* answer for the following items.

10. The speed of a synchronous motor is fixed by the
    a. rotor winding
    b. amortisseur winding
    c. supply voltage
    d. frequency of power supply and number of poles

11. Varying the DC voltage to the rotor field changes the
    a. motor speed
    b. power factor
    c. phase excitation
    d. slip

12. Amortisseur windings are located
    a. in the stator pole faces
    b. in the rotor pole faces
    c. in the controller
    d. leading the power factor

13. DC excitation is applied to the
    a. starting winding
    b. stator winding
    c. rotor winding
    d. amortisseur winding

14. Induction motors and welding transformers require magnetizing current, which causes
    a. lagging power factor
    b. leading power factor
    c. unity power factor
    d. zero power factor

15. A synchronous motor can be used to increase the power factor of an electrical system by
    a. reducing the speed
    b. overexciting the stator field
    c. operating at unity power factor
    d. applying DC to the stator field

UNIT 40

Push-Button Synchronizing

Objectives

After studying this unit, the student should be able to

- Identify terminal markings for a push-button synchronizing controller and synchronous motor.

- Describe two functions of synchronous motor control

- Connect synchronous motors and synchronizing controls.

- Recommend troubleshooting solutions for synchronous motors and controls.

- Describe adjustments that can be made to the circuit of a synchronous motor and control to obtain a power factor of unity or greater.

- Synchronize a synchronous motor.

There are two basic functions of synchronous motor control in starting:

1. *To start the motor as an induction motor.* The motor can be started using any of the typical induction motor starting schemes. These include connecting the motor across the line or the use of autotransformers, primary resistors, or other devices, depending on the size of the motor.

2. *To bring the motor up to synchronous speed by exciting the DC field.* The principal difference between synchronous motor control and induction motor control is in the control of the field. Two voltages are needed, AC for the stator and DC for the rotor.

In Figure 40–1, the synchronous motor is started as an induction motor by pressing the start button. When the motor reaches its maximum speed, the run push button is closed to energize coil (F) and close the DC excitation contacts (F). These contacts, in turn, open the field discharge normally closed contact (F) and energize the DC field, F1 and F2 of the motor. The normally closed contact F opens after normally open F contacts close to avoid a high induced voltage in the field. The stator acts like a transformer primary. The DC field circuit receives an induced current (as does a transformer secondary). When the stator is energized, the rotor circuit must be closed at all times. The overlapping normally closed contacts F keep the loading resistor in the circuit. Without it, dangerous

287

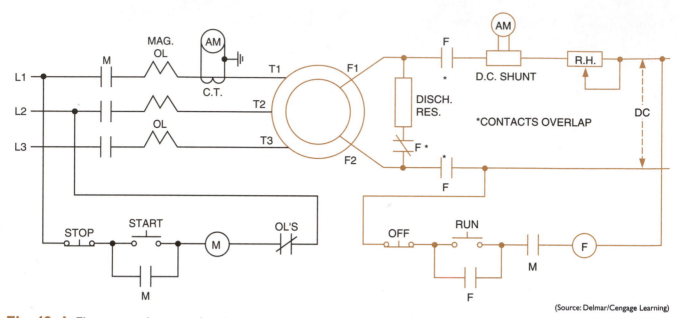

Fig. 40–1 Elementary diagram of push-button synchronizing.

(Source: Delmar/Cengage Learning)

high voltages are induced, which may rupture the insulation on the rotor windings. The induced current in the rotor resistor circuit keeps the induced voltage at a tolerable level.

The discharge resistor also prevents arcing at the contacts. In addition, it dissipates the self-induced voltage from the collapsing DC magnetic field of the rotor windings.

Note in Figure 40–1 that the normally open contacts M in the F coil circuit prevent the DC field from being energized before the motor is started. These interlocking contacts close only

after the AC control is energized. Pushing the stop button on the AC side also opens the DC supply. If the motor does not reach synchronous speed with the first try, the *off* button should be pressed. The run push button can then be reengaged.

The ammeter and rheostat shown in the DC circuit in Figure 40–1 control the excitation current. The unity power factor of the motor can be found by adjusting the rheostat to obtain a minimum reading on the AC ammeter in the stator circuit, Figure 40–2.

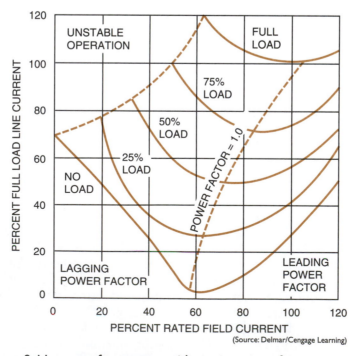

(Source: Delmar/Cengage Learning)

Fig. 40–2 Line current versus field current for a motor with a unity power factor.

Adjustments of the circuit may cause field currents, or line currents, in excess of the motor rating. As a result, electrical instruments must be provided to monitor various circuit values. The operator can then prevent equipment stoppage due to overload tripping. The readings of the instruments must not exceed the rated values shown on the motor nameplate.

The excitation current can be applied manually to excite the DC field to bring the motor up to synchronous speed. For this operation, however, it is the operator who must make the difficult judgment of when the north and south poles of the rotor and stator fields are paired and the motor is ready for synchronous operation. For this reason, the system does not provide automatic motor synchronization. However, it is a comparatively simple and inexpensive method.

Study Questions

1. What are the two basic functions of synchronous motor control?

2. What is the purpose of the interlock (M) in the DC control circuit?

3. If all of the control circuits are AC—and DC is not available—what will be the result of energizing all of the contactor coils?

4. When the rotor and stator fields lock into step, why may the reading of the AC ammeter decrease?

5. Why is manual field application not completely satisfactory?

6. If a synchronizing attempt fails, what procedure should be followed?

7. In Figure 40–2, approximately what is the percentage of current drawn from the line when the field current is adjusted to 100 percent of its capacity at full motor load?

8. Why are the AC ammeter and DC meter necessary in the circuit of Figure 40–1?

9. Why is the operation of the field contacts (F) overlapped in Figure 40–1?

Timed Semiautomatic Synchronizing

Objectives

After studying this unit, the student should be able to

- Identify terminal markings for a timed semiautomatic controller used with a synchronous motor.

- Describe the operation of a timed semiautomatic controller used to bring a motor up to synchronous speed.

- Connect synchronous motors and timed semiautomatic controllers.

- Recommend ways to troubleshoot these motors and controllers.

A synchronous motor may be brought up to synchronous speed with the use of a definite time-delay relay to excite the DC field. This method is shown in the circuit of Figure 41–1.

The timing relay coil (TR) is energized with the main starter coil (M). The instantaneous normally open interlock (M) in the coil circuit (F) is then closed. Both interlock M and the DC contactor coil F must wait for the closing of the delay-in-closing contact TR. After a preset timing period, the rotor has accelerated to the maximum speed possible at this stage. Contact TR then closes to accelerate the rotor until it reaches the point where it synchronizes. The timer setting should be adjusted for the maximum time required to accelerate the motor to the point that it can reach the synchronous speed after contact TR closes.

The attempt to synchronize the motor may not be successful. In this case, the stop button is pressed and the starting cycle is repeated. It is not necessary to bring the rotor to a standstill. Only the timing cycle must be reactivated.

The equipment operator and the electrician should realize that push-button control and timed semiautomatic control of synchronous motors are not guaranteed to be effective on every attempt to obtain synchronous operation of the motor. The timing cycle can be adjusted, however.

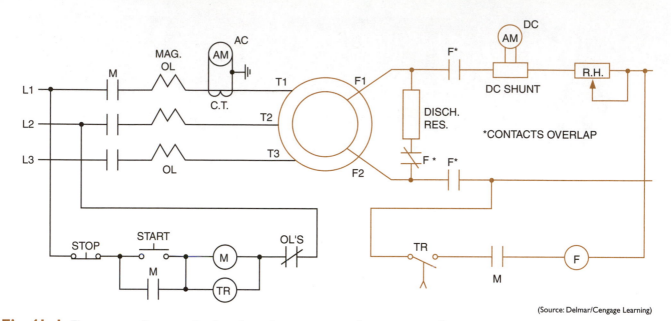

(Source: Delmar/Cengage Learning)

Fig. 41–1 Elementary diagram of a timed, semiautomatic synchronizing installation.

Study Questions

1. What assurance is there that the rotor will lock into step at the synchronous speed with the use of a timing relay?

2. If the motor fails to achieve synchronous operation, what action is necessary?

Synchronous Automatic Motor Starter

Objectives

After studying this unit, the student should be able to

- Describe how an out-of-step relay protects the starting winding of a synchronous motor.

- Describe the action of a polarized field frequency relay in applying and removing DC field excitation on a synchronous motor.

- Connect synchronous motors and controllers that use out-of-step relays and polarized field frequency relays to achieve automatic motor synchronization.

- Recommend troubleshooting solutions for problems.

An automatic synchronous motor starter can be used with a synchronous motor to provide automatic control of the startup sequence. That is, the controller automatically sequences the operation of the motor so that the rotor field is synchronized with the revolving magnetic field of the stator.

There are two basic methods of starting synchronous motors automatically. In the first method, full voltage is applied to the stator winding. In the second method, the starting voltage is reduced. A commonly used method of starting synchronous motors is the across-the-line connection. In this method, the stator of the synchronous motor is connected directly to the plant distribution system at full voltage. A magnetic starter is used in this method of starting.

A polarized field frequency relay can be used for the automatic application of field excitation to a synchronous motor.

ROTOR CONTROL EQUIPMENT

Field Contactor

The field contactor opens both lines to the source of excitation, Figure 42–1. During starting, the contactor also provides a closed field circuit through a discharge resistor. A solenoid-operated field contactor is similar in appearance to the standard DC contactor. However, for this DC operated contactor, the center pole is normally closed. It is designed to provide a positive overlap between the normally closed contact and the two normally open contacts. This overlap is an important feature because it means that the field winding is never open. The field winding of the motor must always be short-circuited through a discharge resistor or connected to the DC line. The coil of the field contactor is operated from the same direct-current source

Fig. 42–1 Magnetic contactor used on synchronous starter for field control. *(Courtesy of Rockwell Automation)*

that provides excitation for the synchronous motor field.

Out-of-Step Relay

The squirrel cage winding, or starting (amortisseur) winding, will not overheat if a synchronous motor starts, accelerates, and reaches synchronous speed within a time interval determined to be normal for the motor. In addition, the motor must continue to operate at synchronous speed. Under these conditions, adequate protection for the entire motor is provided by three overload relays in the stator winding. The squirrel cage winding, however, is designed for starting only. If the motor operates at subsynchronous speed, the squirrel cage winding may overheat and be damaged. It is not unusual for some synchronous motors to withstand a maximum locked rotor interval of only 5 to 7 seconds.

An out-of-step relay (OSR), Figure 42–2, is provided on automatic synchronous starters to protect the starting winding. The normally closed contacts of the relay will open to de-energize the line contactor under the following conditions:

1. The motor does not accelerate and reach the synchronizing point after a preset time delay.
2. The motor does not return to a synchronized state after leaving it.
3. The amount of current induced in the field winding exceeds a value determined by the core setting of the out-of-step relay.

Fig. 42–2 Out-of-step relay used on synchronous starters. *(Courtesy of Rockwell Automation)*

As a result, power is removed from the stator circuit before the motor overheats.

Polarized Field Frequency Relay

A synchronous motor is started by accelerating the motor to as high a speed as possible from the squirrel cage winding and then applying the DC field excitation. The components responsible for correctly and dependably applying and removing the field excitation are a *polarized field frequency relay* (PFR) and a reactor, Figure 42–3.

The operation of the frequency relay is shown in Figure 42–4. The magnetic core of the relay has a direct-current coil (C), an induced field-current coil (B), and a pivoted armature (A) to which contact (S) is attached. Coil C is connected to the source of DC excitation. This coil establishes a constant magnetic flux in the relay core. This flux causes the relay to be polarized. Superimposed on the magnetic flux in the relay core is the alternating magnetic flux produced by the alternating induced rotor field current flowing in coil B. The flux through armature A depends on the flux produced by

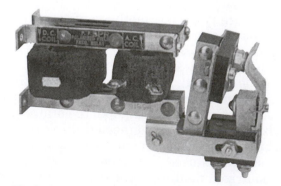

Fig. 42–3 Polarized field frequency relay with contacts in normally closed position. *(Courtesy of Rockwell Automation)*

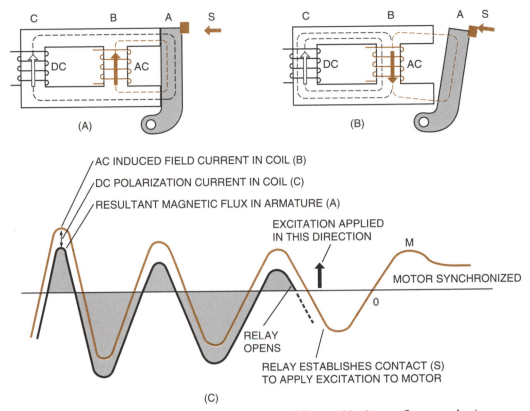

Fig. 42–4 Polarized field frequency relay operation. *(Courtesy of Electric Machinery Company, Inc.)*

AC coil B and DC coil C. Coil B produces an alternating flux of equal positive and negative magnitude each half-cycle. Thus, the combined flux flowing through armature (A) is much larger when the flux from coil B opposes that from coil C. In Figure 42–4(A), the flux from coil B opposes the flux from DC coil C, resulting in a strong flux being forced through armature A of the relay. This condition is shown by the lower shaded loops of Figure 42–4(C). One-half cycle later, the flux produced by coil B reverses and less flux flows through armature A. This is due to the fact that the flux from coil B no longer forces as much flux from coil C to take the longer path through armature A. The resultant flux is weak and is illustrated by the small, upper shaded loops of Figure 42–4(C). The relay armature opens only during the period of the induced field current wave, which is represented by the small, upper loops of the relay armature flux.

As the motor reaches synchronous speed, the induced rotor field current in relay coil B decreases in amplitude. A value of relay armature flux (upper shaded loop) is reached at which the relay armature A no longer stays closed. The relay then opens to establish contact S. DC excitation is then applied at the point indicated on the induced field current wave.

Excitation is applied in the direction shown by the arrow. The excitation is opposite in polarity to that of the induced field current at the point of application. This requirement is necessary to compensate for the time needed to build up excitation. The time interval results from the magnetic inertia of the motor field winding. Because of the inertia, the DC excitation does not become effective until the induced current reverses (point O on the wave) to the same polarity as the direct current. The excitation continues to build up until the motor is synchronized as shown by point M on the curve.

Figure 42–5 indicates the normal operation of the frequency relay. DC excitation is applied to the coil of the relay at the instant the synchronous motor is started. When the stator winding is energized, using either full-voltage or reduced voltage methods, line current is allowed to flow through the three overload relays and the stator winding. Line frequency currents are induced in the two electrically independent circuits of the rotor: (1) the squirrel cage or starting windings and (2) the field windings. The current induced in the field windings flows through the reactor. This device shunts part of the current through the AC coil of the frequency relay, the coil of the out-of-step relay, the field discharge resistor, and finally to the normally closed contact of the field

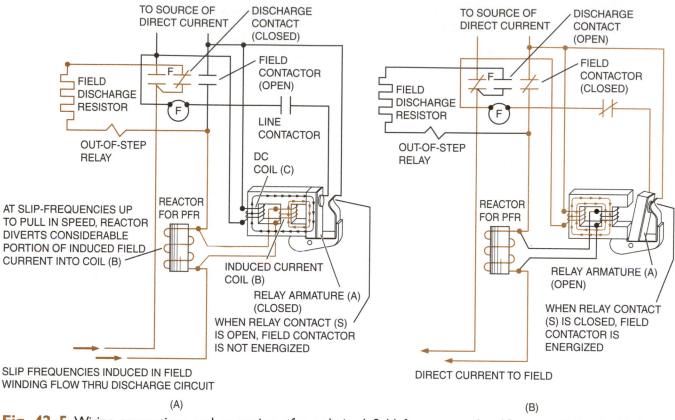

Fig. 42–5 Wiring connections and operation of a polarized field frequency relay. *(Courtesy of Electric Machinery Company, Inc.)*

contactor. The flux established in the frequency relay core pulls the armature against the spacer and opens the normally closed relay contacts, Figure 42–5. As the motor accelerates to the synchronous speed, the frequency of the induced currents in the field windings diminishes. There is, however, sufficient magnetic flux in the relay core to hold the armature against the core. This flux is due to a considerable amount of induced current forced through the AC coil of the frequency relay by the impedance of the reactor at high slip frequency.

At the point where the motor reaches its synchronizing speed (usually 92 to 97 percent of the synchronous speed) the frequency of the induced field current is at a very low value. The reactor impedance is also greatly reduced at this low frequency. Thus, the amount of current shunted to the AC coil is reduced to the point where the resultant core flux is no longer strong enough to hold the armature against the spacer. At the moment that the rotor speed and the frequency and polarity of the induced currents are most favorable for synchronization, the armature is released, the relay contacts close, and the control circuit is completed to the operating coil of the field contactor. DC excitation is applied to the motor field winding, Figure 42–5(B). At the same time, the out-of-step relay and discharge resistor are de-energized by the normally closed contacts of the field contactor.

An overload or voltage fluctuation may cause the motor to pull out of synchronism. In this case, a current at the slip frequency is induced in the field windings. Part of this current flows through the AC coil of the polarized field frequency relay, opens the relay contact, and removes the DC field excitation. The motor automatically resynchronizes if the line voltage and load conditions return to normal within a preset time interval, and the motor has enough pull-in torque. However, if the overload and low-voltage conditions continue so that the motor cannot resynchronize, then either the out-of-step relay or the overload relays activate to protect the motor from overheating.

SUMMARY OF AUTOMATIC STARTER OPERATION

The line diagram in Figure 42–6 shows the automatic operation of a synchronous motor. For starting, the motor field winding is connected

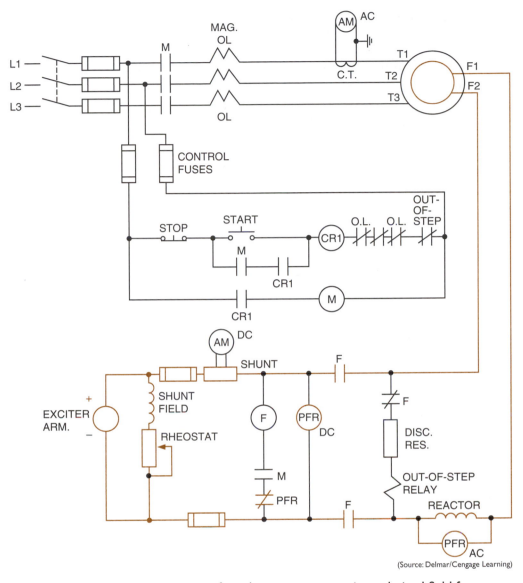

(Source: Delmar/Cengage Learning)

Fig. 42–6 Line diagram for automatic operation of synchronous motor using polarized field frequency relay.

through the normally closed power contact of the field contactor (F), the discharge resistor, the coil of the out-of-step relay, and the reactor. When the start button is pressed, the circuit is completed to the control relay coil (CR1) through the control fuses, the stop button, and contacts of the overload and out-of-step relays. The closing of CR1 energizes the line contactor M, which applies full voltage at the motor terminals with overload relays in the circuit. A normally open contact on CR1 and a normally open interlock on line contactor M provide the hold-in, or maintaining, circuit. The starting and running current drawn by the motor is indicated by an ammeter with a current transformer.

At the moment the motor starts, the polarized field frequency relay (PFR) opens its normally closed contact and maintains an open circuit to the field contactor (F) until the motor accelerates to the proper speed for synchronizing. When the motor reaches a speed equal to 92 to 97 percent of its synchronous speed, and the rotor is in the correct position, the contact of the polarized field frequency relay closes to energize field contactor F through an interlock on line contactor M. The closing of field contactor F applies the DC excitation to the field winding and causes the motor to synchronize. After the rotor field circuit is established through the normally open power contacts of the field contactor, the normally closed contact on this contactor opens the discharge circuit. The motor is now operating at the synchronous speed. If the stop button is pressed, or if either magnetic overload relay is tripped, the starter is de-energized and disconnects the motor from the line.

BRUSHLESS SOLID-STATE MOTOR EXCITATION

An improvement in synchronous motor excitation is the development of the brushless DC exciter. The commutator of a conventional direct-connected exciter is replaced with a three-phase, bridge-type, solid-state rectifier. The DC output is then fed directly to the field winding. A simplified circuit is shown in Figure 42–7. This scheme is similar to that of a brushless generator. The motor mounted exciter receives its excitation and control (normally supplied with the line control unit) from a small rectifier and adjustable autotransformer supplied from the AC power source. Alternating current from the exciter is fed to a rectifier diode assembly.

The DC output of the rectifier is controlled by the static field application system. A rotor-mounted, field-discharge resistor provides a discharge path for the motor induced field current during starting. The frequency-sensitive static field application system replaces the polarized field frequency relay, the reactor, and the field contactor of conventional brush-type motors. It applies field excitation in such a way that maximum pull-in torque is obtained when the motor reaches a synchronizing speed. Field excitation is automatically removed if the motor is out of step. A field monitoring relay (normally supplied with a line control unit) provides instantaneous shutdown or unloading on pull-out and provides out-of-step protection for the squirrel cage winding of the motor.

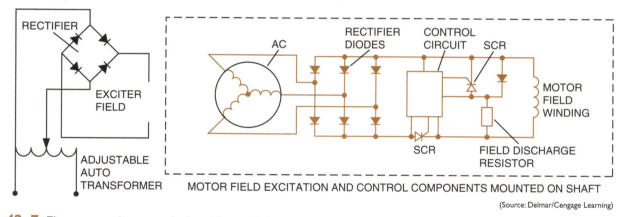

(Source: Delmar/Cengage Learning)

Fig. 42–7 Elementary diagram of a brushless solid-state motor field excitation.

Study Questions

1. What are the two basic methods of automatically starting a synchronous motor?

2. What is an out-of-step relay?

3. Why is an out-of-step relay used on automatic synchronous starters?

4. Under what conditions will the out-of-step relay trip out the control circuit?

5. What is the last control contact that closes on a starting and synchronizing operation?

6. What influence do both of the polarized field frequency relay (PFR) coils exert on the normally closed contact?

7. Why is the PFR polarized with a DC coil?

8. Approximately how much time (in terms of electrical cycles) elapses from the moment the PFR opens to the moment the motor actually synchronizes?

9. How does the AC coil of the PFR receive the induced field current without receiving the full field current strength?

10. Why is a control relay (CR1) used in Figure 42–6?

11. On a brushless rotor, how is DC obtained?

SECTION

9

Direct-Current Controllers

UNIT 43

About DC Motors

Objectives

After studying this unit, the student should be able to

- Cite applications of DC motors.
- Describe the construction and electrical features of DC motors.
- Draw diagrams of DC motor types.
- Reverse the direction of rotation of DC motors.
- Describe base speed, above speed, and below speed.
- Describe motor connection precautions.
- Identify DC motor terminal connections.

APPLICATION

The generation of electrical energy and its conversion to mechanical energy is the basis of our industrial structure. It is a fact that DC motors are used primarily for industrial purposes. By controlling the voltages supplied to the motor windings, a wide range of motor speeds can be obtained. The ability to control the speed of DC motors makes them valuable for factory hoists and cranes, where heavy loads must be started slowly and then accelerated quickly.

DC motors are also used in such precision applications as printing presses and steel mill production operations.

MOTOR CONSTRUCTION

The essential features and parts of DC motors and generators are the same. The DC motor has

field coils, armature coils, and a commutator with brushes, Figure 43–1.

The Field

A strong magnetic field is provided by the field windings of the individual field poles. The magnetic polarity of the field pole system is arranged so that the polarity of any one magnetic field pole is opposite to that of the poles adjacent to it such as north and south. The field structure of a motor has at least two pairs of field poles. However, motors with four or more pairs of field poles are also used.

The Armature

The armature or rotating member of a DC motor is a cylindrical iron assembly mounted directly on the motor shaft, Figure 43–1. Windings are embedded in slots in the surface of the

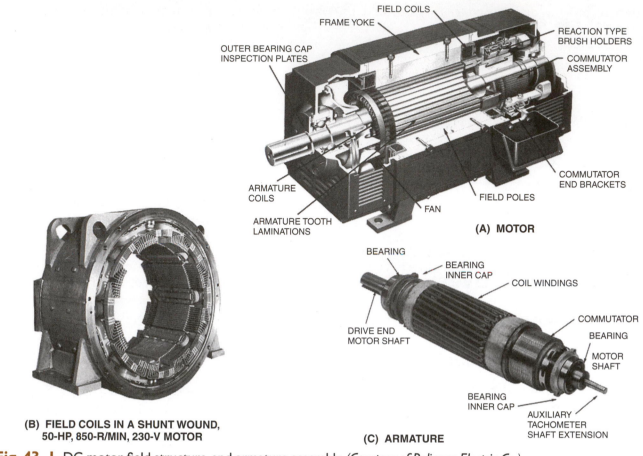

Fig. 43–1 DC motor, field structure, and armature assembly. *(Courtesy of Reliance Electric Co.)*

armature. These windings connect to segments of the commutator. Direct current is fed to the armature windings by carbon brushes, which press against the commutator segments. The commutator changes the direction of the current in the armature conductors as they pass across field poles of opposite magnetic polarity. Continuous rotation in one direction results from these reversals in the armature current.

TYPES OF DC MOTORS

Three types of DC industrial motors are widely used: series, shunt, and compound. The type of motor used is based on the mechanical requirements of the applied load.

As shown in Figure 43–2, the *series* motor field is connected in series with the armature. The wire used for the windings must be large

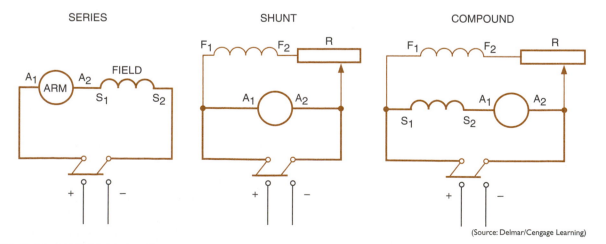

(Source: Delmar/Cengage Learning)

Fig. 43–2 Motor field connections.

enough to pass armature current. The field is labeled S1 and S2.

The field circuit of the *shunt* motor is connected in shunt (parallel) with the armature. Rheostat R is connected in the shunt field circuit (F1, F2). The rheostat is used to increase or decrease resistance in the field circuit; this, in turn, varies the magnetic field flux to control the motor speed. The rheostat is external to the motor.

The *compound motor* has both a shunt and a series field winding.

MOTOR ROTATION

The direction of armature rotation of a DC motor depends on the direction of the current in the field and armature circuits. To reverse the direction of rotation, the current direction in either the field or the armature must be reversed. Reversing the power supply leads does *not* reverse the direction of armature rotation because, in this case, both the field and the armature currents are reversed.

The direction of rotation of a DC motor is determined by facing the commutator end of the motor, which is generally the rear of the motor. Figure 43–3 shows the standard connections for

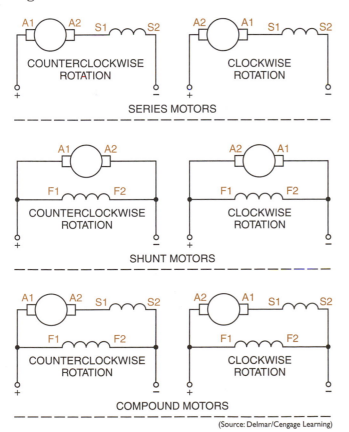

SERIES MOTORS

SHUNT MOTORS

COMPOUND MOTORS

(Source: Delmar/Cengage Learning)

Fig. 43–3 Standard connections to determine rotation of DC motors.

determining the rotation of the series, shunt, and compound motors.

SPEED CONTROL

One of the major advantages of a DC motor is that it can be operated at different speeds. Normal or base speed for a DC motor is obtained by connecting full voltage to both the armature and shunt field. The motor can be operated below normal speed by maintaining rated voltage on the field and reducing the current to the armature. The motor can be operated above normal speed by maintaining rated voltage to the armature and reducing the current flow to the field. Decreasing the field flux causes a decrease in the counter electromotive force (emf) produced in the armature. This permits more armature current to flow and produces a net gain in torque. The motor speed will increase until the motor torque equals the mechanical torque imposed by the load connected to the motor. When a motor will be operated above base speed, the motor manufacturer's specifications should be consulted about the maximum safe operating speed. Do not depend on the controller to protect the motor from excessive speed unless the limit is checked and preset on the controller.

Various methods can be employed to obtain speed control for a DC motor. A method of manual speed control for a compound motor is illustrated in Figure 43–4. Resistance connected in series with the armature allows the motor to be operated below normal speeds. A shunt field resistor connected in series with the shunt field permits the motor to be operated above normal speeds. A field loss relay (FLR) is connected in series with the shunt field. The FLR is used to

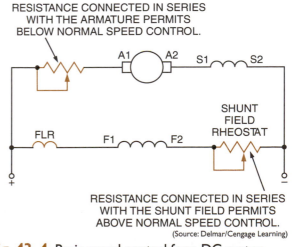

RESISTANCE CONNECTED IN SERIES WITH THE ARMATURE PERMITS BELOW NORMAL SPEED CONTROL.

SHUNT FIELD RHEOSTAT

FLR

RESISTANCE CONNECTED IN SERIES WITH THE SHUNT FIELD PERMITS ABOVE NORMAL SPEED CONTROL.

(Source: Delmar/Cengage Learning)

Fig. 43–4 Basic speed control for a DC motor.

disconnect power to the motor should the current flow through the shunt field be lost or drop below a certain level. If the shunt field should open, the motor will become a series motor and could race to a dangerously high speed. A shunt field rheostat is shown in Figure 43–5.

Most DC speed controllers use solid-state devices such as SCRs to provide speed control for the motor. As a general rule, the shunt field is provided with constant power and the armature circuit is controlled separately, Figure 43–6.

These controllers generally provide current limit protection and speed regulation. The circuit shown in Figure 43–6 employs current transformers on the three-phase lines supplying power to the controller to determine motor current. When the current reaches a certain level, the controller simply will not supply more current to the motor. Speed regulation is generally accomplished by a feedback circuit that senses the speed of the motor. If load is added and the motor speed tries to decrease, the controller will supply more armature current. If load is removed and the motor tries to speed up, the controller will decrease armature current.

Notice that in the circuit in Figure 43–6 the shunt field is connected to a separate power supply. As a general rule the shunt field power remains turned on even when the motor is not operating. The shunt field acts as a heater to keep moisture from forming in the motor. When servicing or disconnecting a DC motor, make certain that all power to the motor has been disconnected.

There are some controllers that provide overspeed control as well as underspeed control. These controllers use variable voltage to supply current to both the armature and shunt field. Motors of this type often have two separate shunt fields. One is connected to variable voltage and the other is connected to a constant voltage source, Figure 43–7. In this way, no matter how

Fig. 43–5 Circular plate rheostat. *(Courtesy of Ward Leonard Electric Co., Inc.)*

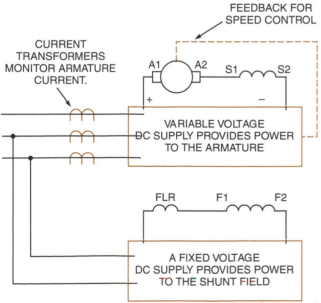

(Source: Delmar/Cengage Learning)

Fig. 43–6 Electronic speed controllers generally provide separate power supplies for the armature and shunt field.

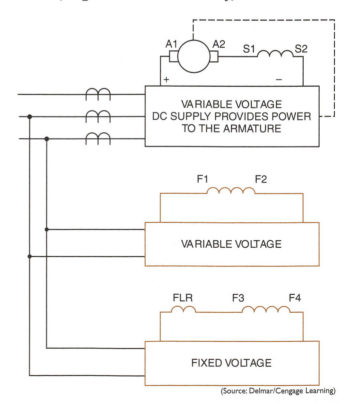

(Source: Delmar/Cengage Learning)

Fig. 43–7 The motor contains two separate shunt field windings.

weak the shunt field connected to variable voltage becomes, a certain amount of magnetic flux is always produced in the pole pieces.

STARTING THE MOTOR

Two factors limit the current taken by the armature from a DC source: (1) the counter emf and (2) the armature resistance. Because there is no counter emf when the armature is at a standstill and its resistance is very low, the current taken by the armature will be abnormally high. As a result, the armature current must be limited by an external resistor or a starting rheostat. This procedure is described in the following units. With solid-state equipment, the current flow is limited by allowing a gradual increase of current strength to start the motor.

Study Questions

1. Draw diagrams of the following DC motors, including the proper terminal markings and field rheostats.
   a. series
   b. shunt
   c. compound

2. How is the direction of rotation of a DC motor reversed?

3. Why is the shunt field circuit never allowed to open while operating?

4. What is the function of the field loss relay?

5. Why is it common practice to leave power connected to the shunt field even when the motor is not operating?

6. Explain why reducing current flow to the shunt field causes the motor speed to increase.

Use of Reduced Voltage for Starting

Objectives

After studying this unit, the student should be able to

- List the two factors that limit the current taken by a motor armature.

- Explain why it is necessary to insert resistance in the proper part of a DC motor circuit on starting.

- Calculate the amount of series resistance required for a DC motor at different speeds during the starting period.

The starting of fairly large DC motors requires that current be increased gradually or that resistance be inserted in series with the armature of the motor.

Two factors that limit the current taken by a motor armature from a given line are the *counter emf* and the *armature resistance*. When the motor is at a standstill, there is no counter emf, nor is there inductive reactance such as in an AC motor. Therefore, the current taken by the armature will be abnormally high unless the DC supply is increased gradually, as with solid-state equipment, or an external current limiting resistor is used.

Figure 44–1 illustrates a shunt motor connected directly across a 250-volt line. The resistance of the armature is 0.5 ohm. The full-load current taken by the motor is 25 amperes and the shunt field current is 1 ampere. The resulting armature current at *full load* is 24 amperes.

The *counter emf* developed by a motor is proportional to the speed of the motor if the field is constant. In addition, the counter emf equals

the applied voltage minus the armature (*IR*) drop. The current through the armature can be found using Ohm's law as follows:

$$I = \frac{E - E_c}{R_a}$$

where: I = armature current, amperes
E = line voltage, volts
E_c = counter emf, volts
R_a = armature resistance, ohms

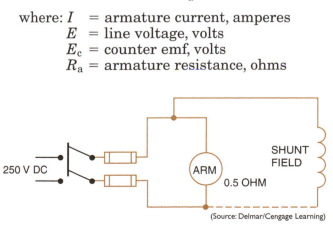

(Source: Delmar/Cengage Learning)

Fig. 44–1 This 25-ampere DC motor will draw 500 amperes from the line on starting without armature resistance.

When there is no starting resistance in the circuit, the current through the armature at the instant of starting, for the shunt motor in Figure 44–1 is

$$I = \frac{250 - 0}{0.5} = 500 \text{ amperes}$$

This starting current is undesirably high. The total starting current is equal to the armature current plus the field current of 1 ampere. The ratio of the starting current to the full-load current is

$$\frac{501}{25} = 20.04$$

The excessive torque and heat produced by this current may harm the motor and the load attached to it. In addition, it may cause the insulation to overheat on starting and the burning of the armature.

These effects can be eliminated by connecting a resistance in series with the armature, Figure 44–2. The resistance reduces the starting armature current to approximately 1.5 times the full-load current value. The resistance is then gradually removed from the circuit.

The value of this series resistance is determined by solving the previous equation for R. Remember that the armature resistance (R_a) must be subtracted:

$$R = \frac{E - E_c}{I} - R_a$$

$$I = 24 \times 1.5 = 36 \text{ A}$$

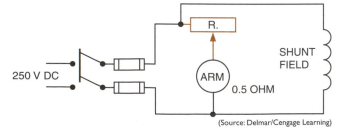

(Source: Delmar/Cengage Learning)

Fig. 44–2 Starting resistors are used to eliminate high starting currents.

When the motor is at a standstill, the series resistance is equal to

$$R = \frac{250 - 0}{36} - 0.5 = 6.44 \text{ ohms}$$

Resistors are also rated in watts. The following formula can be used to find the power requirement of the resistor:

$$I^2R = P$$

$$36 \times 36 \times 6.44 = 8346 \text{ watts}$$

The proper resistor can be selected by knowing the required resistance and the power value in watts.

The required series resistance can be determined for some intermediate speed in the starting period of the motor. In this case, solve for the counter emf at the speed in question and then obtain the resistance from the formula given. Because the counter emf differs from the applied voltage only by the armature (IR) drop, the speed is nearly proportional to the applied voltage. For example, at half speed, the counter emf is equal to

$$\frac{50\% \text{ Speed}}{100\% \text{ Speed}} = \frac{\text{Voltage at Half Speed}}{250}$$

$$\text{Voltage at Half Speed} = 250 \times \frac{50\%}{100\%}$$

$$= 250 \times 1/2 = 125 \text{ volts}$$

The resistance required at half speed is equal to

$$R = \frac{250 - 125}{36} - 0.5 = 2.97 \text{ ohms}$$

The counter emf can be found at the full load:

$$E_c = E - (IR_a)$$
$$E_c = 250 - (24 \times 0.5)$$
$$E_c = 250 - 12$$
$$E_c = 238 \text{ volts}$$

Study Questions

1. Why is a starter necessary on a DC motor?

2. Why is a resistance placed in series with the armature only rather than in series with the entire shunt motor?

3. If an armature has a resistance of 0.5 ohm, what series resistance must be used to limit the current to 30 amperes across a 230-volt circuit at the instant of starting? What is the power rating?

4. What resistance must be used when the motor in question 3 reaches 50 percent of its rated speed?

Across-the-Line Starting

Objectives

After studying this unit, the student should be able to

- Describe across-the-line starting for small DC motors.

- State why a current limiting resistor may be used in the starting circuit for a DC motor.

- Connect across-the-line starters used with small DC motors.

- Recommend troubleshooting solutions for across-the-line starters.

- Draw diagrams for three motor starter control circuits.

Small DC motors can be connected directly across the line for starting because a small amount of friction and inertia is overcome quickly in gaining full speed and developing a counter emf. Fractional horsepower manual starters (discussed in Unit 2) or magnetic contactors and starters (Unit 3) are used for across-the-line starting of small DC motors, Figure 45–1.

Magnetic across-the-line control of small DC motors is similar to AC control or to two- or three-wire control. Some DC across-the-line starter coils have dual windings because of the added load of multiple break contacts and the fact that the DC circuit lacks the inductive reactance that is present with AC electromagnets. Both windings are used to lift and close the contacts, but only one winding remains in the holding position. The starting (or lifting) winding of the coil is designed for momentary duty only. In Figure 45–2, assume that coil M is energized momentarily by the start button. When the starter is closed, it maintains itself through the normally open maintaining contact (M) and the upper winding of the coil because the normally closed contact (M) is now open. Power contacts M close, and the motor starts across the full-line voltage. The double-break power contacts are designed to minimize the effects of arcing. (DC arcs are greater than those due to AC.)

Figure 45–3 shows another control method used to start a DC motor. In this method, a current-limiting resistor is provided to prevent coil burnout. It is used to limit a continuous duty current flow to some coils or when the coils are overheating.

The coil first receives the maximum current required to close the starter. It then receives the minimum current necessary to hold in the contacts and for continuous duty through the current-limiting resistor.

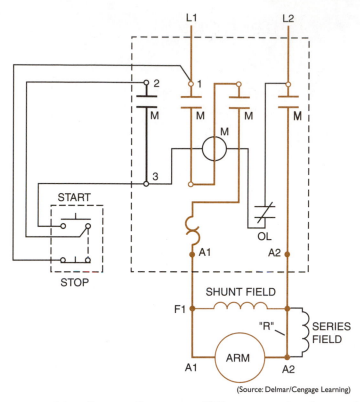

(Source: Delmar/Cengage Learning)

Fig. 45–1 DC full-voltage starter wiring diagram. Connection "R" is removed with use of series field.

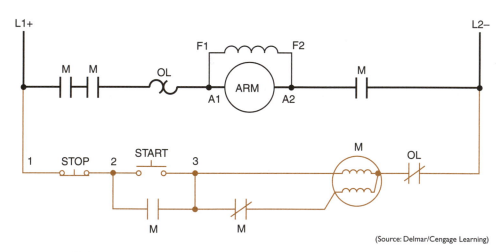

(Source: Delmar/Cengage Learning)

Fig. 45–2 Line diagram of DC motor starter with dual-winding coil.

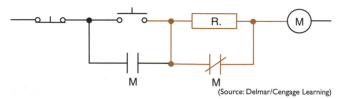

(Source: Delmar/Cengage Learning)

Fig. 45–3 DC starting circuit using current limiting resistor.

Study Questions

1. Why may small DC motors be started directly across the line?

2. When using a coil that is not designed for continuous duty, what may happen if a resistor is not added to the circuit?

3. What is the purpose of a double-break power contact?

Compensating and Definite Time Control Starting

Objectives

After studying this unit, the student should be able to

- Identify terminal markings for a DC compound motor and a counter emf controller.

- Describe the operation of a counter emf controller.

- State the purpose of a relay connected across the armature of a counter emf controller.

- Connect DC motors and counter emf controllers.

- Recommend troubleshooting solutions for problems with counter emf controllers.

- Describe how a definite timer is used to provide time-limit acceleration.

- Describe the use and operating sequence of a field accelerating relay.

- Describe the use and operating sequence of a field failure relay.

COMPENSATING TIME ACCELERATION

This type of controller automatically adjusts the starting time intervals, depending on the load connected to the motor. For example, because a heavy load has more inertia for the motor to overcome, a longer time is required to build up the counter emf. Thus, a longer starting period is required. The following starter is an example.

COUNTER EMF CONTROLLER

When a DC motor starts, the generated counter emf across the armature is low. As the motor accelerates, the counter emf increases. When the voltage across the motor armature reaches a certain value, a relay is actuated to reduce the starting resistance at the right time.

Figure 46–1 shows the schematic connection diagram for a DC motor controller, using the

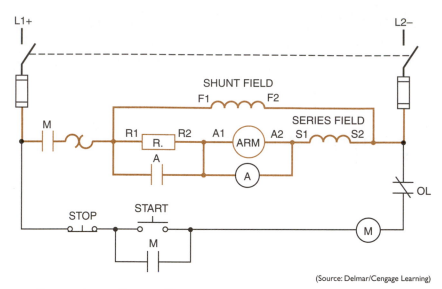

Fig. 46–1 Line diagram of a counter emf controller.

(Source: Delmar/Cengage Learning)

principle of counter emf acceleration. When the start button is pressed, contactor M is energized and the main contact (M) is closed. The current path is then complete from L1 through the resistor (R1-R2), the armature (A1-A2), and the series field (S1-S2) to L2.

To prevent the contactor from opening, the maintaining contact (M) is operated from the coil (M). This means that a shunt is placed around the start button. When the start button is released, M remains energized through the maintaining contacts (M) until the stop button is pressed.

There is a high voltage drop across R1-R2, but a counter emf is not present across the armature. As the motor accelerates, the current is reduced. When the counter emf across the armature reaches a value equal to a certain percentage of the full value, relay A (connected across the armature) closes contact A. Thus, resistor R1-R2 is removed from the circuit and the motor is placed directly on the line.

VOLTAGE-SENSITIVE RELAY

The relay (A) is a voltage-sensitive relay, Figure 46–2. It is commonly used to sense the DC motor armature voltage. This voltage is an indication of motor speed. The voltage-sensitive relay is frequently used in the following applications: hoist controllers; reversing plugging to stop; and applying armature shunt contactors on multistep slowdown and acceleration circuits. Because the relay is adjustable for pickup and dropout, one of its many uses is to control multisteps of acceleration as a motor starter.

Fig. 46–2 DC voltage sensitive relay for DC motor armature voltages. *(Courtesy of Schneider Electric)*

Figure 46–1 shows one relay and one resistor for one step of acceleration. For heavier motors and loads, additional relays and resistors can be added to increase the points of acceleration. The additional relays must be adjusted to ensure that the resistances will be cut out of the circuit at the proper times as the motor accelerates.

FIELD ACCELERATING RELAY

A *field accelerating relay,* or a full field relay, Figure 46–3, is used with starters to control adjustable speed motors. This type of relay provides the full field during the starting period

Fig. 46–3 DC field accelerating relay (left) and DC field failure relay (right). *(Courtesy of General Electric Co.)*

and limits the armature current during sudden speed changes. The coil of the field accelerating relay is connected in series with the motor armature. When there is a current inrush at startup, or when a sudden speed increase causes excessive armature currents, the coil closes the relay contacts. The closed contacts bypass the shunt field rheostat. The full shunt field strength is provided on starting and excessive armature currents are prevented during speed changes.

FIELD FAILURE RELAY

A field failure relay is a single-pole control relay, Figure 46–3. The contacts of the relay are connected in the control circuit of the main contactor. The coil is connected in series with the shunt field. If the shunt field fails, the relay coil is de-energized and the relay contact opens the circuit to the motor starter, disconnecting the motor from the line. The field failure relay is a safety feature that can be added to a starter to prevent the motor from greatly increasing its speed if the field is opened accidentally.

DEFINITE TIME CONTROL STARTING

A timing relay provides a definite timed interval, which is used to control acceleration while a motor is being brought up to speed. The acceleration steps are controlled by resistors inserted in the line ahead of the motor. The definite time limit acceleration provided by the timing relays means that the resistors remain in the line for a fixed time period before the motor is connected across the line. A typical DC reduced voltage magnetic starter (less enclosure) using timing relays is shown in Figure 46–4. A schematic for this starter is also shown.

Time limit acceleration is preferred for some applications. Other methods of acceleration (compensating time, counter emf, and current limiting) have the disadvantage that the motor might not accelerate enough to close the running contactor that bypasses the resistors. In other words, the resistors might remain in the circuit and could burn out. This hazard is eliminated by the use of time limit acceleration, because the running contactor is closed automatically after a definite time interval.

As shown in the typical schematic in Figure 46–5, the closing of the start button energizes coil M. This coil, in turn, instantaneously closes the heavy power contacts (M). The motor is then started through the starting resistor. The double-break contacts (M) are used to minimize the DC arc. The shunt field and field failure relay (FF) are energized at this same moment. The control contacts (FF and M) close and maintain the circuit across the start button. The series coil of the field accelerating relay (FA) receives the maximum current flow and closes contacts FA to shunt the circuit across the shunt field rheostat. Because the maximum field strength is present, there is more counter emf and excessive armature currents are prevented.

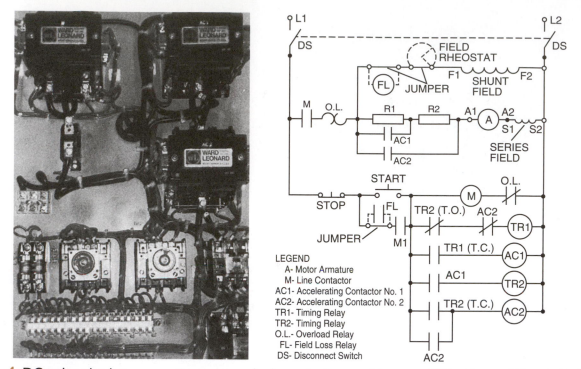

Fig. 46–4 DC reduced voltage magnetic starter and schematic diagram. *(Courtesy of Ward Leonard Electric Co., Inc.)*

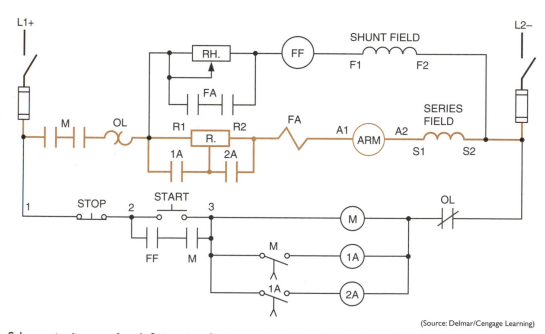

(Source: Delmar/Cengage Learning)

Fig. 46–5 Schematic diagram for definite time limit starting using timing relays, field accelerating relays, and field failure relays.

Main contactor M also activates a normally open, delay-in-closing contact (M). At the end of the time delay of the timing cycle, it energizes the first accelerating relay (1A). Power contact 1A cuts out one step of resistance and also activates the preset timing cycle (1A). The normally open, delay-in-closing contact (1A) energizes the second accelerating relay (2A). This action places the armature across the line. Time limit controllers can be used with series, shunt, or compound wound DC motors. These controllers can drive pumps, fans, compressors, and other classes of machinery when the load varies over a wide range and when automatic or remote control is desired.

Study Questions

1. Why is this type of controller called a counter emf controller?

2. What physical characteristic distinguishes between a *main contactor* and a *relay*?

3. Draw an elementary diagram with three points of acceleration.

4. What protection is available from excessive starting currents in the event the timing relays cut out the starting resistors too rapidly for the connected load?

5. How does a field accelerating relay prevent current excess that may occur because of adjusting the rheostat to change speeds?

6. How do individual timers, which have only light, single-pole control contacts, control the acceleration steps of a large DC motor?

7. How may more steps of acceleration be added to a starter?

8. Why is use made of double-break contacts or contacts connected in series such as contacts FA across the rheostat and M in the armature circuit in Figure 46–4?

Solid-State Adjustable Speed Control

Objectives

After studying this unit, the student should be able to

- Describe power conversion.

- Tell how solid-state control limits current flow on starting.

- Describe methods of feedback for constant-speed control.

- Discuss armature and field current sensing techniques.

- Describe methods of communicating with controllers.

- Perform maintenance on solid-state DC adjustable speed control.

POWER CONVERSION AND CONTROL

DC motors are widely used in industry because of their great versatility. They can be operated at a slow speed with a high torque. They can be run constantly at nameplate speed (base speed) or even above base speed (within safe limits), as shown in Figure 47–1. It is not necessary to have a DC power plant and distribution system due to the wide use of solid-state rectifiers. DC is converted at the motor control site from the AC distribution system.

STARTING THE MOTOR

The following factors limit the current flow taken by the armature from the power source:

1. the counter electromotive force (emf)
2. the armature resistance

Because there is no counter emf when the armature is at a standstill, and its resistance is very low, the current taken by the armature on startup will be abnormally high. As a result, the motor armature is gradually fed an increasing amount of voltage, from zero voltage for start, to full voltage for run. A fully energized field is maintained to generate the counter emf required on run to limit the armature current. Counter emf is gradually built up from start, with an increase in armature speed, to maximum on run.

Motor Speed Control

Motor speed control basics still apply. With full field and armature voltage, the value of

Fig. 47–1 Steel ingot being moved by DC compound-wound motors. *(Courtesy of General Electric Co.)*

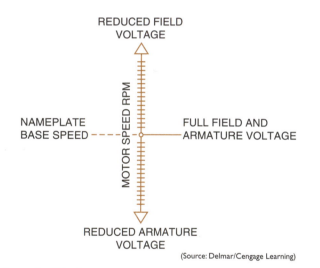

(Source: Delmar/Cengage Learning)

Graph 47–1 DC Drive Speed Control.

revolutions per minute of the motor is the nameplate, or base, speed. When operating the motor below base speed, which is very common, the armature voltage is reduced accordingly. Full field is maintained.

If it is necessary to operate the motor above base speed, the field voltage is reduced and armature voltage is fully applied. Refer to Graph 47–1. The motor nameplate does not give maximum speed limitations. For a safe operating condition, check with the motor manufacturer for the safe maximum speed limit.

DC ADJUSTABLE SPEED CONTROL FROM SINGLE-PHASE AC

A basic DC drive consists of an operator control station that determines when the drive starts and stops. The speed control potentiometer gives a reference signal to the drive controller, as shown in Figure 47–2.

Smaller size DC motors often operate from a single-phase AC source. Rectified and controlled DC is fed to the armature of the motor by silicon controlled rectifiers, or thyristors. This method is similar to the solid-state AC reduced voltage starters shown previously. In this case, however, the SCRs are connected to produce controlled DC from the AC line.

Fig. 47–2 Operator control panel for a DC variable-speed control. *(Courtesy of Reliance Electric Co.)*

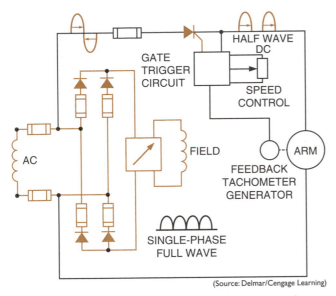

Fig. 47–3 Single-phase AC to DC motor control.

(Source: Delmar/Cengage Learning)

Figure 47–3 shows a simplified single-phase AC circuit rectified to DC supplying a DC motor. The field circuit is a full-wave DC supply (shown graphically) for a greater field strength when required. Motor speed is preset by the operator control. This electronically adjusts for the proper field and armature current strength for the desired speed. The feedback tachometer generator helps to regulate the amount of DC flowing to the armature for high torque, slower-than-base speeds. The tachometer generator output voltage is proportional to its speed. The output voltage is connected to the SCR gate control circuit. A change in speed is detected and immediately corrected to maintain a preset constant speed. Generally, DC motors are run at base speed or below.

INSTALLATION AND MAINTENANCE

Solid-state drives are designed for easy installation. Threaded hublike power entries are provided for top and bottom conduit entry. Enclosure knockouts are also available and wire entry is made directly to incoming marked power terminals. Sufficient space is allowed around power terminals to ensure working room and easy installation of plug-in kits. Refer to Figure 47–4. Keeping these drives operating reliably should not be too difficult. Modular construction makes troubleshooting easier. Components can be removed for repair or replacement. Leads disconnect simply by means of plug connectors. Replacement components are accessible. The visual control panel tells where the problems are.

DC ADJUSTABLE SPEED CONTROL, THREE PHASE

Power Supply

Power conversion cells convert incoming three-phase AC power to the highly controllable DC power used in the armature of the drive

(A)   (B)

Fig. 47–4 (A) Power leads fall straight to incoming termination with ample room for wiring. Note the threaded conduit power entries at the top, which permit quick installation. (B) Modular design allows for easy troubleshooting. *(Courtesy of Reliance Electric Co.)*

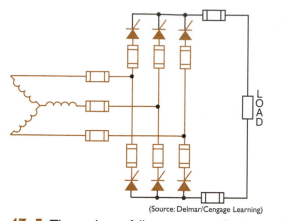

Fig. 47–5 Three-phase, full-wave controlled rectifier with connected load.

(Source: Delmar/Cengage Learning)

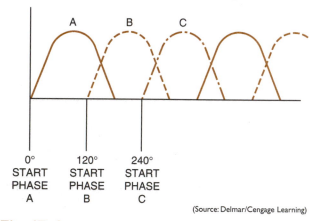

(Source: Delmar/Cengage Learning)

Fig. 47–6 Rectified three-phase power.

motor. Figure 47–5 shows a diagram of a three-phase, full-wave controlled rectifier connected to a load. The controlled rectifier uses SCRs—not uncontrolled diodes as shown in single-phase rectifiers. Note that some single-phase units use SCRs also. Three-phase rectifiers provide smooth DC compared with half-wave single phase. This can be seen in Figure 47–6, looking at the top of the DC ripple (A, B, C). This ripple is further refined by filtering it through electronic circuits containing inductors and capacitors. The power supply takes the incoming three-phase power and transforms it to the proper voltage levels for the support circuitry to power the signal electronics, firing circuits, and excitation modules.

Power Control

Figure 47–7 shows a three-phase power circuit for an adjustable speed drive system. These systems are available in sizes ranging from about 5 to 2500 horsepower. The circuit diagram shows the circuit configuration of the power conversion system and the motor field excitation module. Three-phase power is routed through the fuse-protected legs of this converter and is delivered to the armature. Fuses on each leg provide better protection than line fuses against "shoot-throughs" and reversing inversion faults. SCRs are mounted with proper heat sinks (heat dissipators). Some of this may be seen in Figure 47–8. A main DC contactor isolates the armature when stopped. The motor field may remain energized. If so, its heat helps prevent harmful moisture from forming in the motor due to condensation in a humid or damp environment. Instead of using a tachometer generator feedback for constant-speed control, armature and field current feedback is obtained from shunts, more often used in connecting DC ammeters. These shunts are similar to a resistor. A millivolt drop is read across them, and this information is fed into the speed control circuit. The SCRs then act to control the motor at a constant speed under changing load conditions.

Signal Electronics

A signal electronics module provides the overall drive control. It contains a microprocessor and supporting control electronics mounted on printed wiring boards, Figure 47–9. The same signal electronics module can usually be used regardless of the size of the power electronics (with equipment from the same manufacturer). In this way, the cost of the replacement modules inventory is reduced. It is also convenient for the maintenance personnel, because modules can be switched or interchanged when diagnosing a problem.

The microprocessor is the center of the control system. In addition to providing the SCR firing circuit, control, regulating functions, and drive sequencing, it can communicate with people. It also responds to feedbacks and reference commands, then adjusts both motor field and armature power accordingly. The micro processor also provides diagnostic information. To communicate this information, a number of techniques are available. For example, an LED display panel provides indication of motor faults, converter faults, alarms, drive status, and drive configuration. If more than one fault occurs, each of the faults will activate a light and the first fault will flash. An amber- colored LED is energized on all key circuit boards after the microprocessor control checks each critical circuit. The amber, or white, LED light remains energized as long as the circuit is functional.

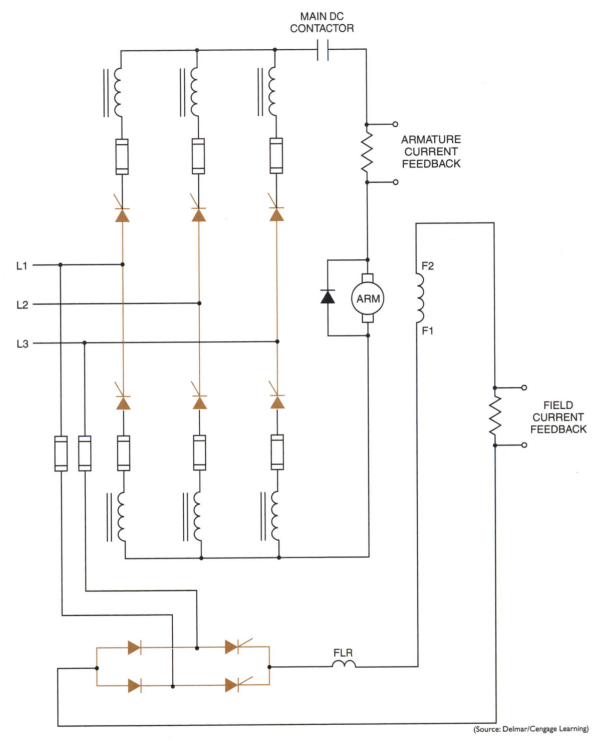

Fig. 47–7 Three-phase AC feeding a DC adjustable speed drive power circuit.

(Source: Delmar/Cengage Learning)

Thyristor cell status lights are energized each time the indicated SCR is fired, thus indicating the operation of the SCR. If a printer is connected, English language and engineering term messages are received from the processor. This information can be used to interrogate the drive unit. However, a keyboard and video monitor are required to do this. These are seen in Figure 47–10.

SELF-TEST DIAGNOSTICS

When power is applied to the drive or when it is requested by an interactive command, each of the circuit boards is sequentially checked by the microprocessor. As each passes the self-test routine, the light for that board is energized and the system progresses to the next board. After the signal electronics have been verified, the microprocessor

Fig. 47–8 SCRs are mounted with proper heat sinks. (*Courtesy of General Electric Co.*)

(Source: Delmar/Cengage Learning)

Fig. 47–10 This wired control module for a drive system is common in the better drives, both AC and DC. The keypad provides a high degree of diagnostics with a 16-character LED feedback in easily understood terms, as well as direct access to the drive.

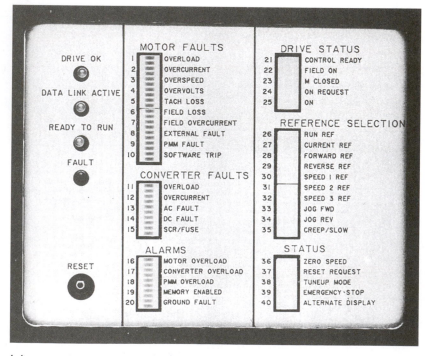

Fig. 47–9 Signal module contains a microprocessor and control electronics mounted on printed wiring boards. (*Courtesy of General Electric Co.*)

fires each of the armature circuit SCRs one at a time. An SCR cell voltage sensor is used to indicate if the gating, fuse, and SCR are operational. If both the signal (control) and the power electronics pass the test, the control is released for drive operation. If the test discovers a problem (failed board, SCR cells, fuses, open wire, and so on), the specific fault is displayed on the lighted panel or is printed out, and the drive is not released for starting until the problem is corrected.

Study Questions

1. Where does the DC motor get its DC power?

2. How does a solid-state device start the motor?

3. Describe methods of feedback control for constant speed.

4. What method is used to sense armature and field current?

5. Describe methods of communication with a machine controller.

6. What might be a first step in troubleshooting a drive control?

Methods of Deceleration

UNIT 48

Plugging

Objectives

After studying this unit, the student should be able to

- Define what is meant by the plugging of a motor.
- Describe how a control circuit using a zero-speed switch operates to stop a motor.
- Describe the action of a time-delay relay in a plugging circuit.
- Describe briefly the action of the several alternate circuits that use the zero-speed switch.
- Connect plugging control circuits.
- Recommend troubleshooting solutions for plugging problems.

Plugging is defined by NEMA as a system of braking, in which the motor connections are reversed so that the motor develops a counter torque, which acts as a retarding force. Plugging controls provide for the rapid stop and quick reversal of motor rotation.

Motor connections can be reversed while the motor is running unless the control circuits are designed to prevent this type of connection. Any standard reversing controller can be plugged, either manually or with electromagnetic controls. Before the plugging operation is attempted, however, several factors must be considered including

1. the need to determine if methods of limiting the maximum permissible currents are necessary, especially with repeated operations and DC motors, and

2. the need to examine the driven machine to ensure that repeated plugging will not cause damage to the machine.

PLUGGING SWITCHES AND APPLICATIONS

Plugging switches, or zero-speed switches, are designed to be added to control circuits as pilot devices to provide quick, automatic stopping of machines. In most cases, the machines will be driven by squirrel cage motors. If the switches are adjusted properly, they will prevent the direction reversal of rotation of the controlled drive after it reaches a standstill following the reversal of the motor connections. One typical use of plugging switches is for machine tools

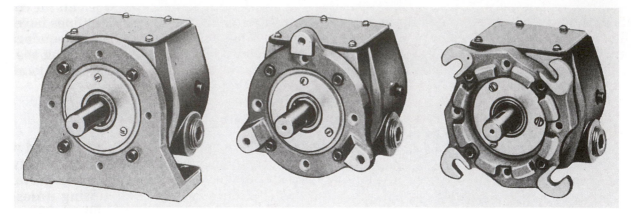

Fig. 48–1 Plugging (zero-speed) switches. Note the mounting methods. *(Courtesy of Rockwell Automation)*

that must stop suddenly at some point in their cycle of operation to prevent inaccuracies in the work or damage to the machine. Another use is for processes in which the machine must stop completely before the next step of work begins. In this case, the reduced stopping time means that more time can be applied to production to achieve a greater total output.

Typical plugging switches are shown in Figure 48–1. The shaft of a plugging switch is connected mechanically to the motor shaft or to a shaft on the driven machine. The rotating motion of the motor is transmitted to the plugging switch contacts either by a centrifugal mechanism or by a magnetic induction arrangement (eddy current disc) within the switch. The switch contacts are wired to the reversing starter, which controls the motor. The switch acts as a link between the motor and the reversing starter. The starter applies just enough power in the reverse direction to bring the motor to a quick stop.

Plugging a Motor to a Stop from One Direction Only

The forward rotation of the motor in Figure 48–2 closes the normally open plugging switch

contact. When the stop button is pushed, the forward contactor drops out. At the same time, the reverse contactor is energized through the plugging switch and the normally closed forward interlock. Thus, the motor connections are reversed and the motor is braked to a stop. When the motor is stopped, the plugging switch opens to disconnect the reverse contactor. This contactor is used only to stop the motor using the plugging operation; it is not used to run the motor in reverse.

Adjustment

The torque that operates the plugging switch contacts will vary according to the speed of the motor. An adjustable contact spring is used to oppose the torque to ensure that the contacts open and close at the proper time regardless of the motor speed. To operate the contacts, the motor must produce a torque that will overcome the spring pressure. The spring adjustment is generally made with screws that are readily accessible when the switch cover is removed.

Care must be exercised to prevent the entry of chips, filings, and hardware into the housing when it is opened. Such material may be attracted to the magnets or hamper spring action. The housing must be carefully cleaned before the cover is removed for maintenance or inspection.

Installation

To obtain the greatest possible accuracy in braking, the switch should be driven from the shaft with the highest available speed that is within the operating speed range of the switch.

The plugging switch may be driven by gears, by a chain drive, or a direct flexible coupling. The preferred method of driving the switch is to connect a direct flexible coupling to a suitable

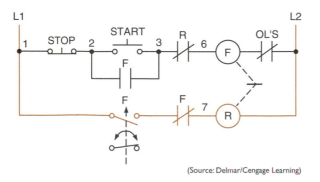

(Source: Delmar/Cengage Learning)

Fig. 48–2 Plugging the motor to stop from one direction only.

shaft on the driven machine. The coupling must be flexible because the centerline of the motor or machine shaft and the centerline of the plugging switch shaft are difficult to align accurately enough to use a rigid coupling. The switch must be driven by a positive means. Thus, a belt drive should not be used. In addition, a positive drive must be used between the various parts of the machine being controlled, especially where these parts have large amounts of inertia.

The starter used for this type of circuit is a reversing starter that interchanges two of the three motor leads for a three-phase motor, reverses the direction of current through the armature for a dc motor, and reverses the relationship of the running and starting windings for a single-phase motor.

Motor Rotation

Experience shows that there is little way of predetermining the direction of the rotation of motors when the phases are connected externally in proper sequence. This is an important consideration for the electrician and the electrical contractor when the applicable electrical code or specifications require that each phase wire of a distribution system be color coded.

If the shaft end of a motor runs counterclockwise rather than in the desired clockwise direction, then the electrician must reconnect the motor leads at the motor. For example, assume that many three-phase motors are able to be connected and the direction of rotation of all the motors must be the same. If counterclockwise rotation is desired, the supply phase should be connected to the motor terminals in the proper sequence, T1, T2, and T3. If the motor does not rotate in the desired counterclockwise direction using these connections, the leads may be interchanged at the motor. Once the proper direction of rotation is established, the remaining motors can be connected in a similar manner if they are from the same manufacturer. If the motors are from different manufacturers, they may rotate in different directions even when all the connections are similar and the supply lines have been phased out for the proper phase sequence and color coded. The process of correcting the rotation may be difficult if the motors are located in a place that is difficult to reach.

Lockout Relay

The zero-speed switch can be equipped with a lockout relay or a safety latch relay. This type of relay provides a mechanical means of preventing the switch contacts from closing unless the motor starting circuit is energized. The safety feature ensures that if the motor shaft is turned accidentally, the plugging switch contacts do not close and start the motor. The relay coil generally is connected to the T1 and T2 terminals of the motor. The lockout relay should be a standard requirement for circuits to protect people, machines, and production processes.

PLUGGING WITH THE USE OF A TIMING RELAY

A time-delay relay may be used in a motor plugging circuit, Figure 48–3. Unlike the zero-speed switch, this control circuit does not compensate for a change in the load conditions on the motor. The circuit shown in Figure 48–3 can be used for a constant load condition once the timer is preset. If the emergency stop button in Figure 48–3(A) is pushed *momentarily* and the normally open circuit is *not* completed, the motor will coast to a standstill. (This action is also true of the normal double contact stop button.) If the emergency stop button is pushed to *complete* the normally open circuit of the push button, then contactor S is energized through the closed contacts (TR and R). Contactor S closes and reconnects the motor leads, causing a reverse torque to be applied. When the relay coil is de-energized, the opening

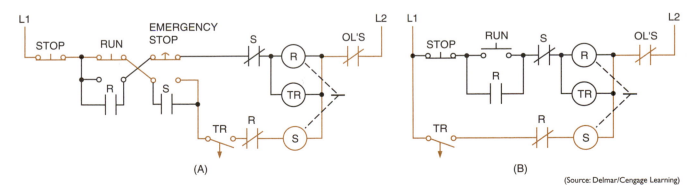

(A)   (B)

(Source: Delmar/Cengage Learning)

Fig. 48–3 Plugging with a time-delay relay.

of contact TR can be retarded. The time lag is set so that contact TR opens at or near the point at which motor shaft speed reaches 0 rpm.

ALTERNATE CIRCUITS FOR PLUGGING SWITCH

The circuit in Figure 48–4 is used for operation in one direction only. When the stop push button is pressed and immediately released, the motor and the driven machine coast to a standstill. If the stop button is held down, the motor is plugged to a stop.

Using the circuit shown in Figure 48–5, the motor may be started in either direction. When the stop button is pressed, the motor can be plugged to a stop from either direction.

The circuit shown in Figure 48–6 provides operation in one direction. The motor is plugged to a stop when the stop button is pressed. Jogging is possible with the use of a control relay.

Figure 48–7 shows a circuit for controlling the direction of rotation of a motor in either direction. Jogging in either the forward or reverse direction is possible if control jogging relays are used. The motor can be plugged to a stop from either direction by pressing the stop button.

The circuit in Figure 48–8 provides control in either direction using a maintained contact selector switch with forward, off, and reverse positions. The plugging action is available from either direction of rotation when the switch is turned to the off position. Low-voltage protection is not provided with this circuit.

The circuit in Figure 48–9 allows motor operation in one direction. The plugging switch is used as a speed interlock. The solenoid, or coil F, will not operate until the main motor reaches its running speed. A typical application of this circuit is to provide an interlock for a conveyor system. The feeder conveyor motor cannot be started until the main conveyor is operating.

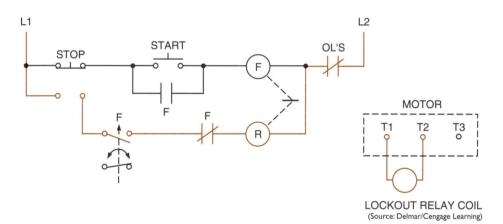

Fig. 48–4 Holding the stop button will stop the motor in one direction.

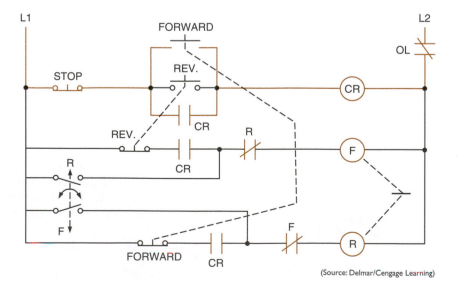

(Source: Delmar/Cengage Learning)

Fig. 48–5 In this circuit, pressing the stop button stops the motor in either direction.

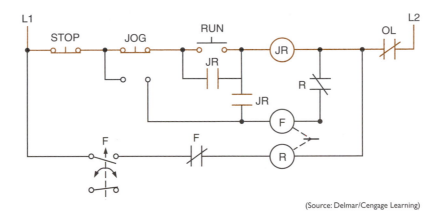

(Source: Delmar/Cengage Learning)

Fig. 48–6 Pressing the stop button will stop the motor in one direction.

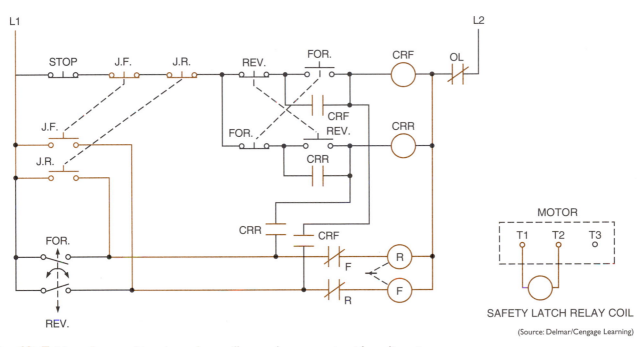

(Source: Delmar/Cengage Learning)

Fig. 48–7 Use of control jogging relays will stop the motor in either direction.

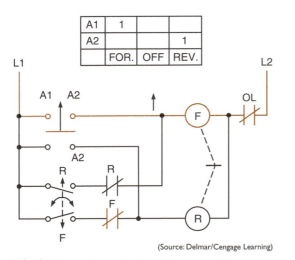

(Source: Delmar/Cengage Learning)

Fig. 48–8 Using the maintained contact selector switch.

ANTIPLUGGING PROTECTION

Antiplugging protection, according to NEMA, is obtained when a device prevents the application of a counter torque until the motor speed is reduced to an acceptable value. An antiplugging circuit is shown in Figure 48–10. With the motor operating in one direction, a contact on the antiplugging switch opens the control circuit of the contactor used to achieve rotation in the opposite direction. This contact will not close until the motor speed is reduced. Then the other contactors can be energized.

Alternate Antiplugging Circuits

The direction of rotation of the motor is controlled by the motor starter selector switch,

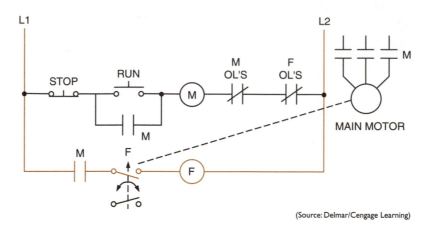

(Source: Delmar/Cengage Learning)

Fig. 48–9 Using the plugging switch as a speed interlock.

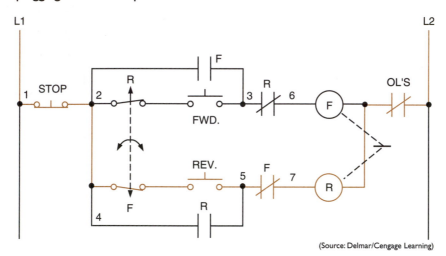

(Source: Delmar/Cengage Learning)

Fig. 48–10 Antiplugging protection; the motor is to be reversed but not plugged.

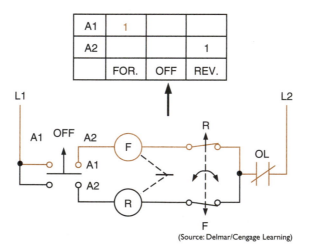

(Source: Delmar/Cengage Learning)

Fig. 48–11 Antiplugging with rotation direction selector switch.

Figure 48–11. The antiplugging switch completes the reverse circuit only when the motor slows to a safe, preset speed. Undervoltage protection is not available.

In Figure 48–12, the direction of rotation of the motor is selected by using the maintained contact, two-position selector switch. The motor is started with the push button. The direction of rotation cannot be reversed until the motor slows to a safe, preset speed. Low-voltage protection is provided by a three-wire, start-stop, push-button station.

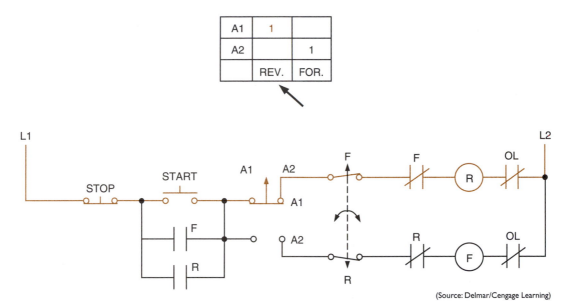

(Source: Delmar/Cengage Learning)

Fig. 48–12 Antiplugging circuit using a selector switch and providing low-voltage protection.

Study Questions

1. In Figure 48–2, what is the purpose of normally closed contact F?

2. Can a time-delay relay be used satisfactorily in a plugging circuit? Explain.

3. In what position must a plugging switch be mounted? Explain.

4. What is the preferred method of connecting the plugging switch to the motor or the driven machine?

5. What happens if the zero-speed switch contacts are adjusted to open too late?

6. What is the purpose of the lockout relay or safety latch relay?

7. What happens if the reverse push button is closed when the motor is running in the forward direction as in Figure 48–5?

8. What alternate methods of stopping are provided by the circuit described in Figure 48–4?

9. If the motor described in Figure 48–6 is plugging to a stop and the operator suddenly wants it to inch ahead or run, what action must be taken?

10. In Figure 48–7, is it necessary to push the stop button when changing the direction of rotation? Explain your answer.

11. In Figure 48–8, what happens to the motor running in the forward direction if the power supply is lost for 10 minutes?

12. In Figure 48–9, how is the feeder motor protected from an overload?

13. What is antiplugging protection?

14. During normal operation, when do the antiplugging switch contacts close?

15. If the supply lines are in the proper phase sequence (L1, L2, and L3) and are connected to their proper terminals on the motor (T1, T2, and T3), will all the motors rotate in the same direction? Why?

UNIT 49

Electric Brakes

Objectives

After studying this unit, the student should be able to

- Describe the general operation of electric brakes, both AC operated and DC operated.

- State the advantages of DC shunt wound brakes over AC-operated brakes.

- Describe the function of typical braking circuits for hoists and cranes.

- Connect motors and braking controllers.

- Recommend troubleshooting solutions for electric brake problems.

Many modern industrial processes require very powerful motors running at increased speeds. These motors must frequently be stopped in a shorter time than can be obtained by disconnecting the power from the motor. When machines require one or more motors and more precise control, and there is less time available to wait for the machines to coast to a standstill, then it is even more important to provide faster stopping methods.

Since the early 1900s, *electric brakes* have been in use. These are also known as *magnetic brakes, friction brakes,* and *mechanical brakes.* An electric brake consists of two friction surfaces, or shoes, which can be made to bear on a wheel on the motor shaft, Figure 49–1. Spring tension holds the shoes on the wheel. Braking is achieved because of the friction between the shoes and the wheel. A solenoid mechanism is used to release the shoes.

Fig. 49–1 Solenoid brake used on machine tools, conveyors, small hoists, and similar devices. *(Courtesy of Eaton Corporation)*

343

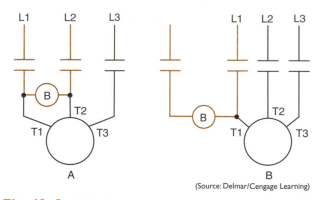

Fig. 49–2 AC brake coil connections for across-the-line starting. Control relays and contactors also are used to control magnetic brakes.

In a magnetically operated brake, the shoes are held in a released position by a magnet as long as the magnet coil is energized. However, if a pilot device interrupts the power, or there is a power failure, then the brake shoes are applied instantly to provide a fast, positive stop.

The coil leads of an AC magnetic brake are normally connected directly to the motor terminals, Figure 49–2. If a reduced-voltage starting method is used, the brake coil should be connected to receive full voltage.

Magnetic brakes provide a smooth braking action. This feature makes them suitable for high-inertia loads. Because these brakes apply and remove the braking pressure smoothly in either direction, they are often used on cranes, hoists, elevators, and other machinery where it is desirable to limit the shock of braking.

This brake is operated by a vertically mounted torque motor. When power is applied to the brake torque motor, the shaft of the motor turns and lifts the operating lever of a ball jack to release the brake. When the brake is fully released, the torque motor is stalled across the line. When power is released from the brake, a heavy compression spring counteracts the jack lever so that the brake is applied quickly.

MAGNETIC DISC BRAKES

Disc brakes can generally be installed wherever shoe brakes are used. Disc brakes can also be installed where appearance and available space prohibit the use of shoe brakes.

Magnetic disc brakes are used on machine tools, cranes, hoists, conveyors, printing presses, saw mills, index mechanisms, overhead doors, and other installations. The control adjustments of the torque and wear of disc brakes are

similar to those for shoe brakes. A disc brake is a self-enclosed unit that is bolted directly to the end bell of the motor. (This end bell operates on the motor shaft.) The braking action consists of pressure released by a solenoid and applied by a spring on the sides of a disc or discs.

DC MAGNETIC BRAKES

The construction of DC magnetic brakes is similar to that of AC shoe brakes. For DC magnetic brakes, however, the operating coils are shunt wound for across-the-line connections and series wound for connecting in series with DC motor armatures.

Shunt-wound brakes pick up at approximately 80 percent of the rated voltage and drop out about 60 percent. Series-wound brakes pick up (release) at about 40 percent of the full-load current of the motor and drop out (set) when the current reaches 10 percent of the full-load value.

The brake linings for DC magnetic brakes are similar to those for AC brakes. The linings are made of molded or woven asbestos with high friction resistance. They are bonded or riveted to steel backplates. Brakes range in size from approximately 1 inch to 30 inches in diameter. Large-diameter shunt-wound brakes are available with a relay and protective resistor to permit operation at high speed. Brakes with smaller diameters pick up and drop out quickly without this added feature. The large-diameter, high-speed brakes release quickly. When the armature closes, the protective resistor is inserted in the circuit. This action reduces the holding current to a low value and keeps the coil cooler. As a result, the brake sets quickly. High-speed brakes are suitable for any duty. They can be operated on alternating current if a rectifier unit is added.

The operation of DC magnetic brakes may be sluggish due to the induction present, unless the manufacturer has attempted to overcome it in order to speed the brake operation. The following list gives advantages of the use of DC brakes for shunt-wound operation (as compared with AC brakes).

- Laminated magnets or plungers are not used.
- Magnet and armature are cast of steel.
- No destructive hammer blow.
- Fast release and fast set.
- No AC chatter.
- Short armature movement.
- No coil burnout due to shoe wear, which affects the air gap.

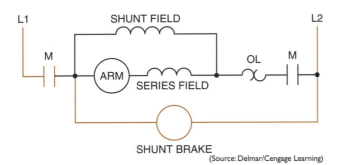

Fig. 49–3 DC shunt brake connections with motor connections.

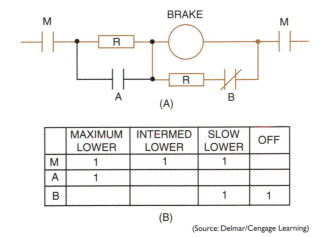

(A)

	MAXIMUM LOWER	INTERMED LOWER	SLOW LOWER	OFF
M	1	1	1	
A	1			
B			1	1

(B)

(Source: Delmar/Cengage Learning)

Fig. 49–4 Varying friction braking.

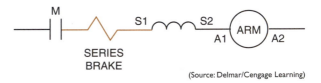

(Source: Delmar/Cengage Learning)

Fig. 49–5 Series brake connected with series motor.

Figure 49–3 shows the connection of a DC brake in a circuit that does not allow armature regeneration through the brake coil. As a result, the brake is prevented from setting.

Figure 49–4(A) shows a shunt brake control circuit. (This circuit is a small portion of a crane hoist elementary diagram.) The target table in Figure 49–4(B) lists the contacts that are closed for different crane-lowering operations. Note how the brake is weakened by the use of resistors in series and in parallel.

Series-wound brakes generally are used with series motors, Figure 49–5, and must have the same full-load current rating as the motor. The brakes must be properly sized for their loads also. Series-wound brakes may be used for heavy hoist applications and for many metal processing applications, such as steel mill drives and conveyors.

Study Questions

1. Why are magnetic brakes used?

2. How do electric brakes operate?

3. When an AC reduced voltage starter is used, how should the electric brakes be wired for release?

4. Why should protective resistors be provided on coils for high-speed shunt brakes?

5. In Figure 49–4, which contacts are closed when the brake coil receives full voltage?

6. What precautions must be taken when matching a series-wound brake to a motor?

7. How are electric brakes adjusted in general?

8. What may cause a de-energized brake to slip under load with maximum spring tension?

UNIT 50

Dynamic and Regenerative Braking

Objectives

After studying this unit, the student should be able to

- Describe what is meant by the process of dynamic and regenerative braking.

- Describe the general series of controller connections that can be made to obtain hoisting, lowering, and braking operations at different speeds.

- List several advantages of dynamic and regenerative braking.

- Describe three methods of providing dynamic braking for light-duty equipment.

- Explain how dynamic braking can be provided for a synchronous motor.

- Connect motors and dynamic braking controllers.

- Recommend troubleshooting solutions for dynamic braking problems.

A motor can be stopped by disconnecting it from the power source. A faster means of stopping a motor is achieved if the motor is reconnected so that it acts as a generator. This method of braking is called *dynamic braking*.

When a DC motor is reconnected so that the field is excited and there is a low-resistance path across the armature, the generator action converts some of the mechanical energy of rotation to electrical energy (as heat in discharge resistors). The result is that the motor slows down

sooner. However, as the motor slows down, the generator action decreases, the current decreases, and the braking lessens. This means that *a motor cannot be stopped by dynamic braking alone.*

A motor driving a crane hoist will let the load down too quickly if it is simply disconnected from the line. Some kind of braking is essential. A slight variation of the dynamic braking method just described works very well for this situation.

A series motor is used because of its characteristics for lifting and moving heavy loads. An electric brake connected in the line keeps the motor shaft from turning when there is no line current. To stop the hoist, therefore, it is necessary only to open the circuit to the motor.

To lower a load, the motor is reconnected as a shunt generator. That is, the series field is connected in series with a resistor. The combination is then connected directly across the line. Other possible connection schemes for series motors may be used.

The armature may be connected to a resistance to obtain dynamic braking as described earlier. It is more satisfactory, however, to connect the armature to the line as well. As a result, some of the mechanical energy is put back in the line as electrical energy. This means that the resistors are not required to handle as much heat. It also saves electrical energy. This process is called *regenerative braking*.

Regenerative braking is available on nearly all DC solid-state adjustable speed drives. The solid-state system provides a very precise controlled stop for those applications in which such control is essential. Unlike the nonregenerative system, which provides a coasting stop, the regenerative system uses the motor's braking force for a more controlled, faster stop. Where such precise control is not essential, and the additional cost of the regenerative system is not economically practical, a nonregenerative system can be used. With a regenerative system, there are twice as many SCR cell stacks within the same panel enclosure, such as those used in the DC solid-state adjustable speed drives.

In an electromagnetic system, the control is moved from the hoisting or lowering positions to the stop position. The load should be slowed by dynamic braking before the mechanical brake takes hold. In this way, wear is reduced on the mechanical brake bands.

The schematic diagrams in Figure 50–1 show how the controller connections can be changed to obtain the hoisting, lowering, and braking operations at different speeds. To simplify these diagrams, the control contactors and other components have not been shown.

For maximum hoist operation, Figure 50–1(A), the DC series motor and the brake receive the full line voltage and the maximum current.

In the intermediate hoist position, Figure 50–1(B), the motor is slowed when a resistor is added in series with the motor. The resistor is not large enough to allow the brake to set or drag, however.

In the slow hoist position, Figure 50–1(C), some current is bypassed around the motor by the parallel resistor. As a result, the motor is slowed and the brake receives enough current to remain open.

In the lower positions, Figures 50–1(D) through 50–1(F), regeneration back into the supply system is possible. The dissipating resistor regulates the field and the brake. The brake is de-energized in the *off* position with maximum braking pressure applied.

Dynamic braking is a simple and safe method of emergency or safety braking. Because the motor is converted to a self-exciting generator to

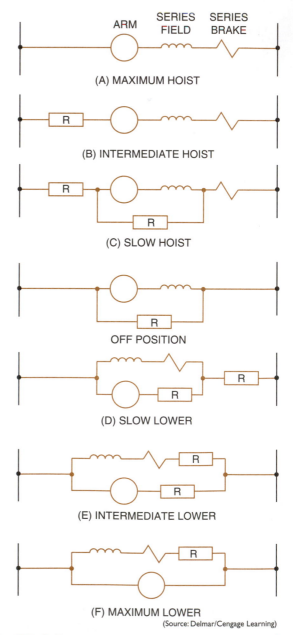

Fig. 50–1 Reconnecting a crane controller for different operations.

(Source: Delmar/Cengage Learning)

provide the slowing action, the braking does not depend on an outside source of power. Machines operating at higher speeds have braking problems not found with lower operating speeds. The need for a more rapid initial slowdown is essential if accidents are to be avoided. Dynamic braking is applied in about one-fifth the time required to set the majority of shunt brakes.

Due to space limitations, some drives cannot be equipped with electric brakes. In other cases, the inertia of the brake wheel is objectionable because it retards acceleration and deceleration.

An advantage of dynamic braking is shown in applications such as outside cranes or ore bridges where wet or icy rails cause the wheels to slip during stopping. When the wheels slip, the dynamic braking decreases, the wheels begin rotating again, and the crane or bridge is stopped in a shorter distance than is possible with locked wheels.

On high-speed drives, it is recommended that dynamic braking be provided to reduce the drive speed to a low value before the brakes are set. An example is an ore bridge trolley where severe braking may endanger the operator.

Dynamic braking can be started by track-type limit switches installed in the end zones of overhead traveling cranes. Cranes and other equipment having this type of dynamic braking can be stopped quickly and automatically when there is a power failure or the overload relays trip, regardless of the position of the master control handle. The braking torque is equally effective in both directions of travel.

DYNAMIC BRAKING FOR LIGHT-DUTY EQUIPMENT

There are many different methods of providing dynamic braking for small production equipment. For the diagram in Figure 50–2, the reduced voltage starting mechanisms are not shown to simplify the diagram. When the stop button is depressed, normally closed contact A completes the braking circuit through the braking resistor. Note that the shunt field must be energized for both deceleration and acceleration. Full field strength must be available for both. If a rheostat is used in the shunt field for speed control, the resistance is cut out manually or automatically. The disconnect switch should be open when the machine is not in service.

Figure 50–3 is a modification of Figure 50–2. The circuit in Figure 50–3 ensures that the braking resistor is not connected across the line. If the series field is used in the braking circuit,

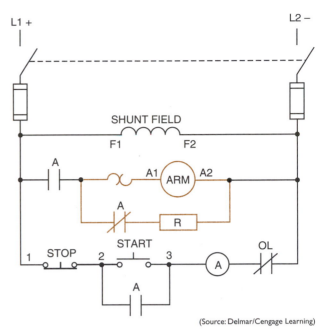

(Source: Delmar/Cengage Learning)

Fig. 50–2 Dynamic braking connections on a motor starter.

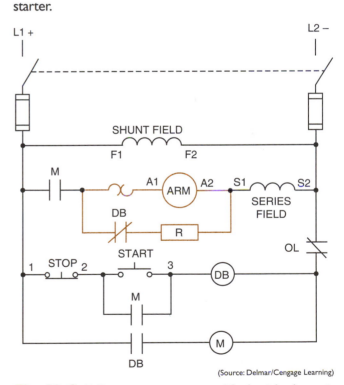

(Source: Delmar/Cengage Learning)

Fig. 50–3 DC motor starter modified with dynamic braking.

the DC compound motor characteristic (differential or cumulative) must be determined. If it is a cumulative compound motor, then the series field current direction must be reversed when it is used in the braking circuit. Automatic operation can be provided for such an action.

Figure 50–4(A) shows a diagram for a braking circuit which uses the starting resistance for braking. This circuit may be satisfactory for a

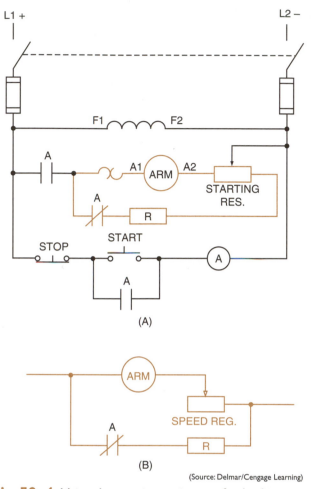

(A)

(B)

(Source: Delmar/Cengage Learning)

Fig. 50–4 Using the starting resistance for braking.

number of situations if the starting and stopping cycle does not exceed the capacity of the resistors. A speed regulator is designed for continuous-duty operation. The fixed resistor (R) in the braking circuit is a current-limiting resistor. It may not be required when a speed regulator is used because of the increased capacity. Figure 50–4(B) shows the control circuit for adjusting the braking speed. Resistor R is a current-limiting resistor, which prevents a short circuit on the armature. The value of R is selected so maximum braking speed is obtained when all resistance is cut out by the speed regulator.

DYNAMIC BRAKING FOR A SYNCHRONOUS MOTOR

Because of the similarity in the construction of a synchronous motor and an alternator (AC generator), a synchronous motor can be reconnected as an alternator to provide faster stopping. The kinetic energy of the rotor and the driven

machine is converted to electrical energy by generator action and then to heat by the dissipating resistors. A method developed by the author for the dynamic braking of a synchronous motor is illustrated in the elementary diagram of Figure 50–5. When the start button is pressed, contactor A is energized and opens the resistor circuits connected to the motor leads in the control panel. Contactor A energizes coil M to maintain the circuit and start the motor across the line. Time-delay relay coil TR is energized and its timing cycle begins. After the motor is brought up to speed by the windings in the rotor, the normally open, delay-in-closing contact (TR) energizes the DC contactor (B) to supply current to the field. The field discharge resistor circuit is opened as well. When the stop button is pressed, the AC supply is removed from the stator. The timer coil, however,

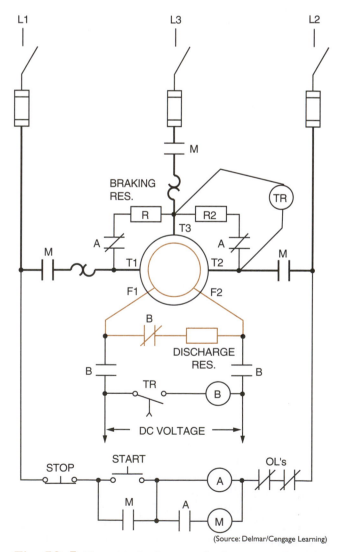

(Source: Delmar/Cengage Learning)

Fig. 50–5 Dynamic braking applied to a synchronous motor.

remains energized so that its contact remains closed. As a result, direct current continues to flow through the rotor. As the rotation continues, the magnetic lines of force of the rotor cut across the stator windings to generate a current. This current keeps the timing relay contact (TD) closed. The contact remains closed as long as the rotor maintains a voltage drop across the braking resistor (R2). Direct current is removed automatically from the rotor when it is almost at a standstill. This timed, semiautomatic method of synchronizing was selected to illustrate this method of braking for reasons of simplicity. With a simple modification, it is readily adaptable to any completely automatic, synchronizing control system.

Study Questions

1. After dynamic braking occurs, why is a mechanical brake generally necessary?

2. What becomes of the energy used to brake the motor dynamically for the two methods described?

3. Referring to Figure 50–1, what will happen if the controller is placed in the off position from a maximum lower position?

4. The brake in Figure 50–1 is not shown in the off position. Why?

5. Why is it necessary to open the disconnect switch when the machine is not in operation (Figure 50–2)?

6. Why may it be necessary to reverse the polarity of the series field of a compound motor if it is used in the dynamic braking circuit?

7. Referring to Figure 50–4(B), why is resistor R necessary?

8. In Figure 50–4(A), why is resistor R necessary?

9. Draw a schematic diagram of a circuit to reverse the polarity of the series field of a compound motor used in the automatic braking circuit when the stop button is pressed (see Figure 50–3 and question 6 above).

10. Referring to Figure 50–5, what keeps the time-delay coil (TD) energized after coil M is de-energized?

11. Design a shunt field circuit for Figure 50–2 to include
    a. a rheostat for motor speed control
    b. automatic removal of the rheostat when the stop button is pushed

UNIT 51

Electric and Electronic Braking

Objectives

After studying this unit, the student should be able to

- Describe the method of operation of a typical electric braking controller.

- List advantages of the use of an electric brake on an AC induction motor.

- Connect electric braking controllers.

- Recommend troubleshooting solutions for electric braking problems.

APPLICATION

As a powerful production machine slowly coasts to a stop, the operator is idle. Work is not being accomplished by either the operator or the machine. This slow stop may be cited as a safety hazard under the occupational safety laws. There is, however, a simple, economical, and effective way to eliminate this type of stop to improve the safety and efficiency of the equipment. An all-electric braking controller provides an effective brake for a standard drive, AC squirrel cage motor. It will stop AC motor-driven equipment as fast as it takes to reach full speed. Operation is smooth, noise-free, and automatic. This type of controller is designed to stop motors driving high-inertia loads such as roving or spinning frames. Electric brakes can be used with conveyors, machine tools, woodworking machines, textile machinery, rubber mills, and processing machinery.

Electric and electronic braking controllers, Figure 51–1, can be installed on new or existing machines. These controllers have several advantages over conventional brakes. For example, it is sometimes difficult to install conventional electromechanical brakes on existing equipment without extensive rebuilding. Mechanical brakes also require extensive mechanical maintenance, adjustment, and periodic replacement of worn parts.

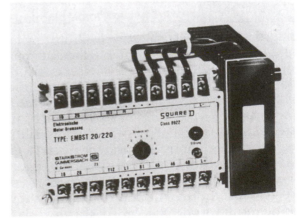

Fig. 51–1 Solid-state braking control for starting and braking. *(Courtesy of Schneider Electric)*

Fig. 51–2 Electronic motor brake. *(Courtesy Ambi-Tech Industries)*

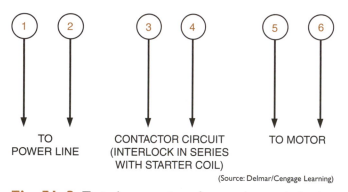

(Source: Delmar/Cengage Learning)

Fig. 51–3 Typical connections for an electronic brake shown in Figure 51–2.

Figure 51–2 shows an electronic braking control that can be added to an existing motor starter.

Typical connections for an electronic brake are shown in Figure 51–3.

OPERATION

The principle of electric braking can be applied to standard AC squirrel cage or wound rotor motors. When a motor is to be stopped, direct current is applied to one or all of the three phases of the motor after the AC voltage is removed. There is now a stationary DC field in the stator instead of an AC rotating field. As a result, the motor is braked quickly and smoothly to a standstill. An electric braking controller provides a smoother positive stop because the braking torque decreases rapidly as the speed approaches zero. Tapped resistors are sometimes used to adjust the braking torque. Controlled SCRs are also used.

Three-phase electric braking improves the efficiency of stopping and reduces the heat buildup in the motor as compared to single-phase braking.

One brake stop is equivalent to a normal start. This means that there must be a reduction in the allowable number of starts that can be made without overheating the motor.

CONSTRUCTION

A self-contained transformer and rectifier provide the direct current required for braking. An adjustable timer is used to provide dependable timing of the braking cycle. A standard reversing starter can be used to supply power to the motor: AC during normal operation and DC during braking.

Terminal blocks are available for motor and control connections. Taps on current-limiting resistors provide a means of adjusting the braking torque over a wide range. Controllers consisting of braking units only can be obtained for use with existing motor starters.

SEQUENCE OF OPERATION OF AN ELECTRIC BRAKING CIRCUIT

Figure 51–4(A) is the schematic diagram of a typical electric braking circuit. When motor starter M is energized, the motor operates as soon as the start button closes. The normally closed contact of the start button ensures that braking contactor B is de-energized before the motor is connected to the AC supply.

While M is energized, its normally closed interlock contact is open and its normally open

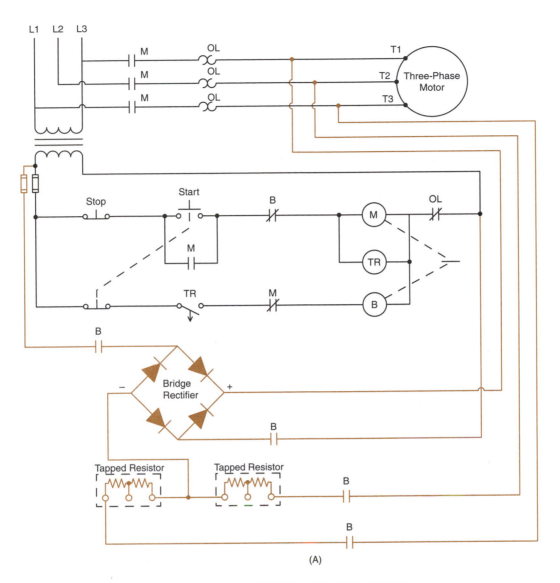

(A)

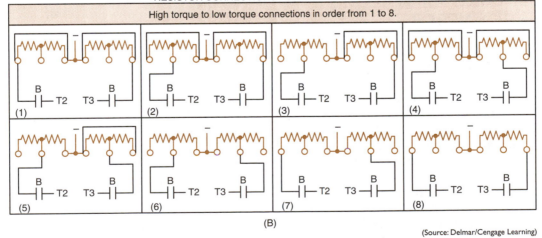

(B)

(Source: Delmar/Cengage Learning)

Fig. 51–4 Typical line diagram and resistor connections for an electric braking circuit.

time-delay contact is closed. When the stop button is pressed, M drops out. When interlock M recloses, braking contactor B is energized. B remains energized until the time-delay contact (M) times open (time delay after de-energized). This timing contact is adjustable and should be set so that contactor B drops out as soon as the motor comes to a complete stop.

Braking contactor B connects all three motor leads to a source of direct current through the rectifier and transformer. Direct current applied to an induction motor polarizes it to a stationary magnetic field and causes it to brake to a stop. Less current is required for the same braking torque on all three phases as compared with applying direct current to one phase. More ampere turns are gained with the additional windings, conserving energy, and reducing motor heating. The braking torque and the speed of braking can be varied by reconnecting the tapped resistors as shown in Figure 51–4(B).

DC and AC supplies must not be connected to a motor at the same time. The motor starter and braking contactor must have adequate interlocks. It is recommended that the motor starter always be included as an integral unit with the brake. If a braking unit only is applied to a separate motor starter, mechanical interlocking is sacrificed. In this case, the start button must be equipped with a normally closed contact. The motor starter must also have extra, normally closed interlock contacts.

ELECTRIC BRAKING FOR A WOUND ROTOR MOTOR

Figure 51–5 illustrates a method of dynamic braking for a wound rotor induction motor developed by the author. The circuit provides three steps of automatic starting for the motor and offers three different braking methods. The three braking options are coast stop, medium stop, and rapid stop. The resistors connected to the rotor circuit used to provide the steps of acceleration for the motor when starting are used to provide the medium braking option when the motor is stopped. As a general rule, the DC voltage applied to the stator winding is kept to about 10 percent of the nameplate voltage of the motor. A current limiting resistor has been added to the DC supply to further control the DC current supplied to the stator winding during braking. Braking current should not exceed the motor's starting current or overheat problems will occur.

The braking speeds for motors depend on motor ratings and the connected load. For example, a 1-horsepower motor with a heavy inertia load such as a flywheel or conveyor can be stopped in about 1 second if necessary. A 125-horsepower, slow-speed motor can be stopped in 2 or 3 seconds if necessary. The amount of stopping force is proportional the strength of the magnetic field developed in the stator winding by the DC supply voltage and the strength of the magnetic field developed in the rotor by the induced voltage. It is like connecting the north and south ends of two magnets together and trying to pull them apart. The force necessary to pull the magnets apart is determined by the magnetic field strength of the two magnets.

The starting sequence for the motor is as follows:

- When the start button is pressed, a circuit is completed to M coil, causing all M contacts to close. The three load contacts close and supply power to the stator winding of the motor. One M auxiliary contact seals the circuit around the start button and the second auxiliary contact provides a current path to timer coil TR1. At this point all the resistors are connected in the rotor circuit and the motor is operating in its lowest speed.
- At the end of TR1's time period, the normally open TR1 contact closes and provides power to the coil of S1 contactor. The two S1 load contacts close and shunt out the first set of resistors in the rotor circuit causing the motor to accelerate to second speed. The normally open S1 auxiliary contact closes and provides power to timer TR2.
- At the end of TR2's time period, the normally open TR2 contact closes and provides power to contactor coil S2. The two S2 load contacts close and short circuit the rotor circuit causing the motor to accelerate to its highest or third speed.

If the coast stop button is pressed, M starter is de-energized and the circuit returns to it normal position. The motor is permitted to coast to a stop because no DC power is applied to the stator winding.

To understand the circuit operation when the medium stop button is pressed, assume that the motor is running. The sequence is as follows:

- When the medium stop button is pressed, a circuit is supplied to B1 contactor.
- The two B1 load contacts close and shunt out the first set of resistors in the rotor circuit.

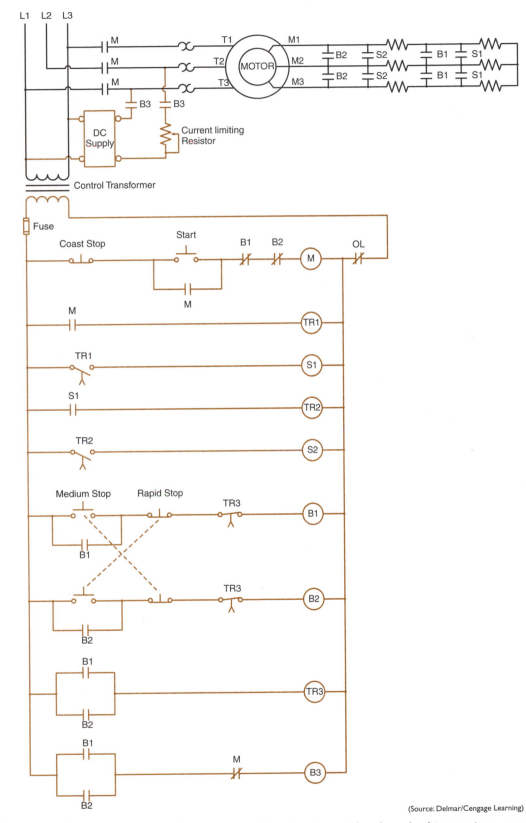

Fig. 51–5 Dynamic braking for a wound rotor motor. The circuit provides three braking options, coast stop, medium stop, and rapid stop.

(Source: Delmar/Cengage Learning)

■ All B1 auxiliary contacts close. One auxiliary contact is connected in parallel with the medium stop button to maintain power to the circuit. A second B1 auxiliary contact provides power to timer coil TR3. TR3 limits the amount of time that DC voltage can be applied to the stator winding. The third B1 auxiliary contact closes and provides power to contactor B3.

■ Contactor B3 closes the two B3 load contacts that supply DC voltage to the stator winding.

The resistors in the rotor circuit limit the amount of induced current in the rotor circuit and therefore, limit the strength of the magnetic field developed in the rotor. It should be noted that if less braking torque is desired, the two B1 load contacts can be disconnected from the circuit. This would permit all the rotor circuit resistors to limit rotor current and further weaken the magnetic field developed in the rotor.

The rapid stop option operates in the same manner as the medium stop option except that the load contacts controlled by contactor B2 directly short the rotor circuit causing maximum rotor current and therefore, maximum magnetic field strength to be developed in the rotor. Some other points concerning the circuit to be considered are:

■ Double acting push buttons permit the medium stop and rapid stop options to be selected after the other one has been selected.

■ Normally closed contacts controlled by contactors B1 and B2 will disconnect motor starter M from the line if either the medium stop or rapid stop option is selected.

■ A normally closed M contact connected in series with coil B3 provides interlock with the motor starter to insure that direct current and three-phase alternating current cannot be connected to the motor at the same time.

The sudden stopping of a motor by mechanical brakes can often harm the motor or load. This is especially true of gear driven loads. Dynamic braking, however, provides a very smooth braking action. Maximum braking force occurs at the beginning of the braking period because the rotor is turning at it greatest speed causing maximum induced current in the rotor. The braking torque will taper off as the motor speed decreases and becomes zero when the motor stops. Dynamic braking cannot be used to hold suspended loads such as in the case with a crane. Dynamic braking can easily be demonstrated by applying direct current to an AC induction motor and turning the motor shaft by hand.

Study Questions

1. Why does three-phase braking reduce motor heating as compared with single-phase braking?

2. Differentiate between the AC and DC magnetic fields applied to an induction motor stator.

3. Electric braking can be applied to what type of AC motor?

4. Why is electric braking smooth and resilient?

5. Braking current is passed through overload heaters; however, the overload control circuit contacts are in the AC starting circuit. How does this protect the motor?

6. How often can a motor be stopped using electric braking?

7. Explain why there is no danger of the motor reversing after it is braked to a stop.

Motor Drives

UNIT 52

Direct Drives and Pulley Drives

Objectives

After studying this unit, the student should be able to

- State the advantages of direct drives and pulley drives.
- Install directly coupled motor drives and pulley motor drives.
- Check the alignment of the motor and machine shafts, both visually and with a dial indicator.
- Install motors and machines in the proper positions for maximum efficiency.
- Calculate pulley sizes using the equation

$$\frac{\text{Drive revolutions per minute}}{\text{Driven revolutions per minute}} = \frac{\text{Driven Pulley Diameter}}{\text{Drive Pulley Diameter}}$$

DIRECTLY COUPLED DRIVE INSTALLATION

The most economical speed for an electrical motor is about 1800 revolutions per minute. Most electrically driven constant-speed machines, however, operate at speeds below 1800 rpm. These machines must be provided with either a high-speed motor and some form of mechanical speed reducer, or a low-speed, directly coupled motor. Electronic controlled drives are available, but the expense sometimes exceeds the need.

Synchronous motors can be adapted for direct coupling to machines operating at speeds from 3600 rpm to about 80 rpm with horsepower ratings ranging from 20 to 5000 and above. It has been suggested that synchronous motors are less expensive to install than squirrel cage motors if the rating exceeds 1 hp. However, this recommendation considers only the first cost. It does not take into account the higher efficiency and better power factor of the synchronous motor. When the motor speed matches the machine input shaft speed, a simple mechanical coupling is used, preferably a flexible coupling.

Trouble-free operation can usually be obtained by following several basic recommendations for the installation of directly coupled drives and pulley, or chain, drives. First, the motor and machine must be installed in a level position. When connecting the motor to its load, the alignment of the devices must be checked more than once from positions at right angles to each other. For example, when viewed from the side, two shafts may appear to be in line. When the

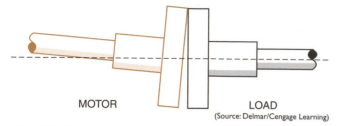

MOTOR   LOAD
(Source: Delmar/Cengage Learning)

Fig. 52–1 Every alignment check must be made from positioning 90 degrees apart, or at right angles to each other.

same shafts are viewed from the top, as shown in Figure 52–1, it is evident that the motor shaft is at an angle to the other shaft. A dial indicator should be used to check the alignment of the motor and the driven machinery, Figure 52–2. If a dial indicator is not available, a feeler gauge may be used.

During the installation, the shafts of the motor and the driven machine must be checked to ensure that they are not bent. Both machine and motor should be rotated together, just as they rotate when the machine is running, and then rechecked for alignment. After the angle of the shafts is aligned, the shafts may appear to share the same axis. However, as shown in Figure 52–3, the axes of the motor and the driven machine may really be off center. When viewed again from a position 90 degrees away

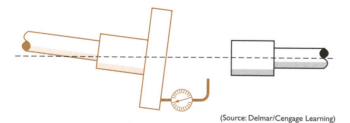

(Source: Delmar/Cengage Learning)

Fig. 52–2 Angular check of direct motor couplings.

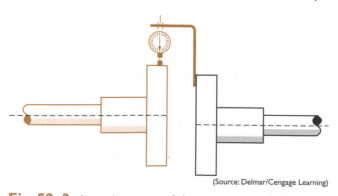

(Source: Delmar/Cengage Learning)

Fig. 52–3 Axis alignment of direct motor couplings.

from the original position, it can be seen that the shafts are not on the same axis.

To complete the alignment of the devices, the motor should be moved until rotation of both shafts shows that they share the same axis when viewed from four positions spaced 90 degrees apart around the shafts. The final test is to check the starting and running currents with the connected load to ensure that they do not exceed specifications.

There are several disadvantages to the use of low-speed, directly coupled induction motors. They usually have a low power factor and low efficiency. Both of these characteristics increase electric power costs. Because of this, induction motors are rarely used for operation at speeds below 500 rpm.

Constant-speed motors are available with a variety of speed ratings. The highest possible speed is generally selected to reduce the size, weight, and cost of the motor. At 5 horsepower, a 1200-rpm motor is almost 50 percent larger than an 1800-rpm motor. At 600 rpm the motor is well over twice as large as the 1800-rpm motor. In the range from 1200 rpm to 900 rpm, the size and cost disadvantages may not be overwhelming factors. Where this is true, low-speed, directly coupled motors can be used. For example, this type of motor is used on most fans, pumps, and compressors.

Laser-Computer Method of Equipment Alignment

Laser light is coherent; that is, it has a single frequency with constant amplitude and wavelength, which gives it precision in many applications. Laser light is directional, traveling in a straight line, rather than radiating outward. Construction jobs where a great deal of alignment is necessary, such as installing conveyor systems, are usually done by the conveyor manufacturer.

The laser beam method is very accurate when performed by qualified people. Once the laser beam system is targeted and set up, it requires periodic checks to make sure it has not been bumped out of position.

Prisms can be used to bend the laser light beam at precise 90-degree angles to ease alignment of components perpendicular to the longitudinal plane. Multiprisms can be used to "see" around obstructions. On some jobs, measurement data are entered into a computer program that models the system. The ways in which a laser can be used for aligning systems are limited only by the ingenuity of the users.

PULLEY DRIVES

Installation

Flat belts, V-belts, chains, or gears are used on motors so that smooth speed changes at a constant rpm can be achieved. For speeds below 900 rpm, it is practical to use an 1800-rpm or 1200-rpm motor connected to the driven machine by a V-belt or a flat belt.

Machine shafts and bearings give long service when the power transmission devices are properly installed according to the manufacturer's instructions.

Offset drives, such as V-belts, gears, and chain drives, can be lined up more easily than direct drives. Both the motor and load shafts must be level. A straightedge can be used to ensure that the motor is aligned on its axis and that it is at the proper angular position so that the pulley sheaves of the motor and the load are in line, Figure 52–4. When belts are installed, they should be tightened just enough to ensure non-slip engagement. The less cross tension there is, the less wear there will be on the bearings involved. Proper and firm positioning and alignment are necessary to control the forces that cause vibration and the forces that cause thrust.

The designer of a driven machine usually determines the motor mount and the type of drive to be used. This means the installer has little choice in the motor location. In many flat belt or V-belt applications, however, the construction or maintenance electrician may be called on to make several choices. If a choice can be made, the motor should be placed where the force of gravity helps to increase the grip of the belts. A vertical drive can cause problems because gravity tends to pull the belts away from the lower sheave, Figure 52–5. To counteract this action, the belts require far more tension than the bearings should have to withstand. The electrician should avoid this type of installation if possible.

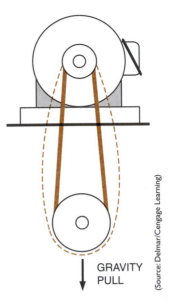

GRAVITY
PULL

(Source: Delmar/Cengage Learning)

Fig. 52–5 Greater belt tension is required in vertical drives where gravity opposes good belt traction. This greater tension generally exceeds the tension the bearings should withstand.

There are correct and incorrect placements for horizontal drives. The location where the motor is to be installed can be determined from the direction of rotation of the motor shaft. It is recommended that the motor be placed with the direction of rotation so that the belt slack is on top. In this position, the belt tends to wrap around the sheaves. This problem is less acute with V-belts than it is with flat belts or chains. Therefore, if rotation is to take place in both directions, V-belts should be used. In addition, the motor should be placed on the side of the most frequent direction of rotation, Figure 52–6.

Pulley Speeds

Motors and machines are frequently shipped without pulleys or with pulleys of incorrect sizes. The drive and driven speeds are given on the

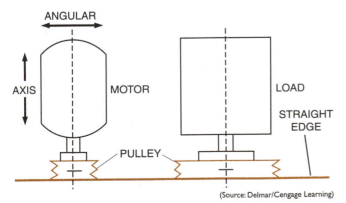

ANGULAR

AXIS MOTOR

LOAD

STRAIGHT EDGE

PULLEY

(Source: Delmar/Cengage Learning)

Fig. 52–4 Using straightedge for angular and axis alignment.

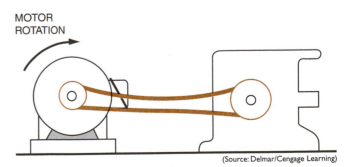

MOTOR ROTATION

(Source: Delmar/Cengage Learning)

Fig. 52–6 Direction of motor rotation can be used to advantage for good belt traction (with slack at top) if the motor is placed in the proper position.

motor and machine nameplates or in the descriptive literature accompanying the machines.

Four quantities must be known if the machinery is to be set up with the correct pulley sizes: the drive revolutions per minute, the driven revolutions per minute, the diameter of the drive pulley, and the diameter of the driven pulley. If three of these quantities are known, the fourth quantity can be determined. For example, if a motor runs at 3600 rpm, the driven speed is 400 rpm, and there is a 4-inch pulley for the motor, the size of the pulley for the driven load can be determined from the following equation:

$$\frac{\text{Drive rpm}}{\text{Driven rpm}} = \frac{\text{Driven Pulley Diameter}}{\text{Drive Pulley Diameter}}$$

$$\frac{3600}{400} = \frac{X}{4} \quad \text{or} \quad \frac{3600}{400} = \frac{X}{4}$$

By cross multiplying and then dividing, we arrive at the pulley size required:

$$4X = 144$$

$$X = \frac{144}{4} = \text{36-inch diameter pulley}$$

If both the drive and driven pulleys are missing, the problem can be solved by estimating a reasonable pulley diameter for either pulley and then using the equation with this value to find the fourth quantity.

Study Questions

1. What are the disadvantages of low-speed, directly coupled induction motors?

2. What type of AC motor can be directly coupled at low rpm with larger horsepower ratings?

3. What three alignment checks should be made to ensure satisfactory and long service for directly coupled and belt-coupled power transmissions?

4. What special tools, not ordinarily carried by an electrician, are required to align a directly coupled, motor-generator set?

5. Induction motors should not be used below a certain speed rpm. What is this speed?

6. What is the primary reason for using a pulley drive?

7. How tightly should V-belts be adjusted?

8. Refer to Figure 52–6 and assume the motor is to rotate in the opposite direction. How can the belt slack be maintained on top?

9. A machine delivered for installation has a 2-inch pulley on the motor and a 6-inch pulley on the load. The motor nameplate reads 180 rpm. At what speed in rpm will the driven machine rotate?

Gear Motors

Objectives

After studying this unit, the student should be able to

- Describe the basic operation of a gear motor.
- State where gear motors are used generally.
- Describe the three standard gear classifications.
- Select the proper gears for the expected load characteristics.

Many industrial machines require power at slow speeds and high torque. Conveyors and concrete mixers are typical examples of machinery with such requirements. At speeds of 780 rpm and less, the following drives may be prudently used: chain drive, belt drive, separate speed reducer coupled to the motor, or gear motor.

The gear motor is a speed-reducing motor that gives a direct power drive from a single unit, Figure 53–1. The gear motor provides an extremely compact, efficient, packaged power drive. A gear motor usually consists of a standard AC or DC motor and a sealed gear train correctly engineered for the load. This assembly is mounted on a single base as a one-package, enclosed power drive. The advantage of this unit is its extreme compactness. A gear motor actually is smaller than a low-speed standard motor of the same horsepower.

The motor-shaft pinion of the gear motor drives the gear or series of gears in an oil bath that

Fig. 53–1 Cutaway view of a speed reducing cycloidal gear motor. Cycloidal gear boxes use a concentric cam and rollers instead of conventional gears. *(Courtesy Sumitomo Machinery Corporation of America)*

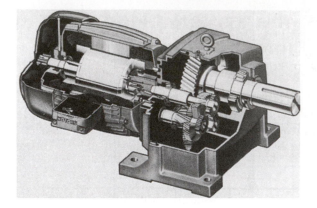

Fig. 53–2 Cutaway view of a speed reducing gear motor (double reduction). (*Courtesy of Emerson Motors*)

is linked with the output shaft, Figure 53–2. This type of arrangement is usually the most economical and convenient way to obtain low speeds of approximately 1 rpm to 780 rpm.

One-unit gear motors are available with the following options: (1) shafts parallel to each other or at right angles, (2) polyphase, single-phase, or DC voltages, and (3) horsepower ratings ranging from approximately ⅛hp to 200 hp.

Gear motors are available in many of the standard motor types such as squirrel cage or wound rotor induction motors, running at either constant or adjustable speeds. The control equipment for the motor is selected the same as for any other motor of the same type.

When selecting a gear motor, an important consideration is the degree of gear service and gear life based on the load conditions to which the motor will be subjected. Gear motors are divided into three classes. Each class uses different gear sizes to handle specific load conditions. Each class gives about the same life for the gears. The American Gear Manufacturer's Association has defined three operating conditions commonly found in industrial service and has established three standard gear classifications to meet these conditions:

Class I For steady loads within the motor rating of 8 hours per day duration, or for intermittent operation under moderate shock conditions.

Class II For 24-hour operation at steady loads within the motor rating, or 8-hour operation under moderate shock conditions.

Class III For 24-hour operation under moderate shock conditions, or 8-hour operation under heavy shock conditions.

For conditions that are more severe than those covered by Class III gears, a fluid drive unit may be incorporated in the assembly to cushion the shock to an acceptable value.

To achieve multiple speeds, separate units are available with a transmission comparable to that of an automobile. These units must be assembled with the motor and the driven machine. Because the amount of power lost in gearing is very small, the multiple drive has essentially constant horsepower. In other words, as the output speed is decreased, the torque is increased. Generally, this means that larger shaft sizes are needed for the output side.

Study Questions

1. Why do many industrial machines use gear motors instead of low rpm induction motors?

2. What is the principal difference in gear motor classifications?

3. In general, what is the maximum speed rpm of gear motors?

4. What is the classification of a gear motor that is subjected to continual reversing during an 8-hour production shift?

5. Why is the output shaft on a gear reduction motor larger than the shaft of a standard motor?

Variable-Frequency Mechanical Drives

Objectives

After studying this unit, the student should be able to

- Describe the operating principle of variable-frequency drives.
- State why variable-frequency drives are used.
- State the advantage of variable-frequency drives.

The speed of an AC squirrel cage induction motor depends on the frequency of the supply current and the number of poles of the motor:

$$\text{RPM} = \frac{60 \times \text{Hertz}}{\text{Pairs of Poles}}$$

A frequency changer may be used to vary the speed of this type of motor.

Several methods can be used for one or more motors to obtain variable frequency from a fixed-frequency power source. A common method is to drive an alternator through an adjustable mechanical speed drive. The voltage is regulated automatically during frequency changes. In this system, an AC motor drives a variable cone pulley, or sheave, which is belted to another variable pulley on the output shaft, Figure 54–1. When the relative diameters of the two pulleys are changed, the speed between the input and the output can be controlled.

As the alternator speed is varied, the frequency varies. In cases where more than one motor is involved, such as in conveyor systems, all of the motors are connected electrically to the power system of the alternator. These motors are generally called slave motors. Each motor thus receives the same frequency and operates at the same speed. If a high-speed motor and a slow-speed gear motor are connected to the same variable-frequency circuit, they will change speeds simultaneously and proportionally, Figure 54–2. To obtain smooth acceleration of the motors attached to the alternator, the frequency is increased. Electric braking with suitable controls can be used in this system.

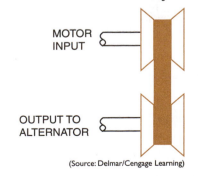

MOTOR INPUT

OUTPUT TO ALTERNATOR

(Source: Delmar/Cengage Learning)

Fig. 54–1 Variable pitch pulleys; mechanical method of obtaining continuously adjustable speed from constant speed shaft.

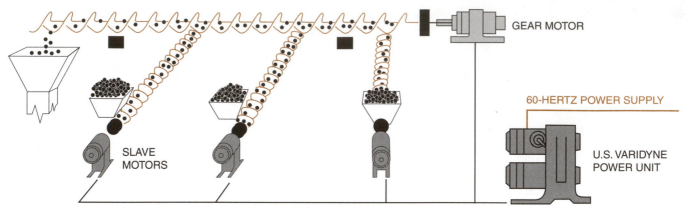

(Source: Delmar/Cengage Learning)

Fig. 54–2 Screw conveyors may be controlled individually by variable drive slave motors. Once the motors are set, they rotate at the same proportional rpm, controlled by the variable-frequency power unit supplying them.

Fig. 54–3 Variable power unit used to control one or more motors at coordinated speeds. *(Courtesy of Emerson Motors)*

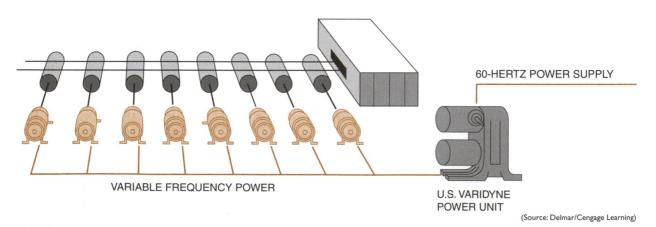

(Source: Delmar/Cengage Learning)

Fig. 54–4 Coordinating many drive motors, all of standard AC, squirrel cage construction.

Variable-frequency power drives are used on many different types of conveyor systems, Figure 54–3. Assembly lines and coordinated screw conveyors for the proportional mixing of ingredients make use of these power drives, Figure 54–2 and Figure 54–4. Variable pitch systems are among the lowest-cost methods of achieving variable speed, both in initial cost and maintenance. Remote control is not an inherent feature. Because the drive uses mechanical means to vary the speed, electrical control signals must be adapted to existing mechanical controls.

Study Questions

1. What is the formula for finding the speed of an AC squirrel cage induction motor?

2. If several motors driving an overhead conveyor system are connected to a variable-frequency supply line and the frequency is changed from 40 hertz to 80 hertz, what effect does this have on the speed of the motors?

3. How is the AC frequency controlled in an alternator?

4. If many feeder conveyor motors of different poles (speed) are connected to a main conveyor and to a variable-frequency power unit, can increased production be achieved by increasing the frequency of the power unit? Explain your answer.

5. What is a *slave motor*?

UNIT 55

AC Adjustable Frequency Drives

Objectives

After studying this unit, the student should be able to

- List factors that determine the speed of the rotating magnetic field.

- Determine synchronous speed for different numbers of poles and frequency.

- Discuss how direct current is converted to alternating current.

- Discuss the characteristics of insulated gate bipolar transistors.

- Discuss how SCRs can be used to change DC voltage into AC voltage.

- List differences between inverter-rated motors and noninverter-rated motors.

Although DC motors are still used in many industries, they are being replaced by variable-frequency drives controlling squirrel cage induction motors. The advantage of a DC motor compared with an AC motor is that the speed of the DC motor can be controlled. Although the wound rotor motor does permit some degree of speed control, it does not have the torque characteristics of a DC motor. DC motors can develop maximum torque at zero rpm. Variable-frequency drives can give the same speed and torque characteristics to AC squirrel cage induction motors. A variable-frequency drive and AC squirrel cage motor are less expensive to purchase than a comparable DC drive and DC motor. Variable-frequency drives and AC motors have less downtime and maintenance problems than DC drives and DC motors.

VARIABLE-FREQUENCY DRIVE OPERATING PRINCIPLES

The operating principle of all AC three-phase motors is the rotating magnetic field. The speed of the rotating field is called *synchronous* speed and is controlled by two factors:

- The number of stator poles per phase
- The frequency of the applied voltage

The chart shown in Figure 55–1 lists the synchronous speed for different numbers of poles

POLES	FREQUENCY IN HZ					
	60	50	40	30	20	10
2	3600	3000	2400	1800	1200	600
4	1800	1500	1200	900	600	300
6	1200	1000	800	600	400	200
8	900	750	600	450	300	150

(Source: Delmar/Cengage Learning)

Fig. 55–1 Speed in rpm for motors with different numbers of poles per phase at different frequencies.

at different frequencies. Variable-frequency drives control motor speed by controlling the frequency of the power supplied to the motor.

Basic Construction of a Variable-Frequency Drive

Most variable-frequency drives operate by first changing the AC voltage into DC and then changing it back to AC at the desired frequency. There are several methods used to change the DC voltage back into AC. The method employed is determined by the manufacturer, age of the equipment, and the size motor that the drive must control. Variable-frequency drives intended to control the speed of motors up to 500 hp generally employ transistors. In the circuit shown in Figure 55–2, a single-phase bridge changes the alternating current into direct current. The bridge rectifier uses two SCRs and two diodes. The SCRs permit the output voltage of the rectifier to be controlled. As the frequency decreases, the SCRs fire later in the cycle and lower the output voltage to the transistors. As the frequency of the voltage is reduced, the inductive reactance of the motor stator winding is also reduced. The voltage applied to the motor must also be reduced to prevent the motor from being damaged by excessive current. A choke coil and capacitor

bank are used to filter the output voltage before transistors Q1 through Q6 change the DC voltage back into AC. An electronic control unit is connected to the bases of transistor Q1 through Q6. The control unit converts the DC voltage back into three-phase alternating current by turning transistors on or off at the proper time and in the proper sequence. Assume, for example, that transistors Q1 and Q4 are switched on at the same time. This permits stator winding T1 to be connected to a positive voltage and T2 to be connected to a negative voltage. Current can flow through Q4 to T2, through the motor stator winding and through T1 to Q1.

Now assume that transistors Q1 and Q4 are switched off and transistors Q3 and Q6 are switched on. Current will now flow through Q6 to stator winding T3, through the motor to T2, and through Q3 to the positive of the power supply.

Because the transistors are turned completely on or completely off, the waveform produced is a square wave instead of a sine wave, Figure 55–3. Induction motors will operate on a square wave without a great deal of problems. Some manufacturers design units that will produce a stepped waveform as shown in Figure 55–4. The stepped waveform is used because it closely approximates a sine wave.

Some Related Problems

The circuit illustrated in Figure 55–2 uses SCRs in the power supply and junction transistors in the output stage. SCR power supplies control the output voltage by chopping the incoming waveform. This can cause harmonics on the line that cause overheating of transformers and motors and can cause fuses to blow and circuit breakers to trip. When bipolar junction transistors are employed as switches, they are generally

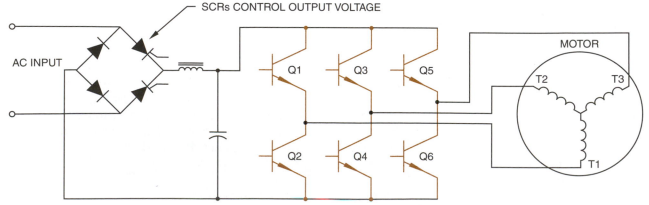

(Source: Delmar/Cengage Learning)

Fig. 55–2 Basic schematic of a variable-frequency drive.

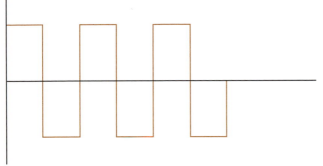

(Source: Delmar/Cengage Learning)

Fig. 55–3 Square wave voltage waveform.

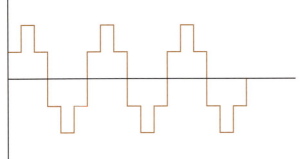

(Source: Delmar/Cengage Learning)

Fig. 55–4 A stepped waveform approximates a sine wave.

(Source: Delmar/Cengage Learning)

Fig. 55–5 Schematic symbol for an insulated gate bipolar transistor.

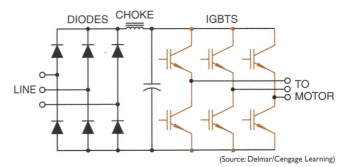

(Source: Delmar/Cengage Learning)

Fig. 55–6 Variable-frequency drives using IGBTs generally use diodes in the rectifier instead of SCRs.

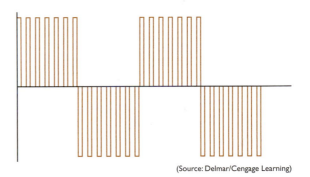

(Source: Delmar/Cengage Learning)

Fig. 55–7 Pulse width modulation is accomplished by turning the voltage on and off several times during each half cycle.

driven into saturation by supplying them with an excessive amount of base-emitter current. Saturating the transistor causes the collector-emitter voltage to drop to between 0.04 and 0.03 volt. This small voltage drop allows the transistor to control large amounts of current without being destroyed. When a transistor is driven into saturation, however, it cannot recover or turn off as quickly as normal. This greatly limits the frequency response of the transistor.

Insulated Gate Bipolar Transistor

Many transistor-controlled variable drives now employ a special type of transistor called an insulated gate bipolar transistor (IGBT). IGBTs have an insulated gate very similar to some types of field effect transistors (FETs). Because the gate is insulated, it has a very high impedance. The IGBT is a voltage-controlled device, not a current-controlled device. This gives it the ability to turn off very quickly. IGBTs can be driven into saturation to provide a very low voltage drop between the emitter and collector, but they do not suffer from the slow recovery time of common junction transistors. The schematic symbol for an IGBT is shown in Figure 55–5.

Drives using IGBTs generally use diodes to rectify the AC voltage into DC, not SCRs, Figure 55–6. The three-phase rectifier supplies a constant DC voltage to the transistors. The output voltage to the motor is controlled by pulse width modulation (PWM). PWM is accomplished by turning the transistor on and off several times during each half cycle, Figure 55–7. The output voltage is an average of the peak or maximum voltage and the amount of time the transistor is turned on or off. Assume that 480 volts, three-phase alternating current is rectified to direct current and filtered. The DC voltage applied to the IGBTs is approximately 630 volts. The output voltage to the motor is controlled by the switching of the transistors. Assume that the transistor is on for 10 microseconds and off for 20 microseconds. In this example the transistor is on for one-third of the time and off for two-thirds of the time. The voltage applied to the motor would be 210 volts (630/3). The speed at

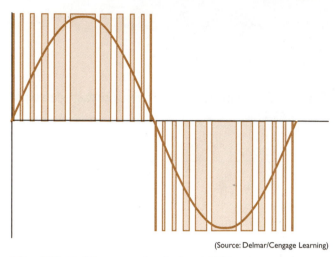

Fig. 55–8 The speed of the IGBT can produce a stepped wave that is similar to a sine wave.

(Source: Delmar/Cengage Learning)

which IGBTs can operate permits pulse width modulation to produce a stepped wave that is very similar to a standard sine wave, Figure 55–8.

Advantages and Disadvantages of IGBT Drives

A great advantage of drives using IGBTs is that SCRs are generally not used in the power supply, and this greatly reduces problems with line harmonics. The greatest disadvantage is that the fast switching rate of the transistors can cause voltage spikes in the range of 1600 volts to be applied to the motor. These voltage spikes can destroy some motors. Line length from the drive to the motor is of great concern with drives using IGBTs. The shorter the line length the better.

Inverter-Rated Motors

Due to the problem of excessive voltage spikes caused by IGBT drives, some manufacturers produce a motor that is "inverter rated." These motors are specifically designed to be operated by variable-frequency drives. They differ from standard motors in several ways:

1. Many inverter rater motors contain a separate blower to provide continuous cooling for the motor regardless of the speed. Many motors use a fan connected to the motor shaft to help draw air through the motor. When the motor speed is reduced, the fan cannot maintain sufficient airflow to cool the motor.

2. Inverter-rated motors generally have insulating paper between the windings and the stator core, Figure 55–9. The high-voltage

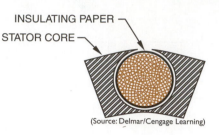

Fig. 55–9 Insulating paper is between the windings and the stator frame.

spikes produce high currents that produce a high magnetic field. This increased magnetic field causes the motor windings to move. This movement can eventually cause the insulation to wear off the wire and produce a grounded motor winding.

3. Inverter-rated motors generally have phase paper added to the terminal leads. Phase paper is insulating paper added to the terminal leads that exit the motor. The high-voltage spikes affect the beginning lead of a coil much more than the wire inside the coil. The coil is an inductor that naturally opposes a change of current. Most of the insulation stress caused by high-voltage spikes occurs at the beginning of a winding.

4. The magnet wire used in the construction of the motor windings has a higher rated insulation than other motors.

5. The case size is larger than most three-phase motors because of the added insulating paper between the windings and the stator core. Also, a larger case size helps cool the motor by providing a larger surface area for the dissipation of heat.

Variable-Frequency Drives Using SCRs and GTOs

Variable-frequency drives intended to control motors over 500 hp generally use SCRs or gate turn-off devices (GTOs). GTOs are similar to SCRs except that conduction through the GTO can be stopped by applying a negative voltage—negative with respect to the cathode—to the gate. SCRs and GTOs are thyristors and have the ability to handle a greater amount of current than transistors. An example of a single-phase circuit used to convert DC voltage to AC voltage with SCRs is shown in Figure 55–10. In this circuit, the SCRs are connected to a control unit that controls the sequence and rate at which the SCRs are gated on. The circuit is constructed so that SCRs A and A′ are gated on at the same time and SCRs

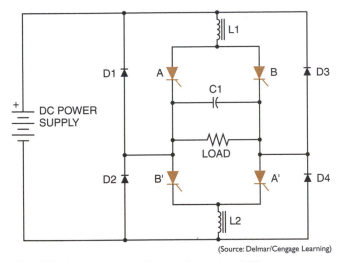

Fig. 55–10 Changing DC to AC using SCRs.

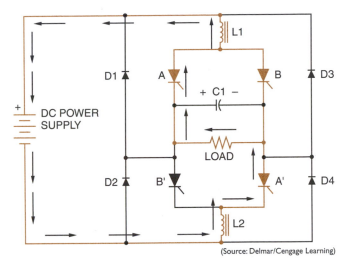

Fig. 55–11 Current flows through SCRs A and A′.

B and B′ are gated on at the same time. Inductors L1 and L2 are used for filtering and wave shaping. Diodes D1 through D4 are clamping diodes and are used to prevent the output voltage from becoming excessive. Capacitor C1 is used to turn one set of SCRs off when the other set is gated on. This capacitor must be a true AC capacitor because it will be charged to the alternate polarity each half cycle. In a converter intended to handle large amounts of power, capacitor C1 will be a bank of capacitors. To understand the operation of the circuit, assume that SCRs A and A′ are gated on at the same time. Current will flow through the circuit as shown in Figure 55–11. Notice the direction of current flow through the load and that capacitor C1 has been charged to the polarity shown. When an SCR is gated on, it can only be turned off by permitting the current flow through the anode-cathode section to drop below a certain level called the holding current level. The SCR will not turn off as long as the current continues to flow through the anode-cathode.

Now assume that SCRs B and B′ are turned on. Because SCRs A and A′ are still turned on, two current paths now exist through the circuit. The positive charge on capacitor C1, however, causes the negative electrons to see an easier path. The current will rush to charge the capacitor to the opposite polarity, stopping the current flowing through SCRs A and A′ permitting them to turn off. The current now flows through SCRs B and B′ and charges the capacitor to the opposite polarity, Figure 55–12. Notice that the current now flows through the load in the opposite direction, which produces alternating current across the load.

To produce the next half cycle of alternating current, SCRs A and A′ are gated on again. The positively charged side of the capacitor will now cause the current to stop flowing through SCRs B and B′, permitting them to turn off. The current again flows through the load in the direction indicated in Figure 55–11. The frequency of the circuit is determined by the rate at which the SCRs are gated on.

Features of Variable-Frequency Control

Although the primary purpose of a variable-frequency drive is to provide speed control for an AC motor, most drives provide functions that other types of controls do not. Many variable-frequency drives can provide the low-speed torque characteristic that is so desirable in DC motors. It is this feature that permits AC squirrel cage motors to replace DC motors for many applications.

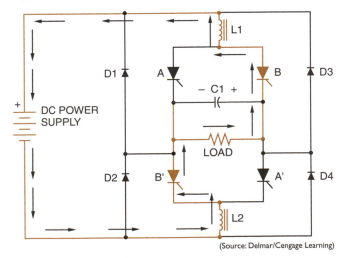

Fig. 55–12 Current flows through SCRs B and B′.

(Source: Delmar/Cengage Learning)

Fig. 55–13 Adjustments are made with potentiometers on some units.

Many variable-frequency drives also provide current limit and automatic speed regulation for the motor. Current limit is generally accomplished by connecting current transformers to the input of the drive and sensing the increase in current as load is added. Speed regulation is accomplished by sensing the speed of the motor and feeding this information back to the drive.

Another feature of variable-frequency drives is acceleration and deceleration control, sometimes called *"ramping."* Ramping is used to accelerate or decelerate a motor over some period of time. Ramping permits the motor to bring the load up to speed slowly as opposed to simply connecting the motor directly to the line. Even if the speed control is set in the maximum position when the start button is pressed, ramping permits the motor to accelerate the load from zero to its maximum rpm over several seconds. This feature can be a real advantage for some types of loads, especially gear drive loads. In some units the amount of acceleration and deceleration time can be adjusted by setting potentiometers on the main control board, Figure 55–13. Other units are completely controlled digitally and the amount of acceleration and deceleration time is programmed into the computer memory.

Some other adjustments that can usually be set by changing potentiometers or programming the unit are as follows.

Current Limit: This control sets the maximum amount of current the drive is permitted to deliver to the motor.

Volts per Hertz: This sets the ratio by which the voltage increases as frequency increases or decreases as frequency decreases.

Maximum Hz: This control sets the maximum speed of the motor. Most motors are intended to operate between 0 and 60 Hz, but some drives permit the output frequency to be set above 60 Hz, which would permit the motor to operate at higher than normal speed. The maximum hertz control can also be set to limit the output frequency to a value less than 60 Hz, which would limit the motor speed to a value less than normal.

Minimum Hz: This sets the minimum speed the motor is permitted to run.

Some variable-frequency drives permit adjustment of current limit, maximum and minimum speed, ramping time, and so on by adjustment of trim resistors located on the main control board. Other drives employ a microprocessor as the controller. The values of current limit, speed, ramping time, and so on for these drives are programmed into the unit, are much easier to make, and are generally more accurate than adjusting trim resistors. A variable-frequency drive that employs IGBTs is shown in Figure 55–14.

(Source: Delmar/Cengage Learning)

Fig. 55–14 Variable-frequency drive.

Study Questions

1. What is the operating principle of all three-phase motors?

2. What factors determine synchronous speed?

3. As the frequency is reduced, what must be done to prevent excessive motor current?

4. What is the advantage of an insulated gate bipolar transistor over a common junction transistor?

5. Explain pulse width modulation.

6. What is the difference between an SCR and a GTO?

7. What type of motor should be used with variable-frequency drives that use IGBTs?

8. Explain ramping.

Magnetic Clutch and Magnetic Drive

Objectives

After studying this unit, the student should be able to

- State several advantages of the use of a clutch in a drive.
- Describe the operating principles of magnetic clutches and drives.
- Distinguish between single- and multiple-face clutches.
- Connect magnetic clutch and magnetic drive controls.
- Recommend troubleshooting solutions for magnetic clutch and drive problems.

ELECTRICALLY CONTROLLED MAGNETIC CLUTCHES

Machinery clutches were originally designed to engage very large motors to their loads after the motors had reached running speeds, Figure 56–1. Clutches provide smooth starts for operations in which the material being processed might be damaged by abrupt starts, Figure 56–2. Clutches are also used to start high-inertia loads, because the starting may be difficult for a motor that is sized to handle the running load. When starting conditions are severe, a clutch inserted between the motor and the load means that the motor can run within its load capacity. The motor will take longer to bring the load up to speed, but the motor and load will not be damaged.

As more automatic cycling and faster cycling rates are being required in industrial production,

electrically controlled clutches are being used more often.

Single-Face Clutch

The single-face clutch consists of two discs: one is the field member (electromagnet) and the other is the armature member. The operation of the clutch is similar to that of the electromagnet in a motor starter, Figure 56–3. The principle of operation is shown in Figure 56–3. However, this is done more frequently electronically using solid-state devices. SCRs are gated to regulate the current flow as in Part (A) and to switch power as in Part (B). When current is applied to the field winding disc through collector (slip) rings, the two discs are drawn together magnetically. The friction face of the field disc is held tightly against the armature disc to provide positive engagement between the rotating drives. When the current is

(Source: Delmar/Cengage Learning)

Fig. 56–1 Magnetic clutches in cement mill service. Note the slip rings for clutch supply.

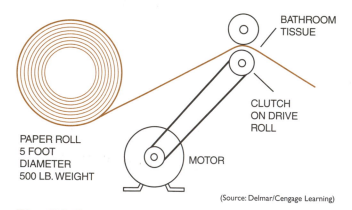

(Source: Delmar/Cengage Learning)

Fig. 56–2 To prevent tearing, a cushioned start is required on a drive roll that winds bathroom tissue off a large roll. Roll is 5 feet in diameter and weighs 500 pounds when full. Pickup of thin tissue must be very gradual to avoid tearing. Application also can be used for filmstrip processing machine.

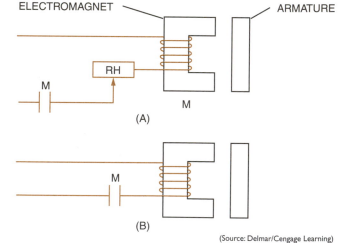

(Source: Delmar/Cengage Learning)

Fig. 56–3 Principle of operation of electrically controlled clutches. (A) Gradual clutch engagement; (B) more rapid clutch engagement.

removed, a spring action separates the faces to provide a definite clearance between the discs. In this manner, the motor is mechanically disconnected from the load.

Multiple-Face Clutch

Multiple-face clutches are also available. In a double-face clutch, both the armature and field discs are mounted on a single hub with a double-faced friction lining supported between them. When the magnet of the field member is energized, the armature and field members are drawn together. They grip the lining between them to provide the driving torque. When the magnet is de-energized, a spring separates the two members and they rotate independently of each other. Double-face clutches are available in sizes up to 78 inches in diameter.

A water-cooled magnetic clutch is available for applications that require a high degree of slippage between the input and output rotating members. Uses for this type of clutch include tension control (windup and payoff) and cycling (starting and stopping) operations in which large differences between the input and the

Fig. 56—4 Electric brake and electric clutch modules. Brake is shown on the center left on driven machine side of the clutch. *(Courtesy of Warner Electric)*

output speeds are required. Flowing water removes the heat generated by the continued slippage within the clutch. A rotary water union mounted in the end of the rotor shaft means that the water-cooled clutch cannot be end coupled directly to the prime mover. Chains or gears must be used.

A combination clutch and magnetic brake disconnects the load from the drive and simultaneously applies a brake to the load side of the drive. Magnetic clutches and brakes are often used as mechanical power-switching devices in module form. Figure 56–4 shows a driveline with an electric clutch (center right) and an electric brake. An application of this arrangement is shown in Figure 56–5. Remember that the quicker the start or stop, the shorter the life of this equipment.

Magnetic clutches are used on automatic machines for starting, running, cycling, and torque limiting. The combinations and variations of these functions are practically limitless.

MAGNETIC DRIVES

The magnetic drive couples the motor to the load magnetically. The magnetic drive can be used as a clutch and can be adapted to an adjustable speed drive.

The electromagnetic (or eddy current) coupling is one of the simpler ways to obtain an adjustable output speed from the constant input speed of squirrel cage motors.

There is no mechanical contact between the rotating members of the magnetic drive. Thus,

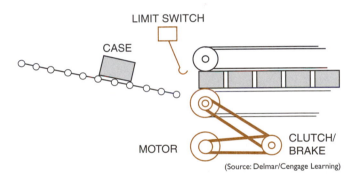

(Source: Delmar/Cengage Learning)

Fig. 56—5 Case sealer is used to hold the top of the carton while the glue is drying. In the application, cartons come down the gravity conveyor and hit the switch in front of the sealer. The clutch on the sealer drive is engaged, moving all cartons in the sealer forward. When the new carton passes the trip switch, the brake is engaged and the clutch is disengaged. Positioning provides even spacing of cartons, ensuring that they are in the sealer for as long a time as possible.

there is no wear. Torque is transmitted between the two rotating units by an electromagnetic reaction created by an energized coil winding. The slip between the motor and load can be controlled continuously, with more precision, and over a wider range than is possible with the mechanical friction clutch.

As shown in Figure 56–6, the magnet rotates within the steel ring or drum. There is an air gap between the ring and the magnet. The magnetic flux crosses the air gap and penetrates the iron ring. The rotation of the ring with relation to the magnet generates eddy currents and magnetic fields in the ring. Magnetic interaction between the two units transmits torque

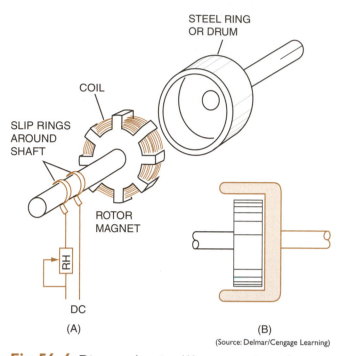

Fig. 56–7 Two magnetic drives driven by 100-hp, 1200-rpm induction motors mounted on top. Machines are used in typical sewage pumping plants. Pumps are mounted beneath the floor of drives. *(Courtesy of Electric Machinery Company, Inc.)*

Fig. 56–6 Diagram showing (A) an open view of a magnetic drive assembly; (B) spider rotor magnet rotates within the ring.

(Source: Delmar/Cengage Learning)

from the motor to the load. This torque is controlled with a rheostat, which manually or automatically adjusts the direct current supplied to the electromagnet through the slip rings.

When the electromagnet drive responds to an input or command voltage, a further refinement can be obtained in automatic control to regulate and maintain the output speed. The magnetic drive can be used with any type of actuating device or transducer that can provide an electrical signal. For example, electronic controls and sensors that detect liquid level, air and fluid pressure, temperature, and frequency can provide the input required.

A tachometer generator provides feedback speed control in that it generates a voltage that is proportional to its speed. Any changes in load condition will change the speed. The resulting generator voltage fluctuations are fed to a control circuit, which increases or decreases the magnetic drive field excitation to hold the speed constant.

For applications where a magnetic drive meets the requirements, an adjustable speed is frequently a desirable choice. Magnetic drives are used for applications requiring an adjustable speed such as cranes, hoists, fans, compressors, and pumps, Figure 56–7.

An electronic microprocessor unit is used to control and regulate the tension applied to strips of metal, paper, film, wire, and cable during the course of processing. It controls the speed and diameter fluctuations of the spool coils. Further, it ensures a constant tension, even at high linear speeds, which is essential to the quality of the finished product.

Study Questions

1. How is the magnetic clutch engaged and disengaged?

2. What devices may be used to energize the magnetic clutch?

3. Which type of drive is best suited for maintaining large differences in the input and output speeds? Why?

4. What is meant by feedback speed control?

5. How is the magnetic drive used as an adjustable speed drive?

DC Variable-Speed Control—Motor Drives

Objectives

After studying this unit, the student should be able to

- Describe the operating principles of DC variable-speed-control motor drives.

- State how above and below DC motor base speeds may be obtained.

- List three advantages of DC variable-speed motor drives.

- List modifications that can be made to packaged variable-speed-control drives to meet specific applications.

- Connect DC variable-speed control motor drives.

- Recommend troubleshooting solutions.

Adjustable speed drives are available in convenient units that include all necessary control circuits, Figure 57–1. These packaged variable speed drives generally operate on AC. They provide a large choice of speeds within given ranges.

Some drive requirements are so exacting that the AC motor drive alone is not suitable. In such cases, the DC motor has characteristics not available with AC motors. The DC motor also possesses many of the characteristics of an AC motor. A DC shunt motor with adjustable voltage control is very versatile and can be adapted to a large variety of applications.

The AC service in industrial plants must be converted to direct current for use with these controls. Smooth and easily controlled direct current is obtained through the use of the Ward Leonard system in which a DC shunt generator with field control is driven by an induction motor.

As described in previous units, electricity can be generated electromagnetically for heavy power consumption only if the following are present:

- A magnetic field
- A conductor
- Relative motion between the two

The motion in the DC variable-speed drive of Figure 57–2 is maintained by the steady driving force of an induction motor. The generator

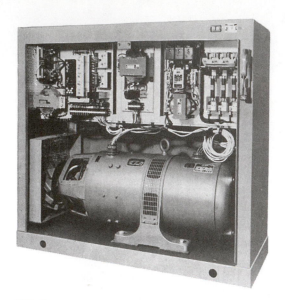

Fig. 57–1 Packaged-type motor-generator with DC variable-speed control system supplied from AC. *(Courtesy of Schneider Electric)*

voltage is increased by increasing the magnetic flux in the generator field. The magnetic flux of the generator field is increased by decreasing the resistance of the field rheostat to allow more current to flow through the shunt winding. Thus, the greater the field current, the stronger is the magnetic field.

The speed and torque of the system shown in Figure 57–2 can be controlled by adjusting the voltage to the field or to the armature, or both. Speeds above the motor base (nameplate) speed are obtained by weakening the motor shunt field with a rheostat or electronically. Speeds below the motor base speed are obtained by weakening the generator field. This, in turn, lowers the generator voltage supplying the DC motor armature. The motor should have a full shunt field for speeds lower than the base speed to give the effect of almost continuous control, rather than step control of the motor speed.

The motor used to furnish power from the line may be a three-phase induction motor, as shown in Figure 57–2, or it may be a DC motor. Once the driving motor is started, it runs continuously at a constant speed to drive the DC generator.

The armature of the generator is coupled electrically to the motor armature as shown. If the field strength of the generator is varied, the voltage from the DC generator can be controlled to send any amount of current to the DC motor. As a result, the motor can be made to turn at many different speeds. Because of the inductance of the DC fields and the time required by the generator to build up voltage, extremely smooth acceleration or deceleration is obtained from zero RPM to speeds greater than the base speed.

The field of the DC generator can be reversed manually or automatically with a resulting reversal of the motor rotation.

The generator field resistance can be changed manually or automatically by the time-delay relays operated by a counter emf across the motor armature.

Electrically controlled variable-speed drives offer a wide choice of speed ranges, horsepower and torque characteristics, means for controlling acceleration and deceleration, and methods of manual or automatic operation. A controlling tachometer feedback signal may be driven by the DC motor driveshaft. This is a refinement of the system to obtain constant speed. The choice of this method depends on the type of application, the speed, and the degree of response required (similar to the magnetic drive automatic control). In addition to speed, the controlling feedback signal may be set to respond to pressure, tension, or some other transducer function.

The circuit used with a motor-generator set to obtain adjustable voltage is shown in Figure 57–3. In this circuit, the exciter has

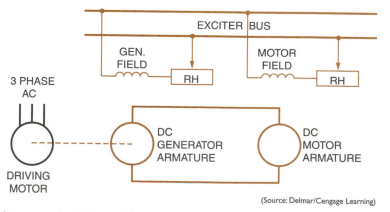

(Source: Delmar/Cengage Learning)

Fig. 57–2 Basic electrical circuit of a DC variable-speed control system.

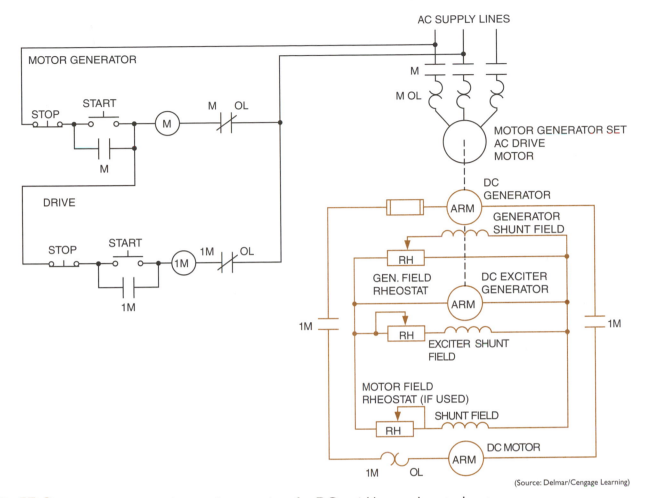

Fig. 57–3 Electrical power and control connections for DC variable-speed control system.

(Source: Delmar/Cengage Learning)

closed connections. An AC motor, usually a squirrel cage motor, drives the DC generator and exciter. The exciter may be operated as a self-excited machine and set to give constant potential output.

The generator field is connected directly across the constant potential exciter. The motor field may be weakened to provide speeds above the base speed after the generator voltage is increased to its maximum value.

The motor-generator set has one inherent advantage over rectifier-type DC supplies, that is, the ability to regenerate.

In Figure 57–3, assume that the generator rheostat is set at its maximum point and the motor is running at base speed. If the generator voltage is decreased by adjusting the rheostat, the motor countervoltage will be higher than the generator voltage and the current reverses. This action results in a reverse torque in the machine and the motor slows down (dynamic braking). This dynamic braking feature is very desirable when used on metal working, textile

and paper processing machines, and for general industry use.

MODIFICATIONS

To meet specific operating requirements, the packaged variable-speed control drives can be furnished with a variety of modifications, including provisions for

- Multimotor drive
- Reversing by reversing the generator field
- Reversing by motor armature control
- Dynamic and regenerative braking
- Motor-operated rheostat operation
- Preset speed control
- Jogging at preset reduced speed
- Current or voltage (speed) regulation
- Timed rate of acceleration and deceleration
- Current-limit acceleration and deceleration

The generator field resistance can be changed manually or automatically by time relays operated by a counter emf across the motor armature. Motor-operated rheostats can also be used in automatic speed control operations.

One or more of these modifications can be combined as required by the specific application. Some control schemes eliminate the standard exciter and provide other methods of energizing the generator and motor fields. Some systems eliminate the motor-generator set entirely and supply the armature and field power through controlled rectifiers directly to the motor from AC lines.

Study Questions

1. What is meant by the base speed of a DC motor?

2. How is above base speed obtained?

3. How is subbase speed for a DC motor obtained?

4. What advantage does the motor-generator type of variable-speed drive have over rectifier-type DC supplies?

5. If the AC driving motor of a self-excited motor-generator set is reversed, does this reverse the direction of rotation of the DC motor? Explain.

6. How does reversing the generator polarity reverse the connected motor or motors?

7. Why should the motor have a full shunt field for below-base speed operation?

8. To increase generator voltage: (a) Should the resistance of the rheostat be increased or decreased? (b) What change in current condition is required in the shunt field? Why?

Programmable and Motion Control

Objectives

After studying this unit, the student should be able to

- Discuss the operation of the input/output modules, the central processing unit, and the operator programming terminal.

- Draw basic diagrams of how the input and output modules function.

- Convert a motor starter elementary drawing to a schematic used for programming a programmable controller.

- Connect some pilot and control devices to input/output modules.

- Discuss installation requirements for a programmable controller.

- Discuss the use of the PLC programmer circuit on and off display.

- Describe some different motors used in automated systems.

- Describe a modern motor control center.

PROGRAMMABLE CONTROLLERS

If you set your alarm clock or clock radio to wake your household in the morning, you have programmed a controller. Programmable controllers range in size from small, versatile controllers that run single machines, to large, complex controllers that oversee entire control networks.

A programmable controller is a solid-state, electronic microprocessor-based control system used widely in industry. The system monitors input signals to detect changes from connected sources such as push buttons, limit switches, and condition sensors. Based on the status of input signals, the controller system reacts by means of user-programmed internal logic. The reaction is to produce output signals to drive external loads such as control relays, motor starters, contactors, and alarms or annunciators, Figure 58–1. Because of their flexibility, accuracy, and economy, more control needs are being met by programmable controllers (PCs) in industry. PCs are available in all sizes. Small ones are equal to as few as five relays, Figure 58–2. The controller shown in Figure 58–2 is specifically designed for

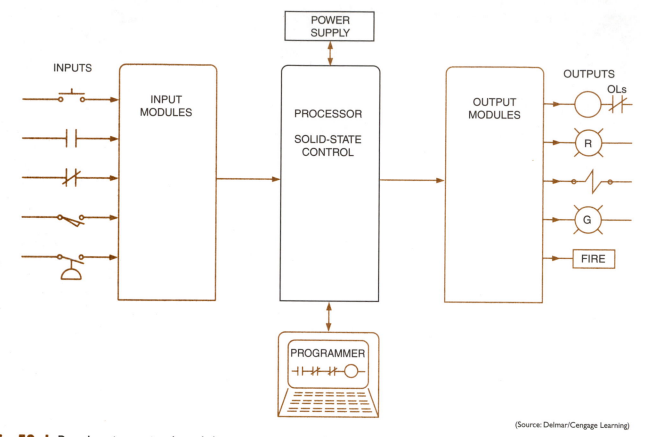

(Source: Delmar/Cengage Learning)

Fig. 58–1 Based on input signals and the program entered into the processor, the output is controlled.

use in automated process control and/or machine control. Large units are equal to thousands of inputs and outputs. This type of control system is widely used in batch processing and special machine control applications. By simply changing an "elementary ladder diagram" stored in the microcomputer logic, program changes are made, thereby changing the process control functions or the machine control applications. The system controls the valves that will open or close and the specific motors that will run at predetermined speeds. Programmable controls are cost competitive with "hard-wired" controls in many applications. They have the added advantage of programmability. Hard-wired systems must be rewired to suit changing control situations. Process manufacturing (food, chemicals, petroleum, and paint, for example) requires different blends for varying brands of a product. PCs are standard components for many completely automated machines being used in industry.

Power Supply

L1 and L2 still mean line one and line two. This AC input is converted to low DC voltages for internal circuit use. The voltage is then filtered and regulated to free the system of any

Fig. 58–2 Small programmable controller with input/output track, designed for use in automatic process control. *(Courtesy Eagle Signal Controls)*

electrical spikes and noise. It might still be necessary to install an isolating transformer. Some PC manufacturers may recommend installing noise suppressors on all magnetic switches being used with their system. Always pay close attention to the manufacturer's installation instructions and recommended operating procedures.

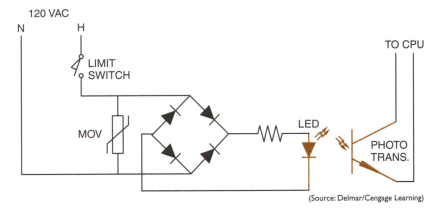

(Source: Delmar/Cengage Learning)

Fig. 58–3 Typical input circuit for a programmable controller.

Input Module

Input modules sense whether the input devices are on or off. This provides the processor module with the input status information that it needs to process a planned program. Typical input devices include pilot devices such as push buttons, limit switches, and the like. Internal circuitry includes optical relay isolation, as well as filtering and electrical surge suppression to guard against possible damage to the central processing unit (CPU) by transients from the user inputs. The input module plugs directly into the processor module. Some input modules may be remotely connected.

The schematic for a typical programmable controller input is shown in Figure 58–3. In this example it is assumed that the input voltage is 120 volts AC. A metal oxide varistor (M.O.V.) is used to suppress voltage spikes that can be produced on the line. A bridge rectifier converts the AC voltage into DC. A current limiting resistor is connected in series with an LED. The LED

activates a photo-transistor. The photo-transistor is connected to the CPU and signals that a voltage is present or not present at that particular input. Notice that communication between the CPU and the outside control device is separated by a light beam. This is known as optical isolation or optoisolation. Optoisolation is used to ensure that no voltage spikes or electrical noise can reach the CPU.

Installation. The CPU must be securely mounted to a panel or cabinet first. The input module is then plugged into the desired input/output (I/O) port of the processor. The module is secured with the screws that are supplied with it. The module is color coded, or marked, to identify it as a 120-volt, AC-specific input module. A corresponding identification is placed on the processor module. If the module is removed for any reason, this identification will help you identify the type of module assigned to this location.

Figure 58–4 shows an input module on the right connected to a control unit (CPU). A typical

Fig. 58–4 Input module wiring. (*Courtesy Rockwell Automation*)

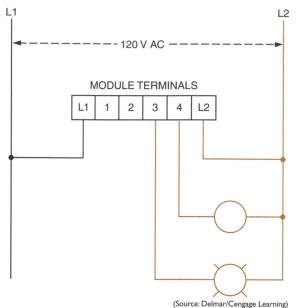

(Source: Delmar/Cengage Learning)

Fig. 58–5 Input module wiring.

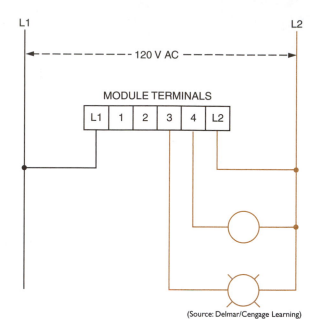

(Source: Delmar/Cengage Learning)

Fig. 58–6 Output module wiring.

electrical ladder connection diagram is shown in Figure 58–5. The wiring for the external push buttons must be brought to the PC cabinet in conduit or to comply with the *NEC®* requirements.

Output Module

Output modules control ON and OFF status of output devices, based on commands received from the processor control unit. Typical output devices include coils for motor starters, contactors, relays, and so on. Some output modules are capable of switching starters through size 5. Module protective circuitry is built-in like the input module. It also plugs directly into the processor unit. Status indicator lamps are available, as with the input module. A typical terminal elementary diagram is shown in Figure 58–6. The physical appearance is similar to that of the input module.

The device used to control the output can vary from one programmable controller to another. Some contain relay-type outputs and others employ solid-state output devices. If the relay-type output is used, the CPU controls the coil of the relay and the contacts connect the load device to the line, Figure 58–7. Notice that the output of the programmable controller does not supply the power to operate the load. It simply acts as a switch or set of contacts to connect the load to power.

The type of solid-state device used in the output is determined by the type and amount of voltage the PC is designed to control. Some programmable controllers are intended to control a

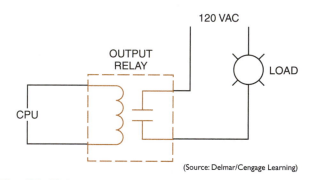

(Source: Delmar/Cengage Learning)

Fig. 58–7 Typical relay output.

voltage of 120 volts AC. Others may be intended to control 24 volts AC or 24 volts DC. Low-voltage controls are generally used in areas where intrinsically safe circuits are desired. Intrinsically safe circuits are low-power circuits that can be used in areas with atmospheres that contain explosive vapors and dusts, such as refineries and chemical plants. The intrinsically safe circuits do not have enough power to ignite the surrounding atmosphere.

If the solid-state output of a programmable controller is intended to control an AC voltage, the output control device will be a triac, Figure 58–8. A triac is a solid-state device similar to an SCR, except the triac is a bidirectional device that will conduct on both half cycles of the waveform. This gives the triac the ability to control an AC current. Triacs are often employed as solid-state AC switches. Notice that the output is optoisolated to prevent any electrical spikes or noise from being able to affect the CPU. The CPU actually turns on an LED. The light produced by the LED activates the gate of a photo triac.

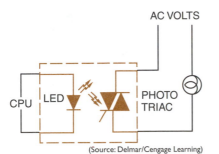

Fig. 58–8 A triac is used to control an AC device.

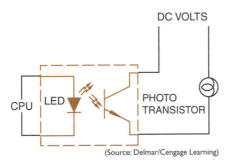

Fig. 58–9 A transistor is used to control a DC device.

If the output is intended to control a DC voltage, the output device will be a photo-transistor, Figure 58–9. Like the triac, the transistor is used as a solid-state switch to connect the load to the power line.

Processor Unit

The processor unit is the principal unit of the controller. It processes and manipulates the information entered in its memory by the operator terminal or programmer. Working with or interfacing with external hard-wired input and output devices is accomplished using the input and output modules. These are plugged into the I/O (input/output) ports of the processor module.

The number of input/outputs, and the capacity to handle them, determines the size of the programmable controller. Even the small processors are a maze of electronic printed circuits, transistors, and other related equipment. Both digital (on/off) and analog signals (such as voltage and current values) are acceptable to the processor. These signals are handled differently internally but can be connected within limits.

The brain of the programmable controller is the processor. It receives all inputs, evaluates and organizes the control activity, and performs logical decisions according to the direction of the memory stored program. It then tells the output what to do. The processor module supplies operating power to the input and output modules and the operator terminal (programmer). Input

and output devices must be separately powered, as may be seen in Figure 58–4 and Figure 58–6.

INSTALLATION

A user's manual is received with a new programmable controller. This manual should be consulted before installing the controller. The environmental conditions will dictate the type of enclosure required. It must be adequate and NEMA approved for a particular application. Consideration should be given to the proper location of a controller disconnect switch. An isolation transformer may also be required.

■ Check to see if the processor module power supply and the input/output devices (I/O modules) require the same AC source. The controller should be adequately grounded. Scrape paint or other nonconductive finish from the panel surface in the bolt mounting areas. This will help ensure good electrical contact for grounding purposes when the mounting bolts are tightened.

■ Check to see if one or more emergency stop switches are to be included or recommended. The location of such switches is also an important consideration.

■ Follow the manufacturer's recommendations for component or accessory spacing within the enclosure. This is to help keep the controller temperature within safe operating limits.

■ Check to see how the wiring should be routed to minimize the effects of electrical noise.

■ Are surge suppressors to be used for inductive loads in series with hard contacts and for other noise-generating equipment?

■ Where many machines or processes are controlled by one processor, it may be economically advantageous to locate I/O modules close to the field devices. This reduces multiple-conductor runs. It also locates the I/O module indicator lights close to the machine or process for easy field monitoring and troubleshooting.

Figure 58–10 shows an I/O track that is wired. The module is plugged into a wired track as can be seen at the upper right. No field wiring need be removed during maintenance or changing modules in this scheme. In checking the user's manual for these few recommendations, other installation suggestions or requirements should be noted before installation.

Fig. 58–10 A wired I/O module. *(Courtesy of Reliance Electric)*

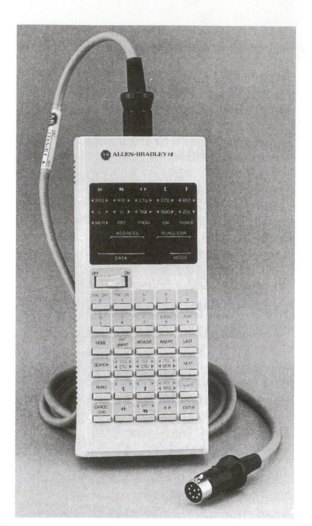

Fig. 58–11 Control logic can be entered into the processor from an operator program terminal. *(Courtesy of Rockwell Automation)*

PROGRAMMER TERMINAL

The programming unit, or operator terminal, permits information to be input to the processor through a keyboard. This input information tells the processor what to do with control loads connected to the output modules. This information includes the step-by-step circuit layout and connections. Figure 58–11 shows a handheld programmer terminal.

Programming may be accomplished either at the processor site or remote to the processor. Many programmers use a standard laptop-type computer loaded with the proper software to develop the program. Once the program has been developed it can be saved to a disk, CD, or other storage device. Special hardware devices that permit the computer to be connected directly to the programmable controller permit the program to be loaded, monitored, or changed using the computer. The actual program is generally done using the standard relay ladder logic format. The program is developed using symbols for normally open or normally closed contacts and coils.

Elementary ladder diagramming ability is essential.

All pilot devices are programmed as normally open or normally closed contacts. There are no keyboard symbols for limit, float, or pressure switches. The processor does not know what is closing a switch, only that it is open or closed. Contacts are assigned numbers to distinguish them.

Because elementary ladder diagrams (refer to Units 4 and 5) follow a logical process, programming a PC should not be difficult. For example, a three-wire, start-stop control circuit is shown in Figure 58–12(A). Figure 58–12(B) shows how this same circuit will appear as it is programmed into the processor, through the program operator terminal. Figure 58–13 shows how these component parts are connected to the input/output (I/O) modules. Note that the starter has been rewired to bring the overload contact to the left control side. The maintaining contact is an "internal relay" of the PC. Note also in Figure 58–12(B) that the normally closed stop button and the normally closed overload contact appear as normally open. Because terminals 1 and 3 are normally closed contacts, power will be fed to the input module, or track. When this occurs to the input module,

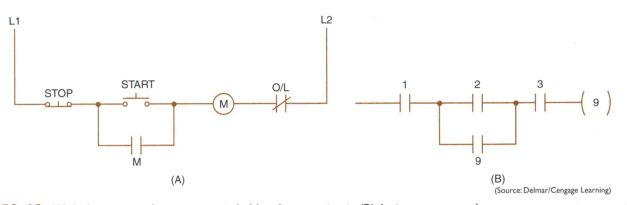

(A)   (B)

(Source: Delmar/Cengage Learning)

Fig. 58–12 (A) A three-wire electromagnetic ladder diagram circuit. (B) As it appears on the program operator terminal.

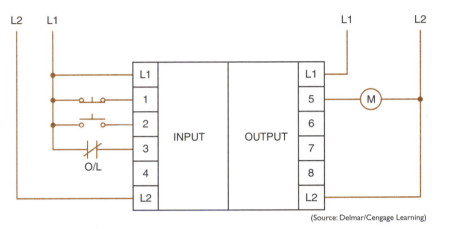

(Source: Delmar/Cengage Learning)

Fig. 58–13 PC connections for start-stop station.

the central processor interprets it as an instruction to change the position of the contact that matches the terminal number of the input module. Contacts 1 and 3 are considered closed by the processor during normal operation, because power is fed through the normally closed contacts. Because a coil will always end the line of a program, overload contact 3 is moved to the left of the coil, where you would place any contact that controls the coil. Note the programmer coil symbol looks like this: ( ). After a circuit is energized, these symbols will assume a crosshatch shading, or other distinctive feature as shown in Figure 58–14. This tells you that the contacts are closed properly and the coil is energized. This feature is also useful for troubleshooting the control system in case of a malfunction.

Figure 58–15 illustrates the actual motor starter connection just described. All I/O module wiring is brought down to the terminal board connections and marked. The operator terminal is plugged into the bottom of the central processor unit. The motor starter is on the student motor control laboratory panel at the right.

Different manufacturers use different methods of programming their controllers. Therefore,

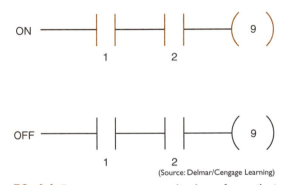

(Source: Delmar/Cengage Learning)

Fig. 58–14 Program operator display of a coil circuit.

Fig. 58–15 PC connected to a motor starter on a student work panel. *(Courtesy Los Angeles Trade-Technical College)*

it is necessary to study and follow the instruction manual that accompanies a programmable controller.

Developing a Circuit

The next circuit to be discussed and constructed is a simple elevator circuit. It is assumed that the elevator is operated by a three-phase motor that runs in forward direction when the elevator goes up, and runs in the reverse direction when the elevator goes down. Limit switches labeled LS1, LS2, and LS3 are used to indicate when the elevator has reached a particular floor; LS1 is for the first floor, LS2 is for the second floor, and LS3 is for the third floor. Pilot lights also are used to indicate when the elevator is at a particular floor. A red light is used for the first floor, a yellow light is used for the second floor, and a green light is used for the third floor.

Developing the Circuit Logic

The control needed for the first and third floors is relatively simple because the elevator must always go down to reach the first floor and always go up to reach the third floor. This control can be accomplished by starting the motor up or down and permitting the appropriate limit switch to stop the motor when it reaches the designated floor. Interlocking is also a necessity to prevent the UP and DOWN motor starters from being energized at the same time. A circuit of this type is shown in Figure 58–16. When the third-floor push button is pressed, the UP starter energizes and the elevator starts up. When it reaches the top, limit switch LS3 opens and stops the motor. At the same time, the normally open LS3 contact closes and turns on the green pilot light to indicate the elevator is on the third floor. When the first-floor push

button is pressed, the DOWN starter is energized and the elevator starts moving in the downward direction. When the elevator reaches the first floor, the normally closed limit switch LS1 opens and stops the motors and the normally open LS1 limit switch closes and turns on the red pilot light.

The addition of the second-floor control, however, greatly increases the complication of the circuit as shown in Figure 58–17. Now that the second floor is added to the circuit, the logic of the control system must be such that if the elevator is located on the first floor and the second-floor push button is pressed, the elevator will proceed up. If the elevator is located on the third floor and the second-floor push button is pressed, the elevator will proceed down. The three limit switches used to detect the location of the elevator must now have more contacts than are generally provided on the switch. Limit switch LS1 must have two normally closed and one normally open contacts, limit switch LS2 must have one normally closed and one normally open contacts, and limit switch LS3 must have two normally closed and one normally open contacts. To supply the necessary contacts, each limit switch will be used to control the coil of a control relay. Limit switch 1 controls the coil of limit switch 1 control relay (LS1CR), limit switch 2 controls the coil of LS2CR, and limit switch 3 controls the coil of LS3CR. Control relays are also used in the remainder of the circuit logic. The first-floor push button, for example, controls the coil of the

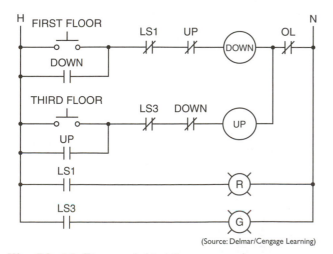

(Source: Delmar/Cengage Learning)

Fig. 58–16 First- and third-floor control.

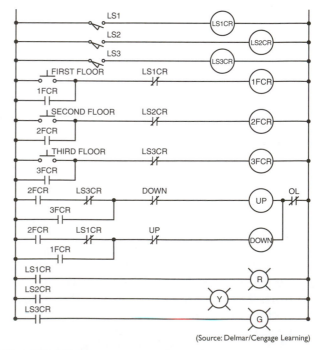

(Source: Delmar/Cengage Learning)

Fig. 58–17 Elevator control circuit.

first-floor control relay (1FCR), the second-floor push button controls the coil of 2FCR, and the third-floor push button controls the coil of 3FCR. These control relays are used to control the action of the two motor starters designated as UP and DOWN.

Changing the Circuit for PC Operation

The relay circuit shown in Figure 58–17 will now be changed so it can be loaded into the memory of the programmable controller. Before the circuit can be developed, the inputs and outputs must be signed. This circuit contains seven inputs and five outputs. Because different types of programmable controllers assign different numbers for input and outputs, it will be necessary to assume input numbers, output number, and internal relay numbers for this particular controller. The chart in Figure 58–18 lists these assigned numbers.

The Overload Relay

In the schematic shown in Figure 58–17, an overload contact is connected to the output side of coils UP and DOWN. Because each line of logic for a programmable controller must end with a coil, the overload contact must be moved to a position behind coils UP and DOWN. Also, both the UP and DOWN coils must be protected by the overload. In this circuit, it will be assumed that the overload relay that controls the action of the overload contact contains two auxiliary contacts instead of one. One contact is normally closed and the other is normally open, Figure 58–19. This type of overload relay is

INPUTS	1 THROUGH 10
OUTPUTS	11 THROUGH 20
INTERNAL RELAY	100 THROUGH 200

(Source: Delmar/Cengage Learning)

Fig. 58–18 Assignment numbers for PC.

becoming very common because it permits the coil or coils of starters to be connected in series with the normally closed contact and the normally open contact can be used to control indicator lights or, in this case, to supply a signal to the input of a programmable controller. In this way, the motor is protected by a hard-wired contact in the event of an overload, and the programmable controller can be informed that an overload has occurred. Before a program can be developed, it is necessary to first determine which control device is connected to which input and which controlled device is connected to which output. The input and output assignments are as follows:

First-floor push button	input 1
Second-floor push button	input 2
Third-floor push button	input 3
LS1 limit switch	input 4
LS2 limit switch	input 5
LS3 limit switch	input 6
NO overload contact	input 7
Motor starter (forward)	output 11
Motor starter (reverse)	output 12
Red pilot light	output 13
Yellow pilot light	output 14
Green pilot light	output 15

In the relay circuit, it was necessary for the limit switches to control the coils of control relays in order to increase the number of contacts operated by each limit switch. Programmable controllers, however, do not have this problem. Any number of contacts can be controlled by a single input, and the contacts can be programmed as normally open or normally closed. Therefore, control relays LS1CR, LS2CR, and LS3CR are not needed in the circuit logic. However, the control relays for the floors, 1FCR, 2FCR, and 3FCR, are needed. Internal relay coil 101 is used for 1FCR, coil 102 is used for 2FCR, and coil 103 is used for 3FCR. The revised control schematic for this circuit is shown in Figure 58–20.

MOTOR CONTROL CENTER

A modern motor control center combines the latest automation technologies, such as programmable controllers, solid-state starters, and AC and DC variable speed drives, in a common enclosure. One center is shown in Figure 58–21. The control center receives incoming power and distributes it to all motors under a controlled and protected process from one location. See Figure 58–22. This center is especially convenient for PC installations with

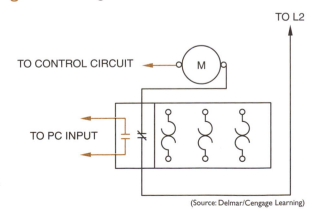

(Source: Delmar/Cengage Learning)

Fig. 58–19 The normally open OL contact connects to programmable controller input.

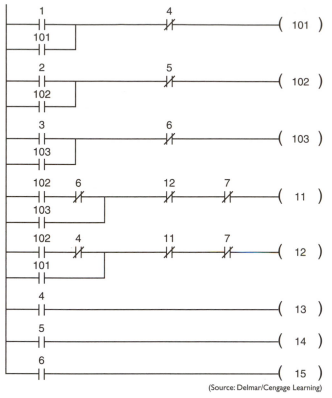

(Source: Delmar/Cengage Learning)

Fig. 58–20 Elevator circuit for programmable controller.

Fig. 58–21 Modern motor control center. *(Courtesy of General Electric Co.)*

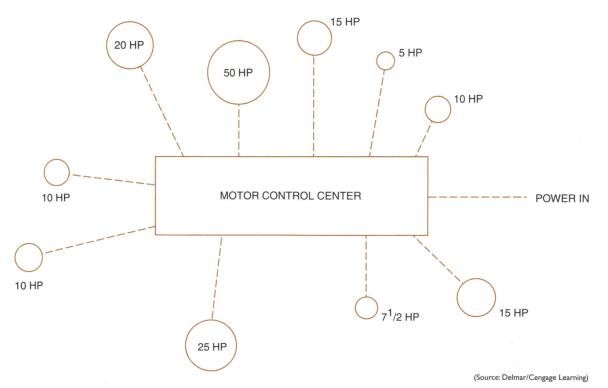

(Source: Delmar/Cengage Learning)

Fig. 58–22 Underground distribution of a motor control center (floor plan). Motor circuits may also be run overhead.

many inputs and outputs. Existing electromagnetic motor control centers are readily adapted to PC control. The wiring is already installed to terminal blocks. The control center is simply rewired to a PC.

MOTION CONTROL

Electric motors do more than turn heavy industrial machinery. *Motion control* is a term usually given to small motors that are used for small incremental movements, for positioning, assembly, or processing. Some are used in industrial robot/automated machines. Many are so small that they are sometimes called electronic motors because they are driven by fine electronic, precision controllers to achieve precise movements. Although the size of such motors may start at about 1/8 horsepower, some range as high as 2 and 5 horsepower. Several types of motors are used with industrial robots and other automated machinery.

Permanent Magnet—DC Operated

Permanent magnet motors have a conventional wound armature with a commutator and brushes and a permanent magnetic field, Figure 58–23. This motor has excellent starting torque with speed regulation just under that of compound motors. Because of the permanent field, motor losses are less and better operating efficiency is obtained. These motors can be dynamically braked and can be reversed at a low voltage (10 percent). They should not be plug reversed with full armature voltage. Reversing current can be no higher than the locked armature current.

Permanent Magnet Synchronous Motor—AC

Permanent magnet synchronous motors are AC motors that eliminate the slip characteristic of induction motors, Figure 58–24. They operate at exactly the synchronous speed of the three-phase power speed, usually 1800 or 3600 rpm. High-density magnets are imbedded in the rotor. Permanent magnet synchronous AC motors are ideally suited for processes requiring accurate speed control or synchronizing of several interconnected operations.

STEPPER MOTORS

The stepper or stepping motor is a precision electromechanical incremental actuator. It uses the digital (on/off) approach for control of industrial machines. It delivers precise incremental motions of great accuracy. Its popularity is broadened by using it with programmable controllers. The stepper motor is DC with a permanent magnet rotor and a wound stator (field). Both rotor and stator are toothed, like small fields in themselves, Figure 58–25. The basic feature here is that the motor will move when energized and come to rest at any number of steps, in either direction. It is adequate for light loads and PC driven applications, and for positioning applications in many forms. It moves and stops according to the digital commands from the PC. The stepper motor converts programmed digital pulse inputs into a series of motions; sequential, fast or slow, forward, and reverse.

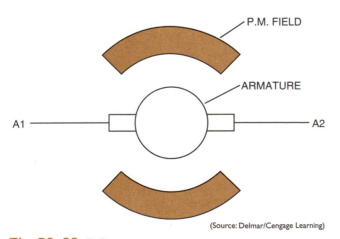

(Source: Delmar/Cengage Learning)

Fig. 58–23 DC permanent magnetic motor.

Fig. 58–24 Permanent magnet synchronous AC motor. *(Courtesy of Reliance Electric)*

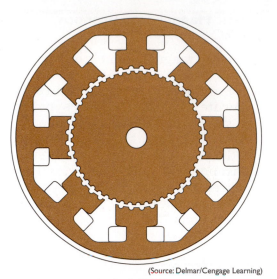

(Source: Delmar/Cengage Learning)

Fig. 58–25 Cross section of a typical stepper motor showing toothed construction of the armature and stator. Signal pulses in serial order advancing sequentially from one field to an adjacent field drive the stepper motor.

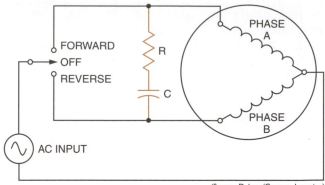

(Source: Delmar/Cengage Learning)

Fig. 58–26 Phase shift circuit converts single phase into two phase.

Fig. 58–27 Stepping motor. *(Courtesy of Danaher Motion)*

Most stepping motors require 400 increments or steps to turn the shaft 360 degrees, resulting in 0.9 degree of rotation per step. They have an accuracy of approximately ±5 percent per step, and the error is not cumulative regardless of the number of steps. Because stepping motors have permanent magnet rotors, they have the ability to provide holding torque when the motor is turned off. If greater holding torque is desired, continuous DC voltage can be applied to the motor. If DC voltage is applied to only one winding, the holding torque will be approximately 20 percent greater than the rate running torque. If DC voltage is applied to both stator and winding, the holding torque will be approximately 150 percent greater than the rate running torque.

It is also possible to operate stepping motors on two-phase alternating current. An eight-pole stepping motor becomes a synchronous motor with a speed of 72 rpm when connected to 60 Hz. The two phases can be formed from single phase with a phase splitting circuit as shown in Figure 58–26. When stepping motors are operated as synchronous motors they can start, stop, or reverse direction of rotation virtually instantly. They will start and obtain full speed within approximately 1½ cycles and stop within 5 to 25 milliseconds. A stepping motor is shown in Figure 58–27. An exploded diagram of a stepping motor is shown in Figure 58–28.

SELF-SYNCHRONOUS (SELSYN) TRANSMISSION SYSTEM

The self-synchronous transmission system is called "selsyn," an abbreviation. It is a system used for transmitting, by electrical means, any signal that can be represented by angular position. The great advantage of this system over other systems of electrical transmission of mechanical motion is that it is self-synchronizing. Selsyn units can control the motion of a remote device at a distant point by controlling its actuating mechanism.

For example, when the transmitter unit is moved to a point (manually, electrically, or mechanically), the receiver will move to this identical point through simple wiring. A good example of this system is probably one of the most common and the oldest uses, a ship's engine order. When the ship captain moves his engine order (transmitter) to a specific desired speed, this is duplicated in the engine room with a pointer rotated by the selsyn receiver, thereby duplicating the captain's direction.

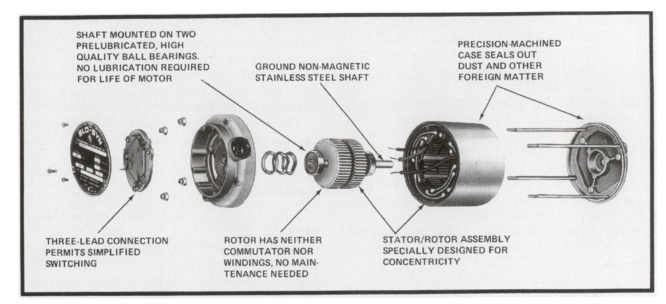

SHAFT MOUNTED ON TWO
PRELUBRICATED, HIGH
QUALITY BALL BEARINGS.
NO LUBRICATION REQUIRED
FOR LIFE OF MOTOR

GROUND NON-MAGNETIC
STAINLESS STEEL SHAFT

PRECISION-MACHINED
CASE SEALS OUT
DUST AND OTHER
FOREIGN MATTER

THREE-LEAD CONNECTION
PERMITS SIMPLIFIED
SWITCHING

ROTOR HAS NEITHER
COMMUTATOR NOR
WINDINGS, NO MAIN-
TENANCE NEEDED

STATOR/ROTOR ASSEMBLY
SPECIALLY DESIGNED FOR
CONCENTRICITY

Fig. 58–28 Exploded diagram of a stepping motor. *(Courtesy of Danaher Motion)*

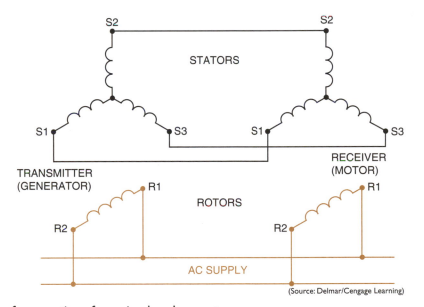

(Source: Delmar/Cengage Learning)

Fig. 58–29 Diagram of connections for a simple selsyn system.

These units look like common small motors and are also called synchros. They are connected together by conductors through necessary switches. See a simple connecting diagram in Figure 58–29.

Operation

The diagram shows interconnected stator, "S" terminals, and rotor, "R" terminals. Note that only the rotors are connected to the AC power source. Assuming both rotors are free to turn, they will take such a position that the voltages induced in the two stators are of balanced magnitude and displacement. Under this condition of stator voltage balance between the two instruments, there is no circulating current in the stator windings. Now, if the rotor of one selsyn unit is displaced by a certain angle and the rotor of the other is held in its original position, the stator voltage balance is altered and a circulating current will flow in the windings. This circulating current reacting on the excitation flux provides a torque tending to turn the rotors of the machines to a position where the induced stator voltages are again equal and opposite. Thus, with both rotors unrestrained, any motion given to the rotor of one synchro unit will be transmitted to and duplicated by the rotor of the second, and we have a system for transmitting mechanical motion for positioning.

If the transmitter rotor is turned, either manually, mechanically, or by some other electrically operated means, the receiver rotor will follow at the same speed and in the same direction as the rotors induce voltages into their three windings of their stators. The self-synchronous alignment of the rotors is the result of voltages induced in the secondary stator windings.

A synchro unit draws but little power from the line and is designed for continuous duty excitation.

It has been the custom to call the selsyn unit, which is used as the transmitter, a synchro (selsyn) generator, and the remote, or receiving, unit a selsyn motor. The one does not generate electrical power nor does the other transform it into mechanical motion. The self-synchronous systems merely transmit mechanical motion. Neglecting friction losses, the mechanical power output of the receiver is exactly equivalent to the mechanical power input to the transmitter. Electrically, the selsyn generator is identical to the motor. Physically, they differ in that the motor has mounted on its rotor shaft an oscillation damper. Thus, there is large dampening effect of the oscillating rotor and its movement is quickly stopped.

Multiple Operation

The system may also be used for multiple operation; that is, several selsyn motors may be operated by a single transmitter. The operation is in general the same as in the simple single transmitter and receiver system.

There are many different kinds and configurations of this basic system.

Application

A good number of factories have installed computer-based systems to keep abreast of growing demands for sophisticated manufacturing technology. Servomechanism devices, as different types of integrated electric motors, activated by electrical or mechanical impulses, automatically operate machines. The human fascination with machines that can move under their own power evolved into the science of robotics—a manufacturing technology currently applied to the production of automobiles, refrigerators, aircraft, and much more. The term *servomechanism* arises from numerous robotic functions that perform desired applications.

Those manufacturers who invested in automated technology have affirmed the robot's ability to support heavy production demands. Thus, the demand for high-technology manufacturing with computer-controlled machines and robots will increase dramatically.

Study Questions

1. Sketch the major components of a programmable controller.

2. Briefly describe a programmable controller.

3. What familiar method of programming is often used for a programmable controller?

4. Discuss some basic installation requirements for a programmable controller.

5. What is the major advantage of the lighted "on" or "off" program entered into the program terminal?

6. What is a modern motor control center?

7. What are the primary characteristics of a stepper motor?

8. What do selsyns look like?

9. In selsyn motors, AC power is connected to what?

10. Explain optoisolation.

11. If a programmable controller is designed to control output devices rated at 120 volts AC, what electronic device is used to connect the load to the line?

12. If a programmable controller is designed to control output devices rated at 24 volts DC, what electronic device is used to connect the load to the line?

Troubleshooting

Motor Startup and Troubleshooting Basics

Objectives

After studying this unit, the student should be able to

- Demonstrate safe working habits.
- Prepare motors for startup.
- Start up newly installed motors.
- Troubleshoot motors using basic sense faculties.
- Separate circuit operation into logical steps for troubleshooting.
- Use industrial control index for troubleshooting tips.
- Seat new motor brushes, AC and DC.
- Test a solid-state diode.
- Change solid-state control modules according to the direction of a visual control panel.

WARNING

High voltage and rotating parts may cause serious or fatal injuries. Installation, operation, and maintenance of electrical machinery should be performed by qualified people!

All maintenance and repairs should be made with good, deliberate safety practices in mind. Fundamental to any job approach is a good safety procedure. Remember, SAFETY FIRST, THE JOB IS SECOND!

MOTOR STARTUP

After new motors and controls are installed, they should be checked for operation under load for an initial period of at least one hour. During this time, the electrician can observe if any unusual noise or hot spots develop. The operating current must be checked against the nameplate ampere rating. This requires skill in the proper connection, setting, and reading of an ammeter. The preferred ammeter is the type

Fig. 59–1 Using an ammeter around a wire.

(Source: Delmar/Cengage Learning)

that has the clamp around the wire, Figure 59–1. The nameplate ampere reading times the service factor (if any) sets the limits of the steady current; this value should not be exceeded.

If the motor has not been installed in a clean, well-ventilated place, clean the area. Good housekeeping, as well as direct accident and fire prevention techniques, must be emphasized at all times.

MOTOR MOUNTS

Motor mounts must be checked to be sure that they are secure and on a firm foundation, Figure 59–2. If necessary, add grout to secure the mounts. Ball-bearing motors can be mounted on sidewalls or the ceiling, vertically or horizontally.

Rotate the end shields to place grease fittings, plugs, or any openings in the best, or most accessible, location. Oil or grease the bearings if necessary.

NAMEPLATE

Check the power supply against the nameplate values; they should agree. Most motors will operate successfully with the line voltage within 10 percent (plus or minus) of the nameplate value; or within 5 percent of the frequency (hertz). Most 220-volt motors can be used on 208-volt network systems but with slightly modified performance. Generally 230-volt motors should not be used on 208-volt systems.

To reconnect a dual-voltage motor to a desired voltage, follow the instructions on the connection diagram on the nameplate. (See also Unit 5, "Interpretation and Application of Simple Wiring and Elementary Diagrams.")

Motor starter overload relay heaters must be installed with the proper size. The motor will not run without them. Sizing information is found inside the control enclosure cover. The starting fuses should be checked in a similar manner. The selection of the correct fuse size must be in accordance with the *NEC©* or with local requirements.

Collector Rings and Commutators

The collector rings on wound rotor motors are sometimes coated with a film at the factory to protect them while in stock and during shipment. The brushes may be fastened in a raised position. Before putting the motor into service, the collector ring surface must be cleaned of the

Fig. 59–2 A motor on a firm, secure foundation. *(Courtesy of Los Angeles Department of Water and Power)*

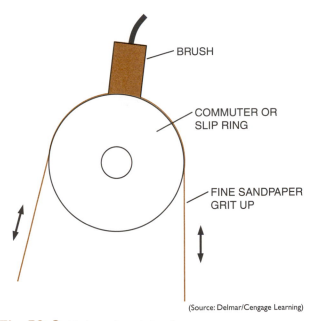

BRUSH

COMMUTER OR
SLIP RING

FINE SANDPAPER
GRIT UP

(Source: Delmar/Cengage Learning)

Fig. 59–3 Fitting a brush in place.

protective coating. The brushes then must be set down on the collector ring or commutator. When the motor is in service, keep the rings clean and maintain their polished surfaces. Chocolate brown is a good operating color for commutators. Ordinarily, the rings will require only occasional wiping with a piece of canvas or nonlinting cloth. Do not let dust, dirt, moisture, or oil accumulate between the collector rings.

Brushes

Brushes should move freely in the holders. At the same time, they must make firm, even contact with the collector rings and commutators. When installing new brushes, fit them carefully to the collector rings. Be sure that the pigtail conductors are securely fastened to, and make good contact with, the brush holders. It may be necessary to "sand in" brushes to fit the contour of commutator or slip rings, Figure 59–3.

LOADS

Electrical

A close check on operating line voltages, both at the incoming line and close to the motor terminals, may reveal imbalances that can be corrected. An imbalanced three-phase electrical system can cause trouble. Avoid unbalanced single-phase loading on a three-phase distribution panel. Check for required surge suppressors for solid-state devices.

Mechanical

Operate the motor without a load to check rotation and free running action. The desired direction of rotation is the required direction of rotation of the driven load. Incorrect rotation direction of some gear boxes, machines, and equipment may cause severe damage. To reverse rotation of a three-phase motor, interchange any two line leads. To reverse rotation of a single-phase motor, follow the connection diagram. The connected loads should be well-balanced to reduce vibration. A smooth-running machine is much more efficient and will cause less trouble later.

TROUBLESHOOTING BASICS

The first step is to shut down the machine and lock it out for repair or adjustment. The expert troubleshooter has mastered basic circuit concepts. The most valuable troubleshooting asset is the ability to apply common sense when analyzing a control operation. Also, good maintenance personnel learn to use sensory functions to diagnose and locate trouble.

- LOOKING may reveal contacts stuck and hung up, thereby creating open circuits.
- LISTENING may indicate loose parts, faulty bearings, excessive speed, and so on.
- SMELLING may indicate burning insulation or a coil failure.
- TOUCHING may reveal excessive motor shaft play, vibration, or abnormal heat.

This seemingly oversimplified procedure to locate a problem has saved many hours of labor. Consider the length of time it would take to become thoroughly familiar with a complicated schematic diagram, compared with locating a few contacts that are stuck—the result of merely LOOKING.

However, finding a problem in an installation is not usually this easy. An orderly, step-by-step approach is required. Circuit operation is separated in logical parts. Circuits and components are then divided into smaller parts to determine their functions, relationships to one another, and the effect that they have on each other in the overall control system operation. Each step leads to the source of the difficulty, finally pinpointing the problem. This procedure may require the use of a voltage tester, ammeter, multimeter, jumper wires, and other tools.

Check the power supply to see if it is on and if it is correct. Test all protective devices. If the coil as shown in Figure 59–4 does not energize,

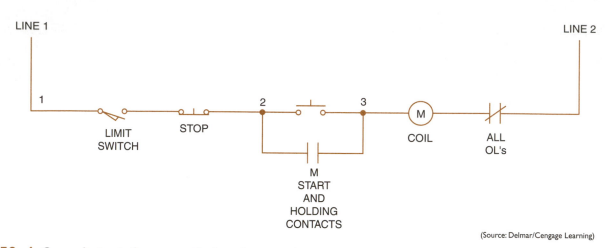

Fig. 59–4 Control circuit for a troubleshooting exercise.

(Source: Delmar/Cengage Learning)

a jumper wire can be placed from L1 to 3 of the control circuit. By jumping across the contacts of the limit switch and push buttons, the circuit operation is separated into logical parts. If the starter coil is now energized, the problem may be in the limit switch, or the stop or start push buttons. Smaller circuits and components are now tested by "jumping" around the limit switch, for example. We may then go to the control station, if necessary. By an orderly process of elimination, by testing all possible fault areas, the problem can be located accurately and efficiently.

Indiscriminate jumping, however, should not be practiced because of the danger of short circuits. For example, a jumper should never be placed across a power-consuming device, such as

a contactor coil; voltage or ohmeter testers are used in this instance. If an ohmeter is used to test a coil for continuity, the power must be off.

Table 59–1 and Table 59–2 are provided as aids to servicing electric control equipment. Use Table 59–1 as an index. Then refer to Table 59–2 for signs and solutions. For example, if the problem involves an AC contactor, look under the heading **Contactors, AC operated.** In the parenthesis next to it are specific number references (1–12). Refer to Table 59–2, items 1 through 12 for the signs and solutions to AC-operated contactor problems.

Because downtime of a machine may be expensive, the best guideline is to select and replace top quality devices designed for specific applications, see Figure 59–5.

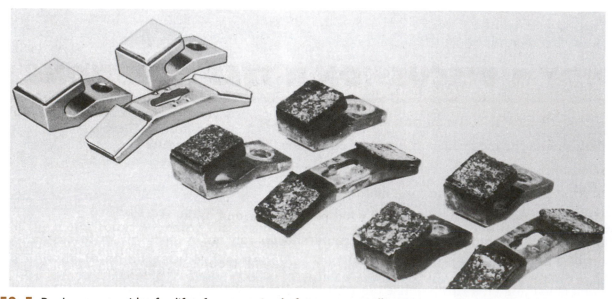

Fig. 59–5 Replacement guides for life of contact tips. Left: new or excellent contact. Center: worn contacts, a signal to watch. Right: excessively worn; change. *(Courtesy of Schneider Electric)*

HOW TO FIND AN INTERMITTENT FAULT

A loose electrical terminal connection or a control pilot device with an intermittent open circuit can cause a motor to stop running. The intermittent fault is one of the most difficult problems encountered during electrical troubleshooting. It need not be frustrating while troubleshooting, if a simple procedure is followed. First check the "LOOK, SMELL, TOUCH, LISTEN" sensory method to diagnose a possible malfunction. Think SAFETY FIRST while doing this procedure.

Next, referring to Figure 59–6, connect a test relay with its normally open contact in series with its coil. Connect two test leads, with alligator clips for convenience, to the wire ends. Connect one wire end to the relay coil and the other wire to the opposite side of the normally open contact. Clip the test relay across the suspected circuit. Close, or energize, the relay manually by pushing the armature

Table 59–1   Industrial Controls Index to Troubleshooting Suggestions

TYPE OF CONTROL	REFERENCE TO CONTROL SIGNS/SOLUTIONS
Electrically Operated Devices	
I. Magnetically Operated Device	
A. Contactors	
1. AC and DC	
a. AC operated	(1–12)
b. DC operated	(5–13)
2. DC	(5–13)
B. Relays	
1. AC	
a. Simple magnetic	(1–12)
b. Timing (mechanical escapement type)	(14, 15)
c. Overcurrent	
(1) Instantaneous	(16, 17)
(2) Inverse-time	(16–19)
2. DC	
a. Simple magnetic	(5–13)
b. Timing	
(1) Mechanical escapement	(14, 15)
(2) Decay-of-flux	(7–13, 20, 21)
(3) Capacitor	(7–13, 22, 23)
c. Overcurrent	
(1) Instantaneous	(16, 17)
(2) Inverse-time	(16–19)
C. Solenoids	
1. AC operated	(24–30)
2. DC operated	(28–31)
D. Valves	
1. AC operated	(27, 32–35)
2. DC operated	(34–36)
E. Brakes	
1. Shoe brake	
a. AC operated	(24–30, 37–41)
b. DC operated	(28–31, 38–41)
2. Disc brake	
a. AC operated	(21, 26, 37–41)
b. DC operated	(12, 31, 38–41)

Table 59–1 Industrial Controls Index to Troubleshooting Suggestions (Continued)

TYPE OF CONTROL	REFERENCE TO CONTROL SIGNS/SOLUTIONS
Electrically Operated Devices	
II. Motor-Operated Devices	
A. Relays	
1. Timing	(11, 15, 42, 43)
2. Induction disc or cup type	(11, 44)
B. Brakes	
1. Thrustor operated	(38–41, 45)
C. Thrustors	(45)
D. Valves	
1. Thrustor operated	(35, 45, 46)
2. Geared	(35, 46)
E. Rheostats	(11, 47, 48)
III. Thermally Operated Devices	
A. Relays	(11, 49–52)
B. Thermostats	
1. Bimetallic	(11, 53)
2. Expanding fluid (bulb and bellows)	(11, 52, 54, 55)
IV. Static Accessories	
A. Resistors	(11, 56–58)
B. Rectifiers (dry type)	(59)
C. Capacitors	(60)
D. Transformers	(61, 62)
E. Fuses	(63, 64)
Mechanically Operated Devices	
I. Manually Operated Devices	
A. Master switches	(11, 52, 65)
B. Drum controllers	(8, 11, 52, 65–67)
C. Push buttons	(11, 52, 65, 68, 69)
D. Selector switches	(11, 52, 65, 68, 69)
E. Knife switches	(70)
F. Manual starters	
1. Full voltage (small sizes)	(8, 11, 49, 50, 65)
2. Reduced voltage	(8, 11, 49, 50, 65)
G. Rheostats	(11, 47, 48)
H. Manual contactors	(8, 11, 65)
II. Otherwise Mechanically Operated Devices	
A. Limit switches	(8, 11, 52, 71)
B. Speed-sensitive switches	(11, 14, 52, 71)
C. Float switches	(11, 14, 52, 65)
D. Flow switches	(11, 14, 52, 65, 72)
E. Pressure switches	(11, 14, 52, 65, 72)

Table 59–2 Industrial Control Troubleshooting Tips

CONTROL SYMPTOMS	POSSIBLE CAUSES AND THINGS TO INVESTIGATE
1. Noisy magnet	Broken pole shader; magnet faces not true as result of wear or mounting strains; dirt on magnet faces
2. Broken pole shader	Heavy slamming caused by overvoltage, weak tip pressure, wrong coil
3. Coil failure	Moisture; overvoltage; high ambient; failure of magnet to seal in on pickup; too rapid duty cycle; corrosive atmosphere; chattering of magnet; metallic dust
4. Wear on magnet	Overvoltage; broken pole shader; wrong coil; weak tip pressure; chattering
5. Blowout coil overheats	Overcurrent; wrong size of coil; loose connections on stud or tip; tip heating (see Table 59-1); excess frequency
6. Pitted, worn, or broken arc-chutes	Abnormal interrupting duty (inductive load); excessive vibration or shock; moisture; improper assembly; rough handling
7. Failure to pick up	Low voltage; coil open; mechanical binding; no voltage; wrong coil; shorted turns; excessive magnet gap
8. Contact-tip troubles	
Filing or dressing	Do not file silver tips. Rough spots or discoloration will not harm efficiency
Interrupting excessively high current	Check for grounds, shorts, or excessive motor currents
Excessive jogging	Install larger device rated for jogging service
Weak tip pressure	Replace contacts and springs; check carrier for damage
Dirt on surfaces	Clean contacts with Freon; reduce exposure
Short circuits or ground fault	Remove fault; be sure fuse/breaker size is correct
Loose connection	Clean, then tighten
Sustain overload	Check for excessive motor current or install larger device
9. Broken flexible shunt	Large number of operations; improper installation (check instructions); extreme corrosive conditions; burned from arcing
10. Failure to drop out	Mechanical binding; gummy substance on magnet faces; air gap in magnet destroyed; contact tip welding; weak tip pressure; voltage not removed
11. Insulation failure	Moisture; acid fumes; overheating; accumulation dirt on surfaces; voltage surges; short circuits
12. Various mechanical failures	Overvoltage; heavy slamming (see 2); chattering; abrasive dust
13. Coil failure	Overvoltage; high ambient; wrong coil; moisture; corrosive atmosphere; intermittent coil energized continuously; holding resistor not cut in

Table 59–2 Industrial Control Troubleshooting Tips (Continued)

CONTROL SYMPTOMS	POSSIBLE CAUSES AND THINGS TO INVESTIGATE
14. Sticking	Dirt; worn parts; improper adjustment; corrosion; mechanical binding
15. Mechanical wear	Abrasive dust; improper application (in general not suited for continuous cycling).
16. Low trip	Wrong coil; assembled wrong
17. High trip	Mechanical binding; wrong or shorted coil; assembled wrong
18. Fast trip	Fluid out or too light; vent too large; high temperature; wrong heaters
19. Slow trip	Fluid too heavy; vent too small; mechanical binding; dirt; low temperature; wrong heaters
20. Too-short time	Dirt in air gap; shim too thick; too much spring and tip pressure; misalignment
21. Too-long time	Shim too thin; weak spring and tip pressure (use brass screws for steel backed shims); gummy substance on magnet faces
22. Too-short time	Same as 20, plus not enough capacitance; not enough resistance
23. Too-long time	Same as 21, plus too much capacitance; too much resistance
24. Noisy magnet	Same as 1, plus low voltage; mechanical overload; load out of alignment
25. Broken pole shader	Heavy slamming caused by overvoltage; mechanical underload; wrong coil; low frequency
26. Coil failure	Same as 3, plus mechanical overload (can't pick up); mechanical underload (slam); mechanical failure
27. Wear on magnet	Overvoltage; underload; broken pole shader; wrong coil; chattering; load out of alignment
28. Failure to pick up	Same as 7, plus mechanical overload
29. Failure to drop out	Mechanical binding; gummy substance on magnet faces; air gap in magnet destroyed; voltage not removed; contacts welded
30. Miscellaneous mechanical failures	Same as 12, plus underload
31. Coil failure	Same as 13, plus mechanical failure of coil
32. Noise	Same as 1, plus water hammer
33. Coil failure	Same as 3, plus handling fluid above rated temperature
34. Failure to open or close	Similar to 28, 29, plus corrosion; scale; dirt; operating above rated pressure

Table 59–2 Industrial Control Troubleshooting Tips (Continued)

CONTROL SYMPTOMS	POSSIBLE CAUSES AND THINGS TO INVESTIGATE
35. Leaks and mechanical failure	Worn seat; solid matter in seat (check strainer ahead of valve)
36. Coil failure	Same as 13, plus handling fluid above rated temperature
37. Noisy magnet	Same as 2, plus improper adjustment (too much pressure or incorrect lever ratio)
38. Excess wear or friction	Abrasive dust; heavy-duty; high inertia load, excess temperature; scarred wheels
39. Failure to hold load	Worn parts; out of adjustment; misapplication; failure to use recommended substitute parts
40. Failure to set	Improper adjustment; mechanical binding; coil not de-energized; worn parts
41. Failure to release	Improper adjustment; coil not energized; mechanical binding; low voltage or current; coil open; shorted turns
42. Failure to time out	Mechanical binding; worn parts; motor damaged; no voltage motor; dirt
43. Failure to reset	Mechanical binding; worn parts; dirt
44. Failure to operate properly	Coils connected wrong; wrong coils; mechanical binding
45. Short life of thrustor	Abrasive dust; dirty oil; water in tank
46. Failure to open or close	Corrosion; scale; dirt; mechanical binding; damaged motor; no voltage
47. Wear on segments or shoes	Abrasive dust; very heavy duty; no lubrication (special materials available)
48. Resistor failure	Overcurrent; moisture; corrosive atmospheres
49. Failure to trip (motor burnout)	Heater incorrectly sized; mechanical binding; relay previously damaged by short-circuit current; dirt; corrosion; motor and relay in different ambient temperatures
50. Failure to reset	Broken mechanism; corrosion; dirt; worn parts; resetting too soon
51. Trips too low	Wrong heater; relay in high ambient (check temperature of motor)
52. Burning and welding of control contacts and shunts	Short circuits on control circuits with too large protecting fuses; severe vibration; dirt; oxidation

Table 59–2 Industrial Control Troubleshooting Tips (Continued)

CONTROL SYMPTOMS	POSSIBLE CAUSES AND THINGS TO INVESTIGATE
53. Arcing and burning of contacts	Misapplied; should handle very little current and have sealing circuit
54. Bellows distorted	Mechanical binding; temperature allowed to overshoot
55. Bulb distorted	Liquid frozen in capillary tube
56. Overheating	Used above rating; running on starter resistor
57. Corrosion	Excess moisture; salt air; acid fumes
58. Breakage, distortion and wear	Overheating; mechanical abuse; severe vibration; shock
59. Breakdown	High temperature; moisture; overcurrent; overvoltage; corrosive atmosphere; mechanical damage
60. Breakdown	Overload; overvoltage, AC on DC capacitor; moisture; high temperature; mechanical damage; continuous voltage on intermittent types
61. Overheating	Overload; overvoltage; intermittent-rating device operated too long
62. Insulation failure	Overheating; overvoltage; voltage surges; moisture; mechanical damage
63. Premature blowing	Extra heating from outside; copper oxide on ferrules and clips (plated ferrules and clips are available); high ambient
64. Too slow blowing	Wrong-sized fuse for application
65. Mechanical failures	Abrasive dust and dirt; misapplication; mechanical damage
66. Short contact life	Jogging; handling abnormal currents; lack of lubrication where recommended; abrasive dirt
67. Flashover	Jogging; short circuits; handling too large motor; moisture; acid fumes; gases; dirt
68. Failure to break arc	Too much current; too much voltage; (usually DC); misapplication; too much inductance
69. Failure to make contact	Mechanical damage; dirt; corrosion; wear allowance gone
70. Heating	Overcurrent; loose connection; spring clips loose or annealed; oxidation; corrosion
71. Mechanical failure	Same as 65, plus excessive operating speed
72. Leaks	Corrosion; mechanical damage; excessive pressure

closed. Now wait for the problem to occur. When it does, the test relay will be de-energized. When the relay is no longer energized, you have found the problem. If a "closed contact" is live on the supply side and dead on the starter coil load side, the problem is in the contact or device located here. While checking the test relay periodically and it is still energized, move it around in the circuit, as shown, until you locate the "troublemaker."

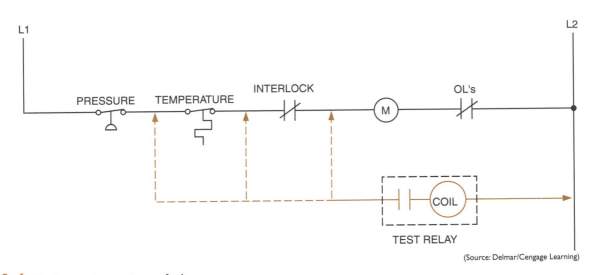

(Source: Delmar/Cengage Learning)

Fig. 59–6 Finding an intermittent fault.

SOLID-STATE EQUIPMENT

Modular construction makes troubleshooting easier. On large controllers, components can be removed for repair or replacement. Leads disconnect simply by means of plug connectors. Figure 59–7 shows an industrial processor controller, which features the convenient drawout construction of all elements, including circuit boards.

With self-diagnostics on some controllers, each of the circuit boards is checked by the microprocessor. As each board passes the self-test routine, this information is displayed on the visual control panel. The panel tells where any problems are located. The malfunctioning modules or circuit boards are then replaced. For electronic equipment, always follow the manufacturer's recommendations about service so as not to void a new equipment guarantee.

With total failures, all fuses should be tested first. If the failure still exists, the power diodes in the converter can be readily tested with an ohmmeter. Remember, these solid-state devices pass current in one direction only. If no continuity is found in either direction on the diode, the diode is open. If continuity exists in both directions, the diode is shorted. In either instance, the diode must be replaced.

Fig. 59–7 Troubleshooting complex systems is simplified by modular construction. This industrial process controller has removable circuit boards to allow for troubleshooting and also for convenient upgrading of capabilities. *(Courtesy of Schneider Electric)*

Study Questions

1. Who shall operate and maintain electrical machinery?
2. Why should the motor be run for a period of time following checkout?
3. What may occur if a jumper is placed across coil M in Figure 59–3?
4. In what position do ball-bearing motors function?
5. Motors will operate satisfactorily at what percent over and under voltage?
6. What basic sensory faculties should a troubleshooter rely on?
7. What is an orderly step-by-step procedure in successful troubleshooting?
8. What might be a hazard if a motor is started in the wrong direction?
9. When a coil does not pick up, what are the probable causes?
10. What are the probable causes of a noisy magnet?
11. Why are some solid-state controllers easy to troubleshoot?
12. How is a solid-state diode tested? Explain.
13. Using a portable relay to locate an intermittent fault, how does the relay close?
14. Why must the relay connections be moved around the circuit?

Troubleshooting Techniques

Objectives

After studying this unit, the student should be able to

- Safely check a circuit to determine if power is disconnected.

- Use a voltmeter to troubleshoot a control circuit.

- Use an ohmmeter to test for continuity.

- Use an ammeter to determine if a motor is overloaded.

It is not a question of *if* a control circuit will eventually fail, but *when* it will fail. One of the main jobs of an industrial electrician is to troubleshoot and repair a control circuit when it fails. In order to repair or replace a fault component it is first necessary to determine which component is at fault. The three main instruments used by an electrician to troubleshoot a circuit are the voltmeter, ohmmeter, and ammeter. The voltmeter and ohmmeter are generally contained in the same meter, Figure 60–1. These meters are called *multimeters* because they can measure several different electrical quantities. Some electricians prefer to use a plunger-type voltage tester because they are not susceptible to ghost voltages. High-impedance voltmeters often give an indication of some amount of voltage, caused by feedback and induction. Plunger-type voltage testers are low-impedance devices and require several milliamperes to operate. The disadvantage of plunger-type voltage testers is that they cannot be used to test control systems that operate on low voltage, such as 24-volt systems.

Ammeters are generally the clamp-on type, Figure 60–2. Both analog and digital meters are in common use. Clamp-on-type ammeters have an advantage in that the circuit does not have to be broken to insert the meter in the line.

SAFETY PRECAUTIONS

It is often necessary to troubleshoot a circuit with power applied to the circuit. When this is the case, safety should be the first consideration. **When de-energizing or energizing a control cabinet or motor control center module, the electrician should be dressed in flame-retardant clothing while wearing safety glasses, a face shield, and a hard hat. Motor control centers employed throughout the industry generally have the ability to release enough energy in an arc-fault situation to kill a person 30 feet away.** Another rule that should always be observed when energizing or de-energizing a circuit is to stand to the side of the control cabinet or module. Do not

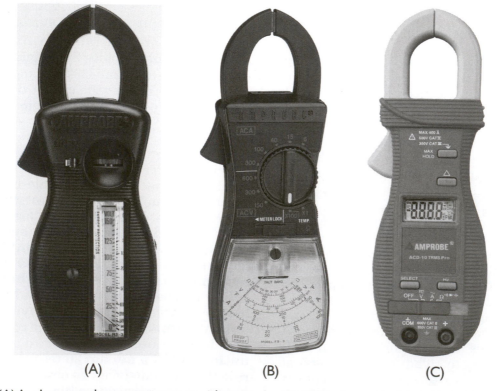

Fig. 60–1 Digital multimeter. *(Courtesy of Advanced Test Products)*

stand in front of the cabinet door when opening or closing the circuit. A direct short condition can cause the cabinet door to be blown off.

After the cabinet or module door has been opened, the power should be checked with a voltmeter to make certain the power is off. A procedure called *check, test, check* should be used to make certain that the power is off.

1. Check the voltmeter on a known source of voltage to make certain the meter is operating properly.
2. Test the circuit voltage to make certain that it is off.
3. Check the voltmeter on a known source of voltage again to make certain that the meter is still working properly.

VOLTMETER BASICS

Recall that one definition of voltage is *electrical pressure*. The voltmeter indicates the amount of potential between two points in much the same way a pressure gauge indicates the pressure difference between two points. The circuit in Figure 60–3 assumes that a voltage of 120 volts exists between L1 and N. If the leads of a voltmeter were to be connected between L1 and N, the meter would indicate 120 volts.

(A)   (B)   (C)

Fig. 60–2 (A) Analog-type clamp-on ammeter with vertical scale. (B) Analog-type clamp-on ammeter with flat scale. (C) Clamp-on ammeter with digital scale. *(Courtesy of Advanced Test Products)*

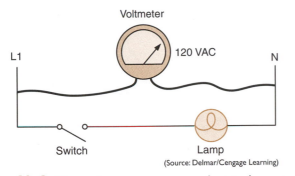

Fig. 60–3 The voltmeter measures electrical pressure between two points.

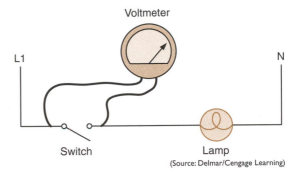

Fig. 60–5 The voltmeter is connected across the switch.

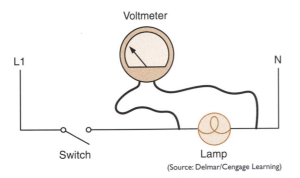

Fig. 60–4 The voltmeter is connected across the lamp.

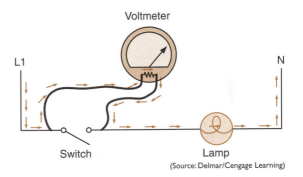

Fig. 60–6 A current path exists through the voltmeter and lamp filament.

Now assume that the leads of the voltmeter are connected across the lamp, Figure 60–4.

Question 1: Assuming that the lamp filament is good, would the voltmeter indicate 0 volts, 120 volts, or some value between 0 and 120 volts?

Answer: The voltmeter would indicate 0 volts. In the circuit shown in Figure 60–4 the switch and lamp are connected in series. One of the basic rules for series circuits is that the voltage drop across all circuit components must equal the applied voltage. The amount of voltage drop across each component is proportional to the resistance of the components and the amount of current flow. Since the switch is open in this example, there is no current flow through the lamp filament and no voltage drop.

Question 2: If the voltmeter were to be connected across the switch as shown in Figure 60–5, would it indicate 0 volts, 120 volts, or some value between 0 and 120 volts?

Answer: The voltmeter would indicate 120 volts. Since the switch is an open circuit, the resistance is infinite at this point, which is millions of times greater than the resistance of the lamp filament. Recall that voltage is electrical pressure. The only current flow through this

circuit is the current flowing through the voltmeter and the lamp filament, Figure 60–6.

Question 3: If the total or applied voltage in a series circuit must equal the sum of the voltage drops across each component, why is all the voltage drop across the voltmeter resistor and none across the lamp filament?

Answer: There is some voltage drop across the lamp filament because the current of the voltmeter is flowing through it. The amount of voltage drop across the filament, however, is so small compared to the voltage drop across the voltmeter, it is generally considered to be zero. Assume the lamp filament to have a resistance of 50 ohms. Now assume that the voltmeter is a digital meter and has a resistance of 10,000,000 ohms. The total circuit resistance is 10,000,050 ohms. The total circuit current is 0.000,011,999 ampere (120/10,000,050) or about 12 microamps. The voltage drop across the lamp filament would be approximately 0.0006 volt or 0.6 millivolt (50 Ω × 12 µA).

Question 4: Now assume that the lamp filament is open or burned out. Would the voltmeter in Figure 60–7 indicate 0 volts, 120 volts, or some value between 0 and 120 volts?

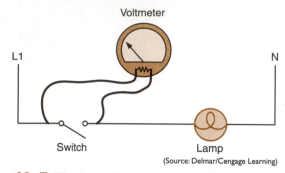

Fig. 60–7 The lamp filament is burned open.

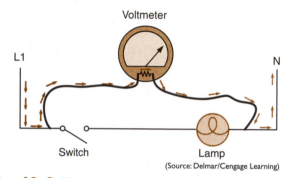

Fig. 60–8 The voltmeter is connected across both components.

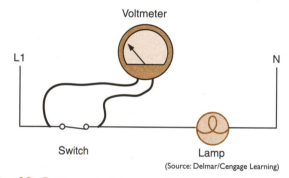

Fig. 60–9 The switch is turned on or closed.

Answer: The voltmeter would indicate 0 volts. If the lamp filament is open or burned out, a current path for the voltmeter does not exist and the ohmmeter would indicate 0 volts. In order for the voltmeter to indicate voltage it would have to be connected across both components so that a complete circuit would exist from L1 to N, Figure 60–8.

Question 5: Assume that the lamp filament is not open or burned out and that the switch has been closed or turned on. If the voltmeter is connected across the switch, would it indicate 0 volts, 120 volts, or some value between 0 and 120 volts, Figure 60–9?

Answer: The voltmeter would indicate 0 volts. Now the switch is closed, the contact resistance is extremely small and the lamp filament now exhibits a much higher resistance than the

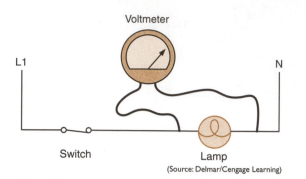

Fig. 60–10 Practically all the voltage drop is across the lamp.

switch. Practically all the voltage drop will now appear across the lamp, Figure 60–10.

TEST PROCEDURE EXAMPLE I

The type of problem determines the procedure to be employed when troubleshooting a circuit. Assume that an overload relay has tripped several times. The first step is to determine what conditions could cause this problem. If the overload relay is a thermal type, a source of heat is the likely cause of the problem. Make mental notes of what could cause the overload relay to become overheated.

1. Excessive motor current.
2. High ambient temperature.
3. Loose connections.
4. Incorrect wire size.

If the motor has been operating without a problem for some period of time, incorrect wire size can probably be eliminated. If it is a new installation, it would be a factor to consider.

Since overload relays are intended to disconnect the motor from the power line in the event that the current draw becomes excessive, the motor should be checked for excessive current. The first step is to determine the normal full-load current from the nameplate on the motor. The next step is to determine the percent of full-load current setting for the overload relay.

Example: A motor nameplate indicates the full-load current of the motor is 46 amperes. The nameplate also indicates the motor has a service factor of 1.00. The *NEC®* indicates the overload should be set to trip at 115 percent of the full-load current. The overload heaters should be sized for 52.9 amperes (46 × 1.15).

The next step is to check the running current of the motor with an ammeter. This is generally accomplished by measuring the motor current at the overload relay, Figure 60–11. The current in each phase should be measured. If the motor

To Load Contacts on Motor Starter

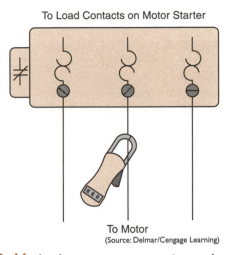

To Motor
(Source: Delmar/Cengage Learning)

Fig. 60–11 A clamp-on ammeter is used to check motor current.

To Load Contacts on Motor Starter

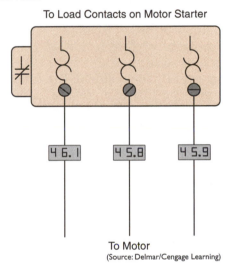

| 4 6. 1 | 4 5.8 | 4 5.9 |

To Motor
(Source: Delmar/Cengage Learning)

Fig. 60–12 Ammeter readings indicate that the motor is operating normally.

is operating properly, the readings may not be exactly the same, but they should be close to the full-load current value if the motor is operating under load, and relatively close to each other. In the example shown in Figure 60–12, phase 1 has a current flow of 46.1 amperes, phase 2 has a current flow of 45.8 amperes, and phase 3 has a current flow of 45.9 amperes. These values indicate that the motor is operating normally. Since the ammeter indicates that the motor is operating normally, other sources of heat should be considered. After turning off the power, check all connections to ensure that they are tight. Loose connections can generate a large amount of heat, and loose connections close to the overload relay can cause the relay to trip.

Another consideration should be ambient temperature. If the overload relay is located in an area of high temperature, the excess heat could cause the overload relay to trip prematurely. If this is the case, bimetal strip type overload relays, Figure 60–13, can often be adjusted

(Source: Delmar/Cengage Learning)

Fig. 60–13 Bimetal strip-type overload relays can be set for a higher value of current.

for a higher setting to offset the problem of ambient temperature. If the overload relay is the solder melting type, it will be necessary to change the heater size to offset the problem, or install some type of cooling device such as a small fan. If a source of heat cannot be identified as the problem, the overload relay probably has a mechanical defect and should be replaced.

Now assume that the ammeter indicated an excessively high-current reading on all three phases. In the example shown in Figure 60–14, phase 1 has a current flow of 58.1 amperes, phase 2 has a current flow of 59.2 amperes, and phase 3 has a current flow of 59.3 amperes. Recall that the full-load nameplate current for this motor is 46 amperes. These values indicate that the motor is overloaded. The motor and load should be checked for some type of mechanical problem such as a bad bearing or possibly a brake that has become engaged.

Now assume that the ammeter indicates one phase with normal current and two phases that have excessively high current. In the example shown in Figure 60–15, phase 1 has a current flow of 45.8 amperes, phase 2 has a current flow of 73.2 amperes, and phase 3 has a current flow of 74.3 amperes. Two phases with excessively high current indicate that the motor probably has a shorted winding. If two phases have a normal amount of current and one phase is excessively high, it is a good indication that one of the phases has become grounded to the case of the motor.

To Load Contacts on Motor Starter

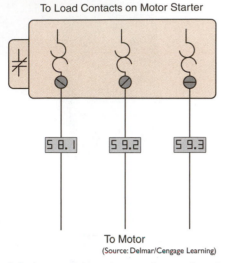

To Motor
(Source: Delmar/Cengage Learning)

Fig. 60–14 Ammeter readings indicate that the motor is overloaded.

To Load Contacts on Motor Starter

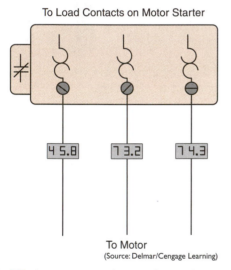

To Motor
(Source: Delmar/Cengage Learning)

Fig. 60–15 Ammeter readings indicate that the motor has a shorted winding.

TEST PROCEDURE EXAMPLE 2

The circuit shown in Figure 60–16 is a reversing starter with electrical and mechanical interlocks. Note that double-acting push buttons are used to disconnect one contactor if the start button for the other contactor is pressed. Now assume that if the motor is operating in the forward direction and the reverse push button is pressed, the forward contactor de-energizes, but the reverse contactor does not. If the forward push button is pressed, the motor will restart in the forward direction.

To begin troubleshooting this problem, make mental notes of problems that could cause this condition:

1. The reverse contactor coil is defective.
2. The normally closed F auxiliary contact is open.
3. The normally closed side of the forward push button is open.
4. The normally open side of the reverse push button does not complete a circuit when pressed.
5. The mechanical linkage between the forward and reversing contactors is defective.

Also make mental notes of conditions that could not cause the problem:

1. The stop button is open. (If the stop button were open, the motor would not run in the forward direction.)
2. The overload contact is open. (Again, if this were true, the motor would not run in the forward direction.)

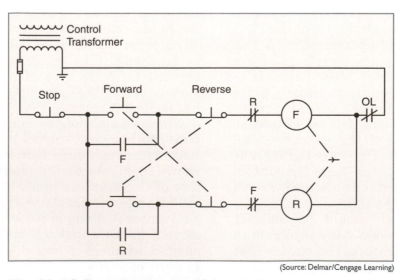

(Source: Delmar/Cengage Learning)

Fig. 60–16 Reversing starter with interlocks.

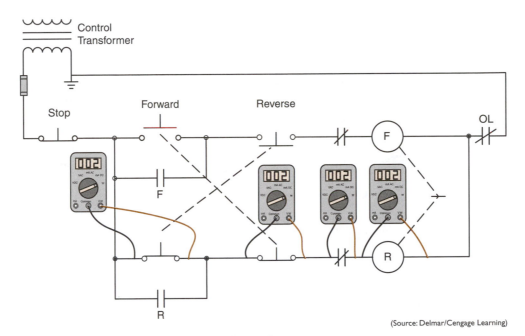

(Source: Delmar/Cengage Learning)

Fig. 60–17 Checking components for continuity with an ohmmeter.

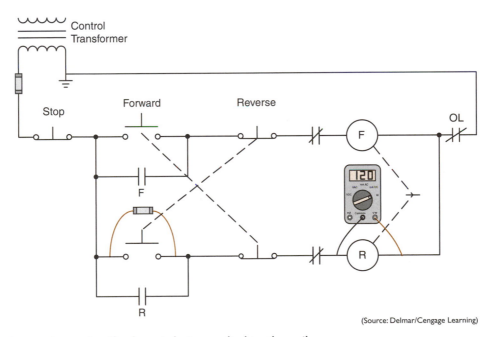

(Source: Delmar/Cengage Learning)

Fig. 60–18 Testing to determine if voltage is being applied to the coil.

To begin checking this circuit, an ohmmeter can be used to determine if a complete circuit path exists through certain components. **When using an ohmmeter make certain that the power is disconnected from the circuit.** A good way to do this in most control circuits is to remove the control transformer fuse. The ohmmeter can be used to check the continuity of the reverse contactor coil, the normally closed F contact, the normally closed section of the forward push button, and across the normally open reverse push button when it is pressed, Figure 60–17.

The ohmmeter can be used to test the starter coil for a complete circuit to determine if the winding has been burned open, but it is generally not possible to determine if the coil is shorted. To make a final determination, it is generally necessary to apply power to the circuit and check for voltage across the coil. Since the reverse push button must be closed to make this measurement, it is common practice to connect a fused jumper across the push button if there is no one to hold the button closed, Figure 60–18. A fused jumper is shown in Figure 60–19. When using a fused jumper, power should be disconnected

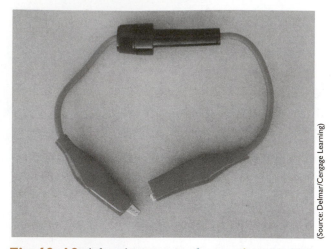

Fig. 60–19 A fused jumper is often used to complete a circuit when troubleshooting.

when the jumper is connected across the component. After the jumper is in position, power can be restored to the circuit. If voltage appears across the coil, it is an indication that the coil is defective and should be replaced or that the mechanical interlock between the forward and reverse contactors is defective.

TEST PROCEDURE EXAMPLE 3

The next circuit to be discussed is shown in Figure 60–20. This circuit permits the motor to be started in any of three speeds with a 5-second time delay between accelerating from one speed to another. Regardless of which speed push button is pressed, the motor must start in its lowest speed and progress to the selected speed. It is assumed that eight pin on-delay timers are used to provide the time delay for acceleration to the next speed.

Assume that when the third speed push button is pressed, the motor starts in it lowest speed. After 5 seconds the motor accelerates to second speed but never increases to third speed. As in the previous examples, start by making a mental list of the conditions that could cause this problem:

1. Contactor S2 is defective.
2. Timed contact TR2 did not close.

3. Timer TR2 is defective.
4. CR2 or S1 contacts connected in series with timer TR2 did not close.

Begin troubleshooting this circuit by pressing the third speed push button and permit the motor to accelerate to second speed. Wait at least 5 seconds after the motor has reached the third speed and connect a voltmeter across the coil of S2 contactor, Figure 60–21. It will be assumed that the voltmeter indicated a reading of 0 volts. This indicates that there is no power being applied to the coil of S2 contactor. The next step is to check for voltage across pins 1 and 3 of timer TR2, Figure 60–22. If the voltmeter indicates a value of 120 volts, it is an indication that the normally open timed contact has not closed.

If timed contact TR2 has not closed, check for voltage across timer TR2, Figure 60–23. This can be done by checking for voltage across pins 2 and 7 of the timer. If a value of 120 volts is present, the timer is receiving power, but contact TR2 did not close. This is an indication that the timer is defective and should be replaced. If the voltage across timer coil TR2 is 0, then the voltmeter should be used to determine if contact CR2 or S1 is open.

Troubleshooting is a matter of progressing logically through a circuit. It is virtually impossible to troubleshoot a circuit without a working knowledge of schematics. You can't determine what a circuit is or is not doing if you don't understand what it is intended to do in normal operation. Good troubleshooting techniques take time and practice. As a generally rule, it is easier to progress backward though the circuit until the problem is identified. For example, in this circuit, contactor S2 provided the last step of acceleration for the motor. Starting at contactor S2 and progressing backward until determining what component was responsible for no power being applied to the coil of S2 was much simpler and faster than starting at the beginning of the circuit and each component.

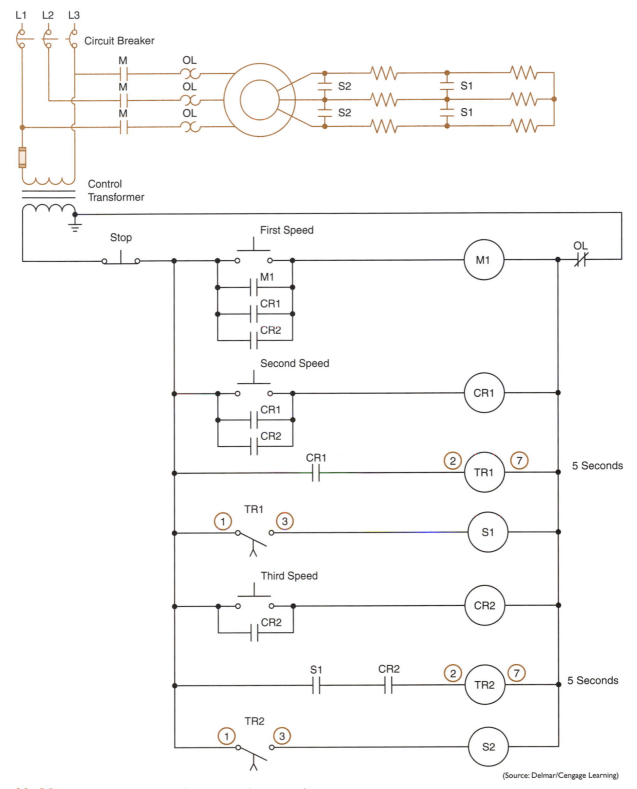

Fig. 60–20 Three-speed control for a wound rotor induction motor.

(Source: Delmar/Cengage Learning)

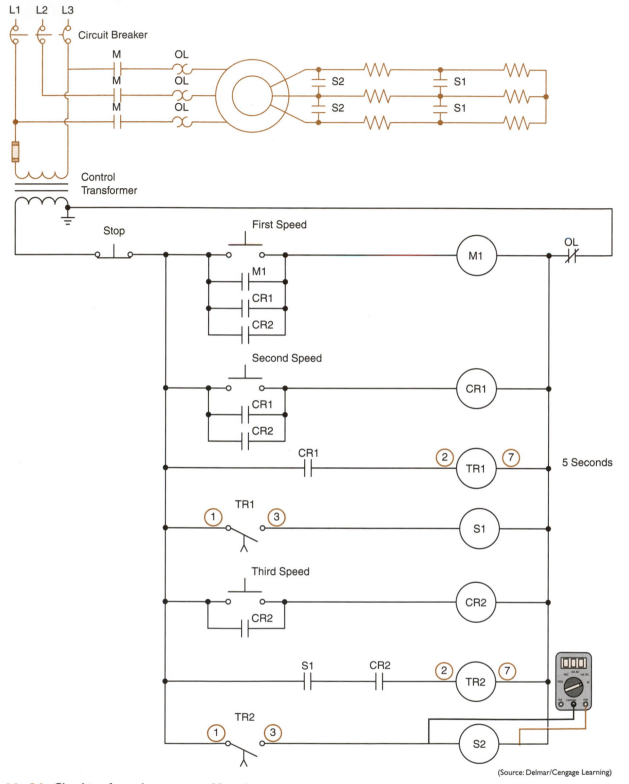

(Source: Delmar/Cengage Learning)

Fig. 60–21 Checking for voltage across S2 coil.

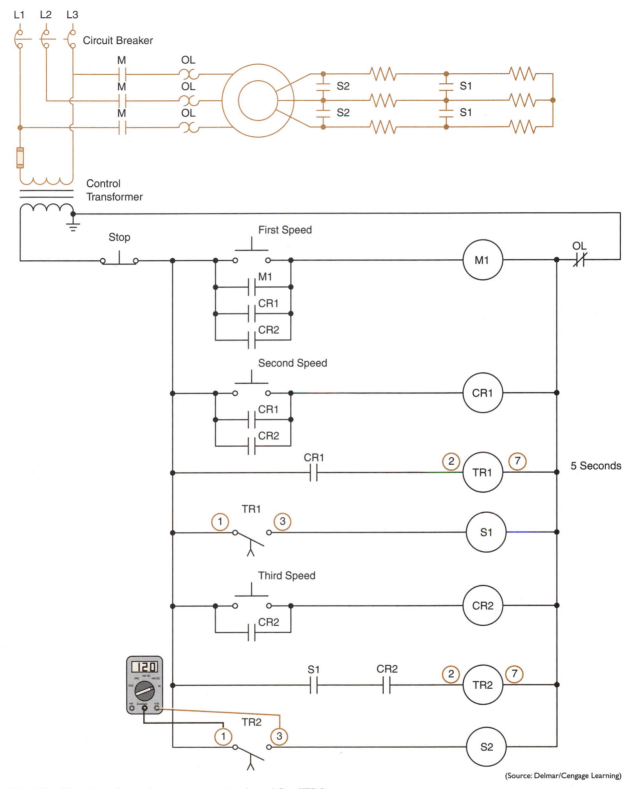

Fig. 60–22 Checking for voltage across pins 1 and 3 of TR2 timer.

(Source: Delmar/Cengage Learning)

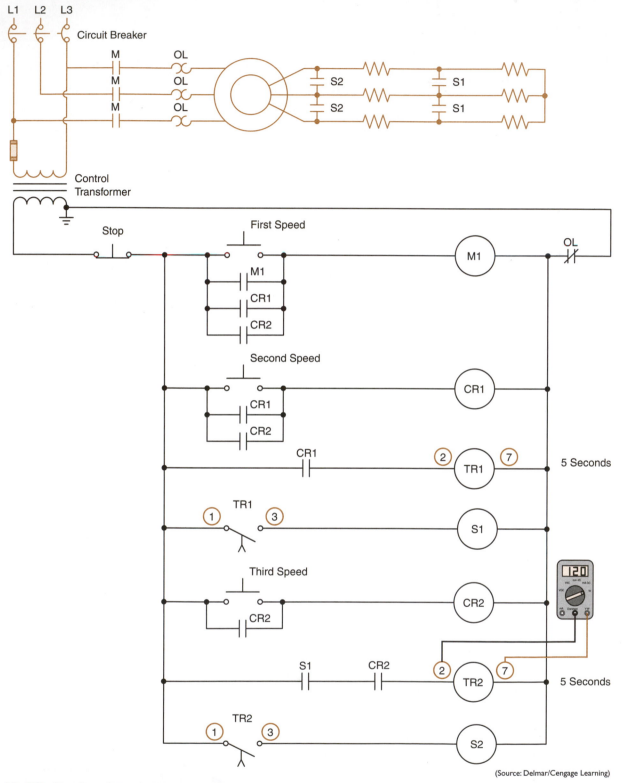

(Source: Delmar/Cengage Learning)

Fig. 60–23 Checking for voltage across TR2 coil.

Study Questions

1. What are the three main electrical test instruments used in troubleshooting?

2. What is the advantage of a plunger-type voltage tester?

3. A motor is tripping out on overload. The motor nameplate reveals a full-load current of 68 amperes. When the motor is operating under load, an ammeter indicates the following: phase 1 = 106 amperes, phase 2 = 104 amperes, and phase 3 = 105 amperes. What is the most likely problem with this motor?

4. A motor is tripping out on overload. The motor nameplate reveals a full-load current of 168 amperes. When the motor is operating under load, an ammeter indicates the following: phase 1 = 166 amperes, phase 2 = 164 amperes, and phase 3 = 225 amperes. What is the most likely problem with this motor?

5. Refer to the circuit shown in Figure 60–16. The motor will not start in either the forward or reverse direction when the start push buttons are pressed. Which of the following could *not* cause this problem?
    a. F coil is open.
    b. The overload contact is open.
    c. The control transformer fuse is blown.
    d. The stop push button is not making a complete circuit.

6. Refer to the circuit shown in Figure 60–16. Assume that the motor is running in the forward direction. When the reverse push button is pressed, the motor continues to run in the forward direction. Which of the following could cause this problem?
    a. The normally open side of the reverse push button is not making a complete circuit when pressed.
    b. R contactor coil is open.
    c. The normally closed side of the reverse push button is not breaking the circuit when the reverse push button is pressed.
    d. There is nothing wrong with the circuit. The stop push button must be pressed before the motor will stop running in the forward direction and permit the motor to be reversed.

7. Refer to the circuit shown if Figure 60–20. When the third speed push button is pressed, the motor starts in first speed but never accelerates to second or third speed. Which of the following could *not* cause this problem?
    a. Control relay CR1 is defective.
    b. Control relay CR2 is defective.
    c. Timer TR1 is defective.
    d. Contactor coil S1 is open.

8. Refer to the circuit shown in Figure 60–20. Assume that the third speed push button is pressed. The motor starts in second speed, skipping first speed. After 5 seconds the motor accelerates to third speed. Which of the following could cause this problem?
    a. S1 contactor coil is open.
    b. CR1 contactor coil is open.
    c. TR1 timer coil is open.
    d. S1 load contacts are shorted.

9. Refer to the circuit shown in Figure 60–16. If a voltmeter is connected across the normally open forward push button, the meter should indicate a voltage value of
    a. 0 volt
    b. 30 volts
    c. 60 volts
    d. 120 volts

10. Refer to the circuit shown in Figure 60–20. Assume that a fused jumper is connected across terminals 1 and 3 of TR2 timer. What would happen if the jumper were left in place and the first speed push button pressed?
   a. The motor would start in its lowest speed and progress to second speed but never increase to third speed.
   b. The motor would start operating immediately in third speed.
   c. The motor would not start.
   d. The motor would start in second speed and then increase to third speed.

MOTOR TYPES AND LINE DIAGRAMS

DC Shunt Motor. Main field winding is designed for parallel connection to the armature; stationary field; rotating armature with commutator; has a no-load constant speed; full speed at full load is less than no-load speed; torque increases directly with load.

DC Series Motor. Main field winding is designed for series connection to the armature; stationary field; rotating armature with commutator; does not have a no-load speed; requires solid direct connection to the load to prevent runaway at no-load; speed decreases rapidly with increase in load; torque increases as square of armature current; main motor for crane hoists; excellent starting torque.

DC Compound Motor. Main field both shunt (parallel) and series; stationary field; rotating armature with commutator; combination shunt and series fields produce characteristics between straight shunt or series DC motor; good starting torque; main motor for DC-driven machinery (mills or presses).

AC Squirrel Cage Motor. Single or three phase; single phase requires a starting winding; three phase, self-starting; stationary stator winding; no electrical connection to short-circuited rotor; torque produced from magnetic reaction of stator and rotor fields; speed a function of supply frequency and number of electrical poles wound

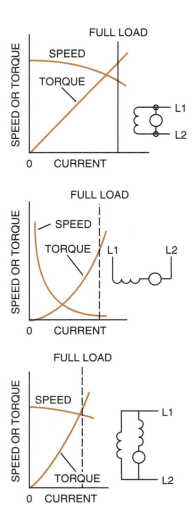

on stator; considered as constant speed even though speed decreases slightly with increased load; good starting on full voltage; high inrush currents during starting torque; rugged construction; easily serviced and maintained; high efficiency; good running power factor when delivering full load; requires motor control for stator windings only.

AC Wound Rotor Induction Motor. Characteristics similar to squirrel cage motor; stationary stator winding; rotor windings terminate on slip rings; external addition of resistance to rotor circuit for speed control; good starting torque; high inrush current during starting on full voltage; low efficiency when resistor is inserted in rotor windings; good running power factor; requires motor controls for stator and rotor circuits.

AC Synchronous Motor. Stationary AC stator windings; rotating DC field winding; no starting torque unless motor has starting winding; generally poor starting torque; constant speed when motor up to speed and DC field winding energized; can provide power factor correction with proper DC field excitation; requires special motor control for both AC and DC windings to prevent the DC field winding from being energized until a specified percent of running speed has been obtained.

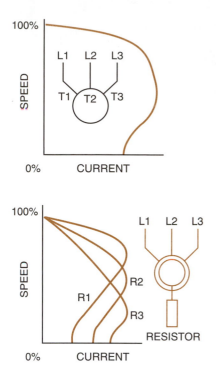

Appendix B

POWER SUPPLIES

All electrical power supplied as AC or DC; primarily AC; generation, transmission, and some distribution of power at high voltage (above 5000 volts) or medium voltage (600 to 5000 volts); most power distribution of voltage for industrial and residential use is 600 volts and under; AC power generally at 60-hertz frequency; AC distribution at use location single or three phase.

Single phase. Two wire, 120 volts, one line grounded; 120/240 volts, three-wire centerline grounded; residential distribution, lighting, heat, fractional horsepower motors and business machines.

Three phase. Three-wire delta 230/460 volts, 580 volts; four-wire wye 208/440, 277/480 volts neutral line grounded; primary industrial power distribution; many motor drives, integral horsepower motors, lighting, heating fractional horsepower motors, and business machines; used as three-phase or single-phase power supply.

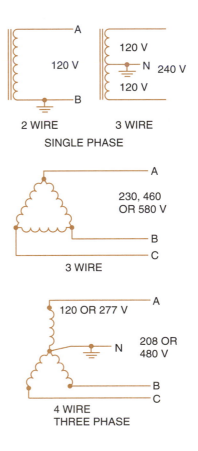

MOTOR CIRCUIT ELEMENTS

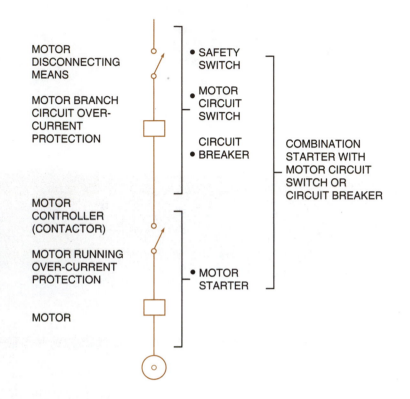

Glossary

Accelerating Relay Any type of relay used to aid in starting a motor or to accelerate a motor from one speed to another. Accelerating relays may function by motor armature current (current limit acceleration), armature voltage (counter emf acceleration), or definite time (definite time acceleration).

Accessory (control use) A device that controls the operation of magnetic motor control. (See also *Master Switch, Pilot Device,* and *Push Button.*)

Across-the-line Method of motor starting that connects the motor directly to the supply line on starting or running. (Also called Full-Voltage Control.)

Alternating Current (AC) Current changing both in magnitude and direction; most commonly used current.

Ambient Temperature The temperature surrounding a device.

Ampacity The maximum current rating of a wire or cable.

Ampere Unit of electrical current.

Analog Signal Having the characteristic of being continuous and changing smoothly over a given range rather than switching on and off between certain levels.

Armature Resistance Ohmic resistance of an armature.

Armortisseur Winding Squirrel cage winding used to start a synchronous motor.

ASA American Standards Association.

Automatic Self-acting, operating by its own mechanism when actuated by some triggering signal; for example, a change in current strength, pressure, temperature, or mechanical configuration.

Automatic Starter A self-acting starter that is completely controlled by master or pilot switches or other sensing devices; designed to control automatically the acceleration of a motor during the acceleration period.

Auxiliary Contacts Contacts of a switching device in addition to the main circuit contacts; auxiliary contacts operate with the movement of the main contacts.

Blowout Coil Electromagnetic coil used in contactors and starters to deflect an arc when a circuit is interrupted.

Brake An electromechanical friction device to stop and hold a load. Generally electric release spring applied—coupled to motor shaft.

Branch Circuit That portion of a wiring system that extends beyond the final overcurrent device protecting the circuit.

Breakdown Torque (of a motor) The maximum torque that will develop with the rated voltage applied at the rated frequency, without an abrupt drop in speed. (ASA)

Busway A system of enclosed power transmission that is current and voltage rated.

Capacitor-Start Motor A single-phase induction motor with a main winding arranged for direct connection to the power source and an auxiliary winding connected in series with a capacitor. The capacitor phase is in the circuit only during starting. (NEMA)

Circuit Breaker Automatic device that opens under abnormal current in carrying circuit; circuit breaker is not damaged on current interruption; device is ampere, volt, and horsepower rated.

Compound Motor A motor that has both a shunt and a series field winding.

Contact A conducting part that acts with another conducting part to complete or to interrupt a circuit.

Contactor A device to establish or interrupt an electric power circuit repeatedly.

Control Circuit A circuit used to control the operation of some device.

Controller A device or group of devices that governs, in a predetermined manner, the delivery of electric power to apparatus connected to it.

Controller Function Regulate, accelerate, decelerate, start, stop, reverse, or protect devices connected to an electric controller.

Controller Service Specific application of controller. General purpose: standard or usual service. Definite purpose: service condition for specific application other than usual.

Counter emf Counter voltage generated in a turning armature.

Current Relay A relay that functions at a predetermined value of current. A current relay may be either an overcurrent relay or an undercurrent relay.

Dashpot Consists of a piston moving inside a cylinder filled with air, oil, mercury, silicon, or other fluid. Time delay is caused by allowing the air or fluid to escape through a small orifice in the piston. Moving contacts actuated by the piston close the electrical circuit.

Deceleration Reducing speed.

De-energize Remove power from.

Definite-Purpose Motor Any motor designed, listed, and offered in standard ratings with standard operating characteristics or mechanical construction for use under service conditions other than usual or for use on a particular type of application. (NEMA)

Definite Time (or Time Limit) Definite time is a qualifying term indicating that a delay in action is purposely introduced. This delay remains substantially constant regardless of the magnitude of the quantity that causes the action.

Device A unit of an electrical system that is intended to carry but not utilize electrical energy.

Digital The representation of data in the form of pieces, bits, or digits.

Diode A two-element electronic device that permits current to flow through it in only one direction.

Direct Current (DC) A continuous nonvarying current in one direction.

Disconnecting Means (Disconnect) A device, or group of devices, or other means whereby the conductors of a circuit can be disconnected from their source of supply.

Drum Controller Electrical contacts made on a surface of rotating cylinder or sector; contacts made also by operation of a rotating cam.

Drum Switch A switch having electrical connecting parts in the form of fingers held by spring pressure against contact segments or surfaces on the periphery of a rotating cylinder or sector.

Duty Specific controller functions. Continuous (time) duty: constant load, indefinite long time period. Short time duty: constant load, short or specified time period. Intermittent duty: varying load, alternate intervals, specified time periods. Periodic duty: intermittent duty with recurring load conditions. Varying duty: varying loads, varying time intervals, wide variations.

Dynamic Braking Using the motor as a generator, taking it off the line, and applying an energy dissipating resistor to the armature.

Eddy Currents Circular induced currents contrary to the main currents; a loss of energy that shows up in the form of heat.

Electric Brakes Mechanical brakes operated by an electric solenoid.

Electric Controller A device, or group of devices, that governs, in some predetermined manner, the electric power delivered to the apparatus to which it is connected.

Electrical Interlocking Accomplished by control circuits in which the contacts in one circuit control another circuit.

Electronic Control Control system using gas and/or vacuum tubes or solid-state devices.

Electronic or Solid-State Overload Relay A relay that employs a solid-state component to sense motor current.

Enclosure Mechanical, electrical, and environmental protection for control devices.

Energize Apply power to.

Eutectic Alloy Metal with low and sharp melting point; used in thermal overload relays; converts from a solid to a liquid state at a specific temperature; commonly called solder pot.

Excessive Starting Starting a motor too many times within a certain period of time.

Feeder The circuit conductor between the service equipment, or the generator switchboard of an isolated plant, and the branch circuit over-current device.

Feeler Gauge A precision instrument with blades in thicknesses of thousandths of an inch for measuring clearances.

Field Accelerating Relay Used with starters to control adjustable-speed motors.

Filter A device used to remove the voltage/current ripple produced by a rectifier.

Frequency Number of complete variations made by an alternating current per second; expressed in Hertz. (See *Hertz*.)

Friction Brakes Brakes that convert kinetic energy into thermal energy to slow or stop an object.

Full Load Maximum amount of power that can be safely delivered by an electrical or mechanical device.

Full-Load Torque (of a Motor) The torque necessary to produce the rated horsepower of a motor at full-load speed.

Full-Voltage Control (Across-the-line) Connects equipment directly to the line supply on starting.

Fuse An overcurrent protective device with a fusible member that is heated directly and destroyed by the current passing through it to open a circuit.

General-Purpose Motor Any open motor having a continuous 40C rating and designed, listed, and offered in standard ratings with standard operating characteristics and mechanical construction for use under usual service conditions without restrictions to a particular application or type of application. (NEMA)

Hall Effect Sensors Proximity sensor that detects the presence of a magnetic field.

Heat Sink Metal used to dissipate the heat of solid-state components mounted on it; usually finned in form.

Hertz International unit of frequency equal to one cycle per second of alternating current.

High-Voltage Control Formerly, all control above 600 volts. Now, all control above 5000 volts. See *Medium Voltage* for 600- to 5000-volt equipment.

Horsepower Measure of the time rate of doing work (working rate).

Impedance Total opposition to current flow in an electrical circuit.

Input Power delivered to an electrical device.

Instantaneous A qualifying term indicating that no delay is purposely introduced in the action of the device.

Integral Whole or complete; not fractional.

Interface A circuit permitting communication between the central processing unit (of a programmable controller) and a field input or output.

Interlock To interrelate with other controllers; an auxiliary contact. A device is connected in such a way that the motion of one part is held back by another part.

Inverse Time A qualifying term indicating that a delayed action is introduced purposely. This delay decreases as the operating force increases.

Jogging (Inching) Momentary operations; the quickly repeated closure of the circuit to start a motor from rest for the purpose of accomplishing small movements of the driven machine.

Jumper A short length of conductor used to make a connection between terminals or around a break in a circuit.

Limit Switch A mechanically operated device that stops a motor from revolving or reverses it when certain limits have been reached.

Line Diagram A type of electrical diagram.

Load Center Service entrance; controls distribution; provides protection of power; generally of the circuit breaker type.

Local Control Control function, initiation, or change accomplished at the same location as the electric controller.

Locked Rotor Amount of inrush current when power is applied to a motor.

Locked Rotor Current (of a motor) The steady-state current taken from the line with the rotor locked (stopped) and with the rated voltage and frequency applied to the motor.

Locked Rotor Torque (of a motor) The minimum torque that a motor will develop at rest for all angular positions of the rotor with the rated voltage applied at a rated frequency. (ASA)

Lockout A mechanical device that may be set to prevent the operation of a push button.

Logic A means of solving complex problems through the repeated use of simple functions that define basic concepts. Three basic logic functions are AND, OR, and NOT.

Low Voltage Less than normal voltage.

Low-Voltage Protection (LVP) Magnetic control only; nonautomatic restarting; three-wire control; power failure disconnects service; power restored by manual restart.

Low-Voltage Release (LVR) Manual and magnetic control; automatic restarting; two-wire control; power failure disconnects service; when power is restored, the controller automatically restarts the motor.

Magnetic Brake Friction brake controlled by electromagnetic means.

Magnetic Contactor A contactor operated electromagnetically.

Magnetic Controller An electric controller; device functions operated by electromagnets.

Magnetic Overload Relay Overload relays that use magnetism to sense motor currents.

Maintaining Circuit Contacts connected in parallel with a start push button.

Maintaining Contact A small control contact used to keep a coil energized, actuated by the same coil usually. Holding contact; pallet switch.

Manual Controller An electric controller; device functions operated by mechanical means or manually.

Master Switch A main switch to operate contactors, relays, or other remotely controlled electrical devices.

Mechanical Brakes See *Friction Brakes.*

Medium-Voltage Control Formerly known as high voltage; includes 600- to 5000-volt apparatus; air break or oil-immersed main contactors; high interrupting capacity fuses; 150,000 kVA at 2300 volts; 250,000 kVA at 4000–5000 volts.

Memory Part of a programmable controller where instructions and information are stored.

Metal Oxide Varistor (M.O.V.) A solid-state, voltage-sensitive resistor.

Microprocessor A small computer.

Microswitch A type of small limit switch.

Motor Device for converting electrical energy to mechanical work through rotary motion; rated in horsepower.

Motor Circuit Switch Motor branch circuit switch rated in horsepower; capable of interrupting overload motor current.

Motor Control Control used to operate a motor.

Motor-Driven Timer A device in which a small pilot motor causes contacts to close after a predetermined time.

Multispeed Starter An electric controller with two or more speeds; reversing or nonreversing; full or reduced voltage starting.

NEMA National Electrical Manufacturers Association.

NEMA Size Electric controller device rating; specific standards for horsepower, voltage, current, and interrupting characteristics.

Noise A condition that interferes with the desired voltage, or signal, in a circuit. Noise can produce erratic operation.

No-Load Speed Speed of a motor with no load applied to it.

Nonautomatic Controller Requires direct operation to perform function; not necessarily a manual controller.

Nonreversing Operation in one direction only.

Normally Open and Normally Closed When applied to a magnetically operated switching device, such as a contactor or relay, or to the contacts of these devices, these terms signify the position taken when the operating magnet is de-energized. The terms apply only to non-latching types of devices.

Off Delay A type of timer that starts its time when power is turned off.

On Delay A type of timer that starts its time when power is turned on.

Open Transition Starting A method of wye-delta starting.

Optoisolation A method of controlling the operation of relays using LEDs and photosensitive components to separate two circuits with a light beam.

Optoisolator A device used to connect sections of a circuit by means of a light beam.

Output Devices Elements such as solenoids, motor starters, and contactors that receive input.

Overload Protection The result of a device that operates on excessive current, but not necessarily on short circuit, to cause and maintain the interruption of current flow to the device governed. NOTE: Operating overload means a current that is not in excess of six times the rated current for AC motors, and not in excess of four times the rated current for DC motors.

Overload Relay Running overcurrent protection; operates on excessive current; not necessarily protection for short circuit; causes and maintains interruption of device from power supply. Overload relay heater coil: coil used in thermal overload relays; provides heat to melt eutectic alloy. Overload relay reset: Push button used to reset thermal overload relay after relay has operated.

Palm-Operated Push Button Mushroom push button.

Panelboard Panel, group of panels, or units; an assembly that mounts in a single panel; includes buses, with or without switches and/or automatic overcurrent protection devices; provides control of light, heat, power circuits; placed in or against wall or partition; accessible from the front only.

Part Winding Starting Starting a motor with part of the stator winding connected to reduce starting current.

Permanent-split Capacitor Motor A single-phase induction motor similar to the capacitor start motor except that it uses the same capacitance, which remains in the circuit for both starting and running. (NEMA)

Permeability The ease with which a material will conduct magnetic lines of force.

Phase Relation of current to voltage at a particular time in an AC circuit. Single phase: A single voltage and current in the supply. Three phase: Three electrically related (120-degree electrical separation) single-phase supplies.

Phase-Failure Protection Provided by a device that operates when the power fails in one wire of a polyphase circuit to cause and maintain the interruption of power in all the wires of the circuit.

Phase-Reversal Protection Provided by a device that operates when the phase rotation in a polyphase circuit reverses to cause and maintain the interruption of power in all the wires of the circuit.

Phase Rotation Relay A relay that functions in accordance with the direction of phase rotation.

Photodetectors Proximity sensor consisting of a light source and light sensor used in a variety of industrial applications.

Pilot Device Directs operation of another device. Float switch: A pilot device that responds to liquid levels. Foot switch: A pilot device operated by the foot of an operator. Limit switch: A pilot device operated by the motion of a power-driven machine; alters the electrical circuit with the machine or equipment.

Plugging Braking by reversing the line voltage or phase sequence; motor develops retarding force.

Polarized Field Frequency Relay A relay used in the starting of a synchronous motor.

Pole The north or south magnetic end of a magnet; a terminal of a switch; one set of contacts for one circuit of main power.

Poor ventilation Insufficient airflow.

Potentiometer A variable resistor with two outside fixed terminals and one terminal on the center movable arm.

Pressure Switch A pilot device operated in response to pressure levels.

Printed Circuit A board on which a predetermined pattern of printed connections has been formed.

Pull-up Torque (of AC motor) The minimum torque developed by the motor during the period of acceleration from rest to the speed at which breakdown occurs. (ASA)

Push Button A master switch; manually operable plunger or button for an actuating device; assembled into push-button stations.

Ramping Changing speed of a motor at a controlled rate.

Rectifier A device that converts AC into DC.

Regenerative Braking Braking a DC motor by letting it become a generator.

Relay Operated by a change in one electrical circuit to control a device in the same circuit or another circuit; rated in amperes; used in control circuits.

Remote Control Controls the function initiation or change of an electrical device from some remote point or location.

Remote Control Circuit Any electrical circuit that controls any other circuit through a relay or an equivalent device.

Residual Magnetism The retained or small amount of remaining magnetism in the magnetic material of an electromagnet after the current flow has stopped.

Resistance The opposition offered by a substance or body to the passage through it of an electric current; resistance converts electrical energy into heat; resistance is the reciprocal of conductance.

Resistor A device used primarily because it possesses the property of electrical resistance. A resistor is used in electrical circuits for purposes of operation, protection, or control; commonly consists of an aggregation of units.

Armature Regulating Resistors Used to regulate the speed or torque of a loaded motor by resistance in the armature or power circuit.

Dynamic Braking Resistors Used to control the current and dissipate the energy when a motor is decelerated by making it act as a generator to convert its mechanical energy to electrical energy and then to heat in the resistor.

Field Discharge Resistors Used to limit the value of voltage that appears at the terminals of a motor field (or any highly inductive circuit) when the circuit is opened.

Plugging Resistors Used to control the current and torque of a motor when deceleration is forced by electrically reversing the motor while it is still running in the forward direction.

Starting Resistors Used to accelerate a motor from rest to its normal running speed without damage to the motor and connected load from excessive currents and torques, or without drawing undesirable inrush current from the power system.

Rheostat A resistor that can be adjusted to vary its resistance without opening the circuit in which it may be connected.

Ripple An AC component in the output of a DC power supply; improper filtering.

Rotor The rotating member of an AC motor.

Safety Switch Enclosed manually operated disconnecting switch; horsepower and current rated; disconnects all power lines.

SCR Silicon controlled rectifier; a semiconductor device that must be triggered on by a pulse applied to the gate before it will conduct.

Secondary Resistor Starters Starters that insert resistance in series with a motor to limit current.

Selector Switch A master switch that is manually operated; rotating motion for actuating device; assembled into push-button master stations.

Semiautomatic Starter Part of the operation of this type of starter is nonautomatic while selected portions are automatically controlled.

Semiconductors See *Solid-State Devices.*

Semimagnetic Control An electric controller whose functions are partly controlled by electromagnets.

Sensing Device A pilot device that measures, compares, or recognizes a change or variation in the system it is monitoring; provides a controlled signal to operate or control other devices.

Sequence Chart A chart that lists the sequence of operation of identified contacts in a circuit. Also known as Target Table.

Sequence Control Forcing a control system to operate in a predetermined manner.

Series Components that are connected in such a manner that there is only one current path.

Service The conductors and equipment necessary to deliver energy from the electrical supply system to the premises served.

Service Equipment Necessary equipment, circuit breakers or switches and fuses, with accessories mounted near the entry of the electrical supply; constitutes the main control or cutoff for supply.

Service Factor (of a general-purpose motor) An allowable overload; the amount of allowable overload is indicated by a multiplier, which, when applied to normal horsepower rating, indicates the permissible loading.

Shaded Pole Motor A single-phase induction motor provided with auxiliary short-circuited winding or windings displaced in magnetic position from the main winding. (NEMA)

Shunt Parallel.

Signal The event, phenomenon, or electrical quantity that conveys information from one point to another.

Slip Difference between rotor rpm and the rotating magnetic field of an AC motor.

Snubber A circuit that suppresses transient spikes.

Solder Pot See *Eutectic Alloy.*

Solenoid A tubular, current-carrying coil that provides magnetic action to perform various work functions.

Solenoid-and-Plunger A solenoid provided with a bar of soft iron or steel called a plunger.

Solid-State Devices Electronic components that control electron flow through solid materials such as crystals; for example, transistors, diodes, integrated circuits.

Special-Purpose Motor A motor with special operating characteristics or special mechanical construction, or both, designed for a particular application and not falling within the definition of a general-purpose or definite-purpose motor. (NEMA)

Speed Regulators Devices or circuits that control motor speed.

Split-Phase A single-phase induction motor with auxiliary winding, displaced in magnetic position from, and connected in parallel with, the main winding. (NEMA)

Squirrel Cage Motor An AC motor that contains a squirrel cage rotor.

Squirrel Cage Rotor A type of rotor that contains a squirrel cage winding.

Starter A controller designed for accelerating a motor to normal speed in one direction of rotation. NOTE: A device designed for starting a motor in either direction of rotation includes the additional function of reversing and should be designated as a controller.

Startup The time between equipment installation and the full operation of the system.

Static Control Control system in which solid-state devices perform the functions. Refers to no moving parts or without motion.

Stator The winding of an AC motor to which power is applied.

Surge A transient variation in the current and/or potential at a point in the circuit, unwanted, temporary.

Switch A switch is a device for making, breaking, or changing the connections in an electric circuit.

Switchboard A large single panel with a frame or assembly of panels; devices may be mounted on the face of the panels, on the back, or both; contains switches, overcurrent, or protective devices; instruments accessible from the rear and front; not installed in wall-type cabinets. (See *Panelboard.*)

Synchronous Rotor The rotor of a synchronous motor.

Synchronous Speed Motor rotor and AC rotating magnetic field in step or unison.

Tachometer Generator Used for counting revolutions per minute. Electrical magnitude or impulses are calibrated with a dial gage reading in rpm.

Target Table See *Sequence Chart.*

Temperature Relay A relay that functions at a predetermined temperature in the apparatus protected. This relay is intended to protect some other apparatus such as a motor or controller and does not necessarily protect itself.

Temperature Switch A pilot device operated in response to temperature values.

Terminal A fitting attached to a circuit or device for convenience in making electrical connections.

Thermal Overload Relays Overload relays that sense motor current with heater elements.

Thermal Protector (as applied to motors) An inherent overheating protective device that is responsive to motor current and temperature. When properly applied to a motor, this device protects the motor against dangerous overheating due to overload or failure to start.

Time Limit See *Definite Time.*

Timer A pilot device that is also considered a timing relay; provides adjustable time period to perform function; motor driven; solenoid actuated; electronic.

Torque The torque of a motor is the twisting or turning force that tends to produce rotation.

Transducer A device that transforms power from one system to power of a second system; for example, heat to electrical.

Transformer Converts voltages for use in power transmission and operation of control devices; an electromagnetic device.

Transient Temporary voltage or current that occurs randomly and rides an AC sine wave. (See also *Surge.*)

Transient Suppression A method of eliminating electrical noise and voltage spikes with electronic devices.

Transistor An active semiconductor device which can be used for rectification and amplification by placing the proper voltage on its electrodes.

Trip Free Refers to a circuit breaker that cannot be held in the on position by the handle on a sustained overload.

Troubleshoot To locate and eliminate the source of trouble in any flow of work.

Undervoltage Protection The result of when a device operates on the reduction or failure of voltage to cause and maintain the interruption of power to the main circuit.

Undervoltage Release Occurs when a device operates on the reduction or failure of voltage to cause the interruption of power to the main circuit but does not prevent the reestablishment of the main circuit on the return of voltage.

USASI United States of America Standards Institute.

Voltage Relay A relay that functions at a predetermined value of voltage. A voltage relay may be either an overvoltage or an undervoltage relay.

Voltage Unbalance When the voltages of a multiphase system have different values.

VOM Volt-ohm-milliammeter; a test instrument designed to measure voltage, resistance, and milliamperes.

Wiring Diagrams A diagram that shows a pictorial of components with connecting wires.

Wound Rotor A type of three-phase motor.

Index

C